Drugs: Photochemistry and Photostability

Drugs
Photochemistry and Photostability

Edited by

A. Albini
Dell' Università Di Pavia, Italy

E. Fasani
Dell' Università Di Pavia, Italy

THE ROYAL
SOCIETY OF
CHEMISTRY
Information
Services

Based on the proceedings of the 2nd International Meeting on Photostability of Drugs held in Pavia, Italy on 4–17 September 1997.

Special Publication No. 225

ISBN 0-85404-743-3

A catalogue record for this book is available from the British Library

Published by The Royal Society of Chemistry,
Thomas Graham House, Science Park, Milton Road,
Cambridge CB4 4WF, UK

For further information see our web site at www.rsc.org

Printed and bound by MPG Books Ltd, Bodmin, Cornwall, UK.

Preface

That many drugs, just as non-pharmaceutically active compounds, are photoreactive has been long known. As an example, Pasteur noticed the photolability of quinine in 1846[1] and industry-sponsored studies on the photochemistry of drugs were already systematically carried out in the twenties.[2] However, until recently the matter has received only limited attention, mainly on the assumption that by using the appropriate opaque container no significant decomposition could have taken place.

As a result, the available knowledge is quite sparse. All Pharmacopoeias mention that some drugs have to be protected from light, but one cannot rely upon such qualitative (and incomplete) information. The number of reports in specialised journals is growing, but remains low.

The situation has changed recently, however, and this is due to several causes.

First, more sensitive analytical methods are now available and the standard of purity required has become more and more stringent. Thus, even traces of (photochemically formed) impurities must be revealed. This has led to the formulation by ICH of internationally accepted Guidelines for Drug Photostability (see p. 66), which have been implemented since January 1998.

Second, there have been cases of promising drugs which have been discarded late in the development process due to a too high photolability. The development of a new drug is very expensive and this calls for more attention to the photochemical properties of a molecule early in the development, or for a way to predict the photostability of a new molecule.

Third, significant phototoxic effects have been ascertained for several drugs in common clinically, and in general there is now more attention to the phototoxic effects of drugs (as well as of cosmetic products and sunscreens). Here again, control of the photobiological effects demands that the photohemistry of the active molecule is known.

The awareness of this situation has led to the organisation of two international meetings, the first one in Oslo in June 1995, the latter in Pavia in September 1997. Both have been attended by scientists of different affiliations (industries, regulatory agencies, universities) and of different specialisations (pharmaceutical techniques, pharmaceutical chemistry, photochemistry, photophysics, biology, toxicology). The need for a close collaboration between such different areas has been recognised.

This book is based on the communications presented at the Pavia meeting, and is organised as follows.

1. Introductory part. This includes an overview on the photochemistry of drugs and on some related problems (dependence on conditions, protection of photolabile drugs) by the editors, the text of the ICH Guidelines on Photostability, and an introduction to medicinal chemistry with attention to the kinetics of photochemical processes by *Beijerbergen van Henegouwen*.

2. Photochemistry of drugs. Photochemistry of drug families, *viz.* antimalarials (*Tønnesen*), diuretic drugs (*Moore*), antimycotics (*Thoma*), phenothiazines (*Glass*), anti-inflammatory drugs (*Monti*), coumarins (*Zobel*), sunscreens (*Allen*), Leukotriene B4 antagonists (*Webb*). The photosensitising properties by some drugs are treated by *De Guidi* and *Tronchin*.

3. Photostability of drugs. Methods for implementing the ICH guidelines (*Drew*) and a discussion of their application (*Helboe*); the choice of lamps (*Piechocki*) and in general of the appropriate conditions for carrying out photostability studies (*Boxhammer* and *Forbes*); the choice of the actinometer (*Favaro* and *Bovina*).

It is hoped that these contributions may help to determine on a sound basis the significance of drug photostability for the pharmaceutical industry and also help to serve as support for phototoxicity studies.

Thanks are due to Mr F. Barberis and Misses M. Di Muri, M. Parente and F. Stomeo for their help in preparing the manuscripts.

A. Albini and E. Fasani   Pavia, March 1998

1. L. Pasteur, *Comp. Rend.*, 1853, **37**, 110.
2. J. Piechocki, p. 247.

Contents

Photochemistry of Drugs:
An Overview and Practical Problems

Angelo Albini and Elisa Fasani
Department of Organic Chemistry
University of Pavia
v.le Taramelli 10, I-27100 Pavia, Italy

1 INTRODUCTION

Absorption of light (UV or visible) by the ground state of a molecule (S_0) generates electronically excited states, either directly (the singlet states) or after intersystem crossing from the singlet manifold (the triplet states). Alternatively, triplet states may be generated by energy transfer from another excited state (a sensitiser). In both multiplicities, very fast internal conversion leads to the lowest states (S_1 and T_1 respectively). These states, although still quite short lived (typical lifetime $\tau < 10^{-8}$ s for S_1 and $< 10^{-6}$ s for T_1) live long enough that a chemical reaction competes with decay to the ground state.

Electronically excited states are electronic isomers of the ground state, and not surprisingly show a different chemistry. These, however, can be understood with the same kind of reasoning that is used for ground state chemistry, taking into account that the very large energy accumulated in excited states makes their reactions much faster (in the contrary case, there would be no photochemistry at all, in view of the short lifetime of the key intermediates). As an example, ketones are electrophiles in the ground state due to the partial positive charge on the carbon atom. The reaction with nucleophiles occurs. In the $n\pi^*$ triplet excited state electrons are differently distributed, and the important thing is now the presence of an unpaired electron on the non-bonding orbital localised on the oxygen atom. This makes atom transfer to that atom so fast a process ($k \approx 10^6 s^{-1}$, many orders of magnitude faster than any reaction of ground state molecules) that it competes efficiently with the decay of such a state.

On the basis of such principles, the many photochemical reactions now known have been rationalised. This is shown in many fine books of photochemistry,[1-5] which demonstrate both the dramatic development of this science in the last decades and the high degree of rationalisation that has been reached. The photoreactions of drugs[6] obviously can be discussed in the same way, and G. M. J. Beijersbergen van Henegouwen (p. 74) pointed out some key points that one should take into account. It is therefore generally possible to predict the photochemical behaviour of a new drug, as of any other molecule, or at least to point out the most likely alternatives.

More exactly, as it has been pointed out by Grenhill in a recent review,[7] it is possible to indicate some molecular features that are likely to make a molecule liable to photodecomposition, even if it is difficult to predict the exact photochemical behaviour of a specific molecule. This is due to the fact that competition between the chemical reaction(s) and physical decay to the ground state depends in a complex way on the structure (and on conditions). Thus both the efficiency of a photochemical reaction and product distribution may vary significantly even among closely related compounds and further depend on conditions.

At any rate, several chemical functions are expected to introduce photoreactivity (see Scheme 1). These are:

a. The carbonyl group. This behaves as an electrophilic radical in the $n\pi^*$ excited state. Typical reactions are reduction via intermolecular hydrogen abstraction and fragmentation either via α-cleavage ("Norrish Type I") or via intramolecular γ-hydrogen abstraction followed by C_α-C_β cleavage ("Norrish Type II").

b. The nitroaromatic group, also behaving as a radical, and undergoing intermolecular hydrogen abstraction or rearrangement to a nitrite ester.

c. The *N*-oxide function. This rearranges easily to an oxaziridine and the final products often result from further reaction of this intermediate.

d. The C=C double bond, liable to *E/Z* isomerisation as well as to oxidation (see case g).

e. The aryl chloride, liable to homolytic and/or to heterolytic dechlorination

f. Products containing a weak C-H bond, e.g. at a benzylic position or α to an amine nitrogen. These compounds often undergo photoinduced fragmentations via hydrogen transfer or electron-proton transfer.

g. Sulphides, alkenes, polyenes and phenols. These are highly reactive with singlet oxygen, formed through photosensitisation from the relatively harmless ground state oxygen.

Such functions are present in a very large fraction, if not the majority, of commonly used drugs. Thus, many drug substances, and possibly most of them, are expected to react when absorbing light. However, photodegradation of a drug is of practical significance only when the compound absorbs significantly ambient light (λ>330 nm), and even in that case the photoreaction may be too slow to matter, particularly if concentrated solutions or solids are considered. It is important to notice that most information about photoreactions available in the literature refers to the conditions where such processes are most easily observed and studied, viz. dilute solutions in organic solvents, whereas what matters for drug photostability are (buffered) aqueous solutions or the solid state. Under such different conditions the photoreactivity of a drug may be dramatically different. To give but one example, benzophenone triplet - probably the most thoroughly investigated excited state - is a short-lived species in organic solvents, e. g. τ ca 0.3 μs in ethanol, and is quite photoreactive via hydrogen abstraction under such conditions, and in general in an organic solution. However,

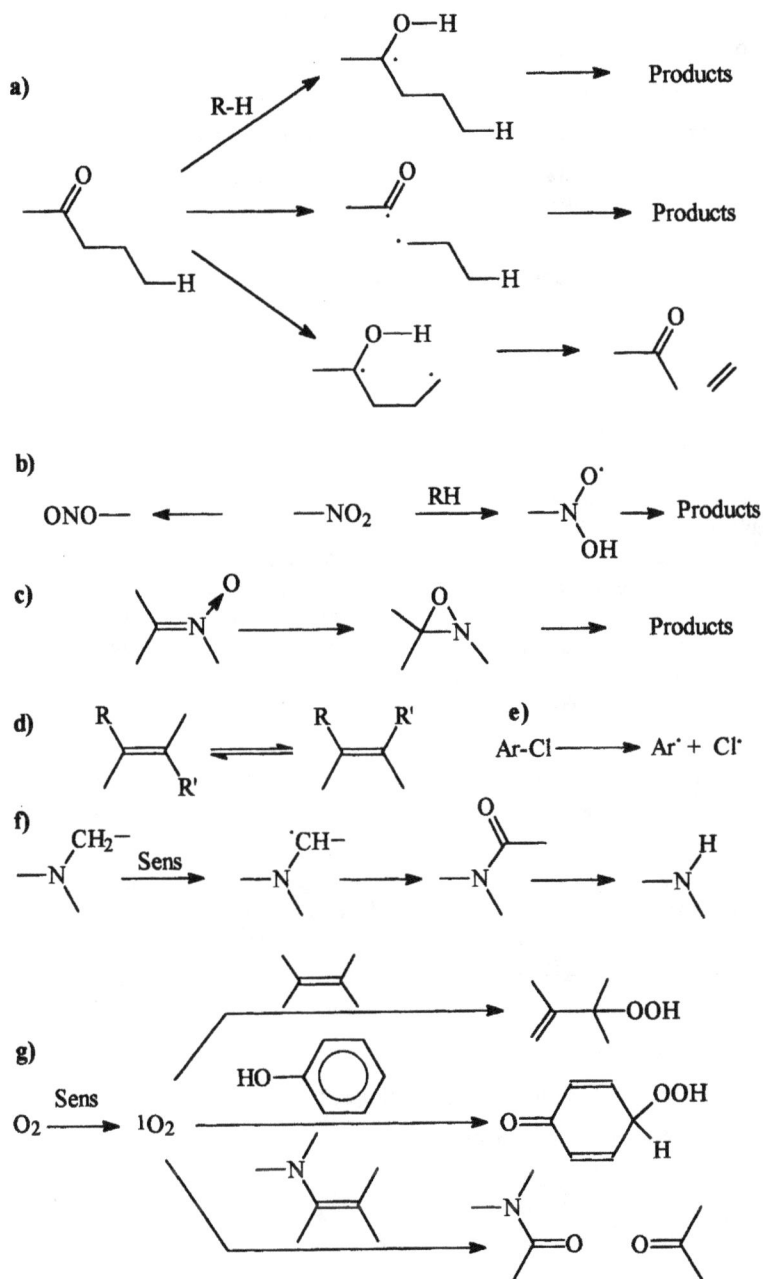

Scheme 1

the lifetime of this species increases by two orders of magnitude in water, where benzophenone is almost photostable.

The present chapter has the following aims:

a. to offer an overview of reported photochemical reactions of drugs (see Sec. 2).

b. to discuss practical problems related with drug photoreactivity, such as the dependence on the physical state of the drug or drug preparation and the quantitative assessment of drug photostability (see Sec. 3).

c. to make reference to the possible ways for protecting a drug against photoreactions (see Sec. 4).

The ICH Guidelines on Drug Photostability are enclosed as an Appendix.

2 PHOTOREACTIONS OF DRUGS

Information on drug photoreactivity is probably not sufficient among practitioners of pharmaceutical chemistry. Reports about this topic have been growing in number in the last years, but they are scattered in a variety of journals (oriented towards chemistry, pharmaceutical sciences and techniques, pharmacology, biology and medicine), thus possibly not reaching all interested readers. Furthermore, both the approach used (ranging from the simple assessment of the photolability to detailed product or mechanistic studies) and the experimental conditions used (e.g. radiation source) are quite various, and thus care is required when extending the results obtained with a drug to different conditions (let alone for predicting the reactivity of related substrates).

Several more or less extended reviews about the photochemistry of drugs are available in the literature,[6-16] and an extensive compilation of reference groups by compound name has been published by Tønnesen.[17]

It is hoped that the present review may help to give a better "feeling" of the type of photochemical reactions occurring with drugs. Due to limitation of the available space the overview presented here is intended to be exemplificative rather than exhaustive. The drugs are grouped according to the following broad therapeutic categories:

- anti-inflammatory, analgesic and immunosuppressant drugs;

- drugs acting on the central nervous system;

- cardiovascular, diuretic and hemotherapeutic drugs;

- gonadotropic steroids and synthetic estrogens;

- dermatologicals;

- chemotherapeutic agents;

- vitamins.

2.1 Anti-inflammatory, Analgesic and Immunosuppressant Drugs

2.1.1 Non-steroidal Anti-inflammatory and Analgesic Drugs. A variety of 2-aryl- (or heteroaryl-) propionic (or acetic) acid derivatives are used as anti-inflammatory agents. Most of these are photoreactive and have some phototoxic action. As a consequence, their photochemistry has been intensively investigated.[18-20] The main process in aqueous solution is decarboxylation to yield a benzyl radical, a general reaction with α-arylcarboxylic acid ("photo-Kolbe"reaction).[21] Under anaerobic conditions, benzyl radicals undergo dimerisation or reduction (and in an organic solvent abstract hydrogen).[22] In the presence of oxygen, addition to give a hydroperoxy radical and the corresponding alcohol and ketone (the latter in part from secondary oxidation of the former) takes place (Scheme 2). A further path leading to the oxidised products may involve singlet oxygen.[19, 23]

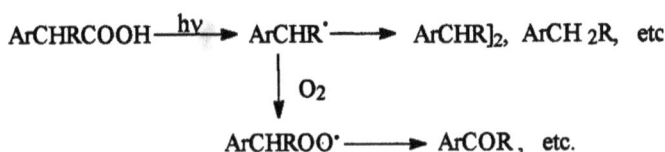

Scheme 2

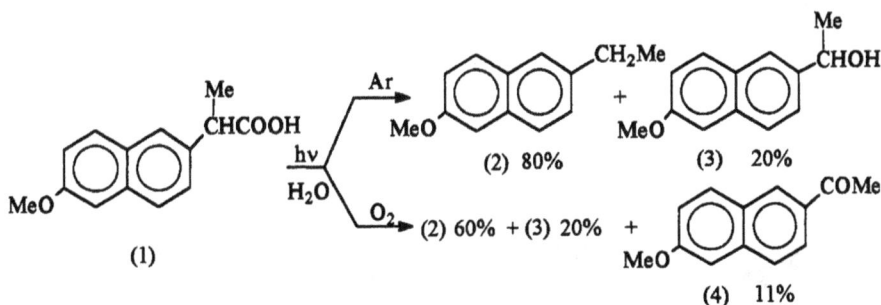

Scheme 3

The results from the irradiation of naproxen (1) in water are shown in Scheme 3,[111, 83] and a related chemical course is followed with several drugs pertaining to this group, such as ibuprofen (5),[24] butibufen (6),[25] flurbiprofen (7),[24] ketoprofen (8),[20, 26, 27] suprofen (9),[28] benoxaprofen (10),[19, 22, 25, 29] tiaprofenic acid (11)[30] (Scheme 4) and ketorolac tromethamine (12) (Scheme 5).[31] The triplet state is responsible for initial decarboxylation. Some detailed mechanistic studies have been carried out;[26, 29] in the case of ketoprofen, as an example, it has been shown that the fast decarboxylation of the triplet in water (τ_T 250 ps, quantum yield 0.75) may involve an adiabatic mechanism via internal charge transfer and, in part, ionisation.[26]

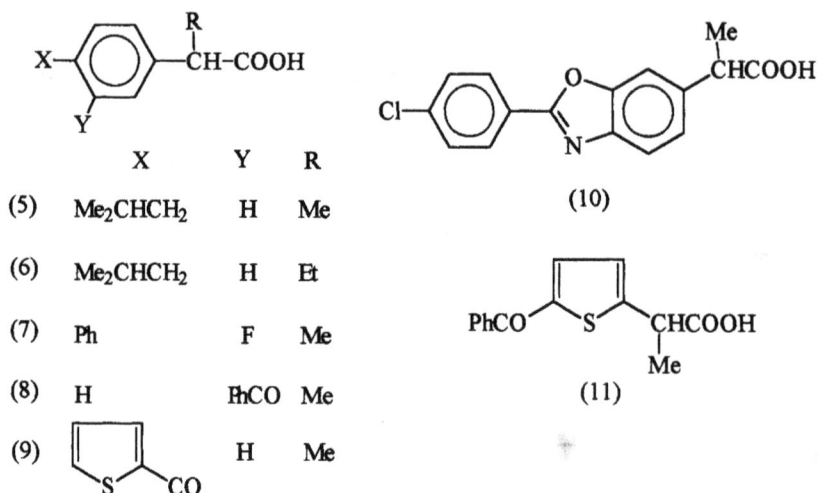

	X	Y	R
(5)	Me$_2$CHCH$_2$	H	Me
(6)	Me$_2$CHCH$_2$	H	Et
(7)	Ph	F	Me
(8)	H	PhCO	Me
(9)	[thiophene-CO]	H	Me

Scheme 4

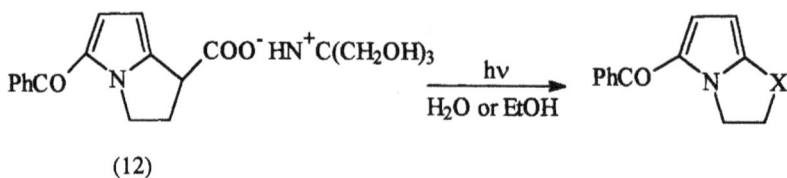

$$X = CH_2, CHOH, CHO_2H, CO$$

Scheme 5

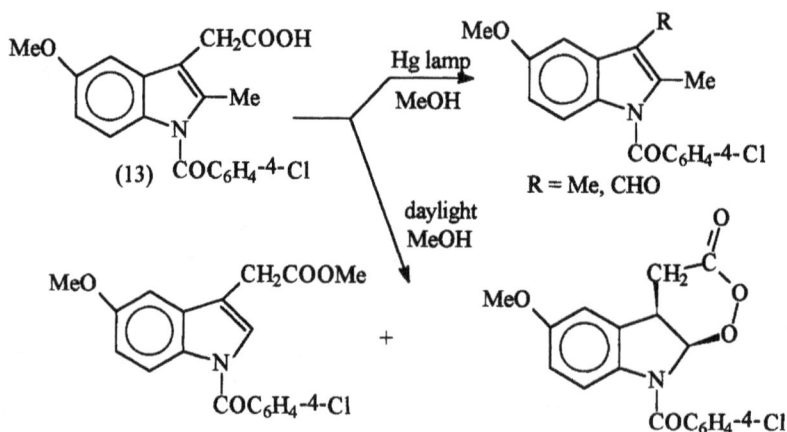

Scheme 6

Indomethacin (13) is quite photostable in the solid state (7.5% decomposition after 72 h irradiation)[32] but reacts in solution.[18] In methanol the usual decarboxylation is the main process[33, 34] when mercury lamps are used, while daylight irradiation leads to products conserving the carboxyl group which have been rationalised as arising via the acyl radical (Scheme 6).[35]

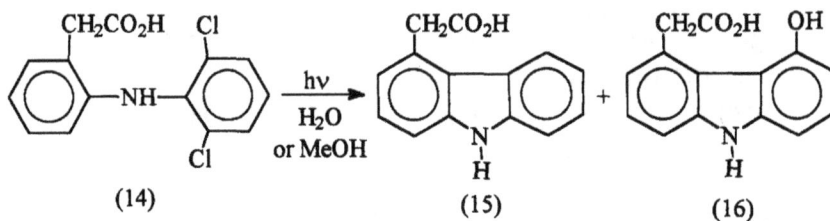

Scheme 7

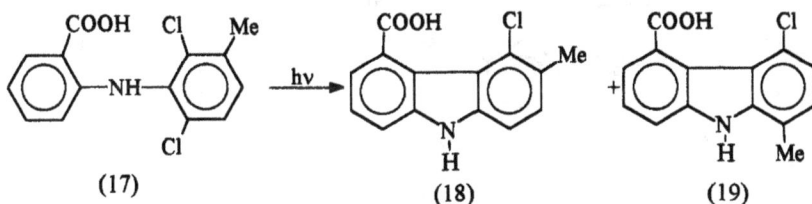

Scheme 8

In the case of the related drug 2-(2,6-dichlorophenylamino)phenylacetic acid (diclofenac, 14), on the other hand, dechlorination - as stated above, one of the general photochemical reaction of aromatics - is the dominant process. Sequential loss of both chlorine atoms is followed by ring closure, reasonably via radical addition, to yield the carbazole-1-acetic acids (15) and (16) as the main products (Scheme 7).[36] It may be noticed that 2-(2,6-dichloro-3-methylphenylamino)benzoic acid, also an anti-inflammatory agent (meclofenamic acid, 17), likewise undergoes photochemical dechlorination and ring closure to the carbazoles (18) and (19) (Scheme 8).[37]

Photoreactivity has been reported also for some anti-inflammatory and analgesic drugs different from arylacetic acids. Thus, benzydamine (20) (irradiation of the hydrochloride in methanol leads to hydroxylation in position 5 as well as well as to Fries type *O-N(2)* chain migration, to yield products (21) and (22) respectively, see Scheme 9).[38] Benorylate (23) likewise undergoes a Fries rearrangement to give (24) which then further rearranges thermally to product (25) (see Scheme 10).[39] The photo-Fries rearrangement is a general reaction with aromatic esters and amides, and occurs via a radical mechanism, rather than via the ionic mechanism of the thermal reaction. 5-Aminosalicylic acid (26), used for the treatment of chronic inflammatory bowel diseases, undergoes light-accelerated oxidation and polymerisation (Scheme 11).[40]

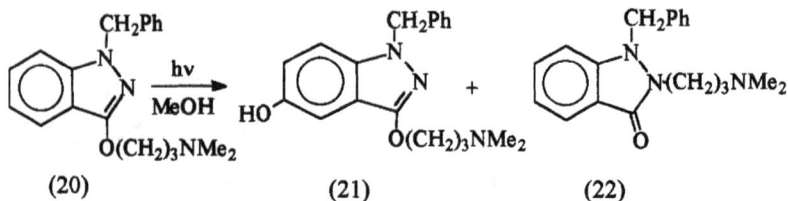

Scheme 9

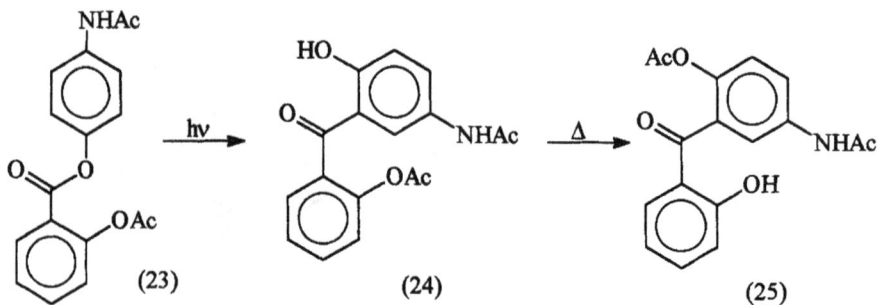

Scheme 10

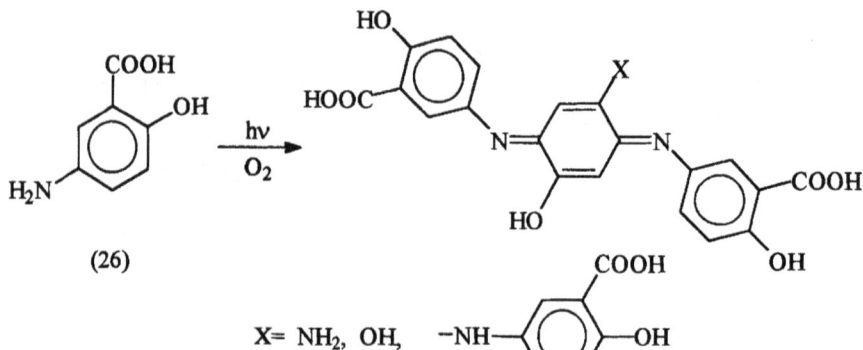

X= NH₂, OH,

Scheme 11

The narcotic analgesic methadone hydrochloride (27) reacts when irradiated by UV light both in aqueous solution[41] and in the solid state.[42] The processes observed (fragmentation and cyclisation, see Scheme 12) are a typical manifestation of the radical-like character of the nπ* state of ketones (α- and β-cleavage, see Scheme 1, a). However, this drug is photostable in an isotonic solution when exposed to ambient light.[43]

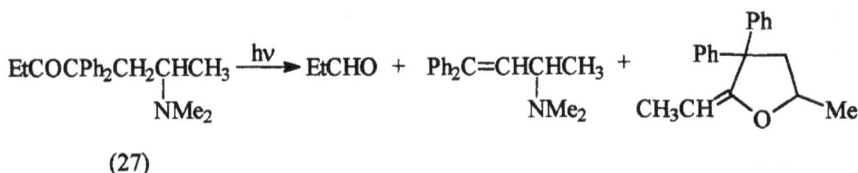

(27)

Scheme 12

The enkefalinase inhibitor thiorfan (28), a new generation analgesic, is quite sensitive to oxidation and is converted to the disulphide; this reaction is accelerated by light.[44]

$$PhCH_2CH(SH)CONHCH_2COOH \quad (28)$$

2.1.2 Pyrazolone Analgesic and Antipyretic Drugs. The largely used drugs of this structure are photoreactive, and cleavage of the pyrazole ring occurs in most cases.[45, 46] Typical reactions are shown below for the case of aminopyrine (29) (Scheme 13) and for that of phenylbutazone (30) (Scheme 14).

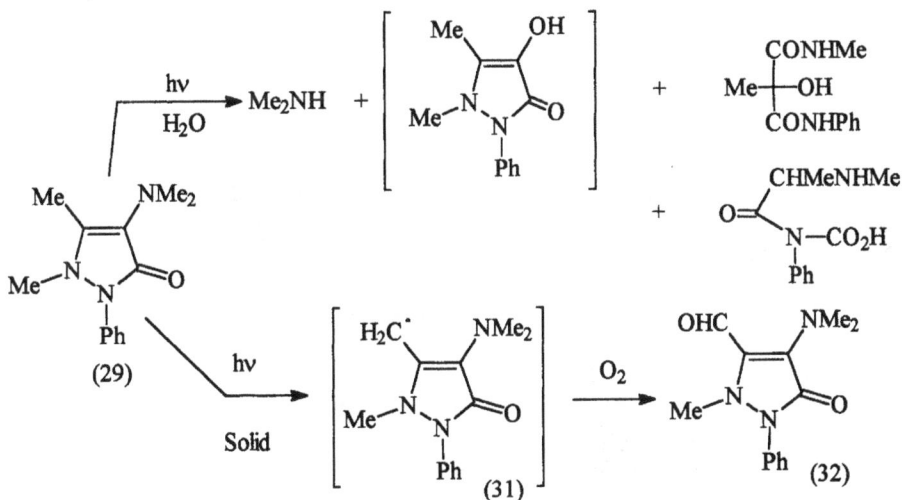

Scheme 13

Comparative studies in aqueous solutions[47] showed that aminopyrine is the most reactive derivative,[45] and in general 4-amino substituted pyrazolones react faster than 4-alkyl[48, 49] derivatives. In the presence of oxygen, photolysis is accompanied by a photo-oxidation reaction.[50] The above order of photoreactivity for pyrazolones remains the same in the solid state.[51] However, in the latter case different processes may be involved, as with aminopyrine for which the main reaction in the solid is type I (i.e. involving addition of ground state oxygen to a radical) photo-oxidation of the methyl group in position 5. This has been attributed to the small distance between the methyl group of one molecule and the carbonyl group of a neighbouring molecule in the lattice. This makes hydrogen abstraction easy and the resulting radical (31) adds oxygen to finally yield (32) (Scheme 13).[52]

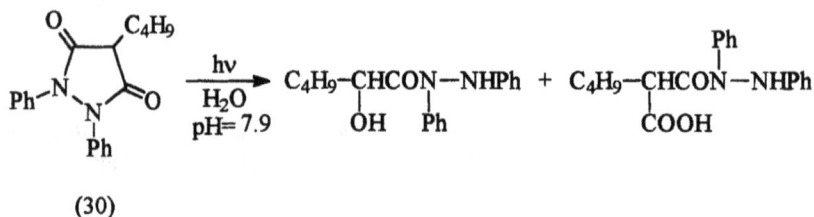

(30)

Scheme 14

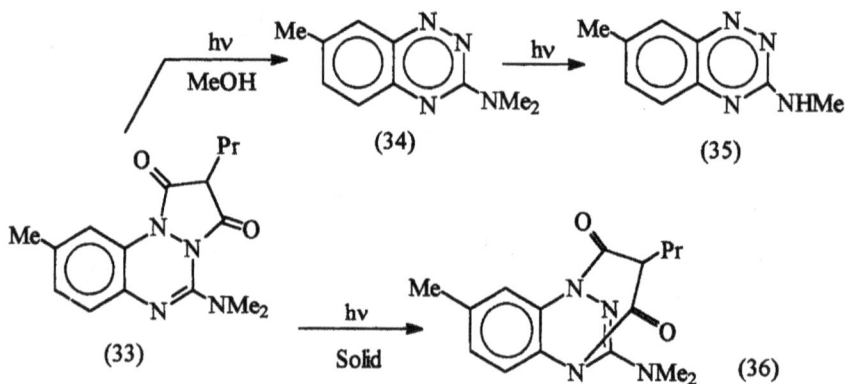

(34)

(35)

(33)

(36)

Scheme 15

Azapropazone (apazone, 33), a fused pyrazole derivative, undergoes cleavage of the five-membered ring to give a benzotriazine (34) by irradiation in methanol; the product then undergoes *N*-dealkylation to give (35). In the solid state a different process, 1,3-sigmatropic migration of one of the acyl groups, occurs and leads to isomeric (36) (Scheme 15).[53]

2.1.3   Immunosuppresant and Anti-histaminic Drugs. The immunosuppressant drug azathioprine (37) undergoes fragmentation of the C-S bond to give 6-mercaptopurine (38) and 1-methyl-4-nitro-5-hydroxyimidazole (39) as well as a cyclisation reaction   suggested to give (40) (Scheme 16).[54] Among drugs with anti-histaminic action, terfenadine (41) undergoes oxidation (main process) and dehydration at the benzylic position, to give products (42) and (43) respectively, upon irradiation in aqueous solution (Scheme 17),[55] and diphenylhydramine (44) suffers progressive *N*-dealkylation.[56] The thiazine derivative promethazine (45) is *N*-dealkylated to phenothiazine (46) and this in turn oxidised to the sulfone (47) and to 3*H*-phenothiazin-3-one (48) (Scheme 18).[57]

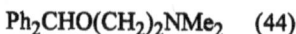

$Ph_2CHO(CH_2)_2NMe_2$ (44)

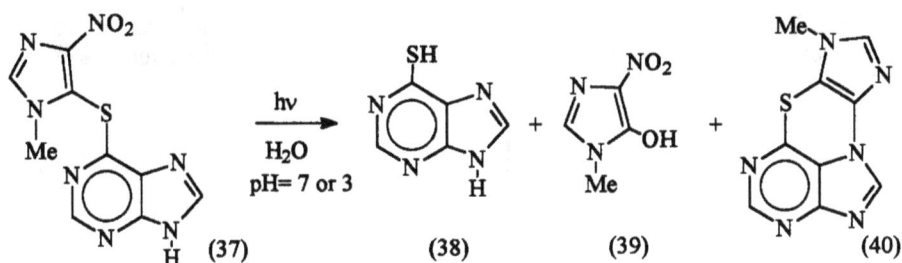

Scheme 16

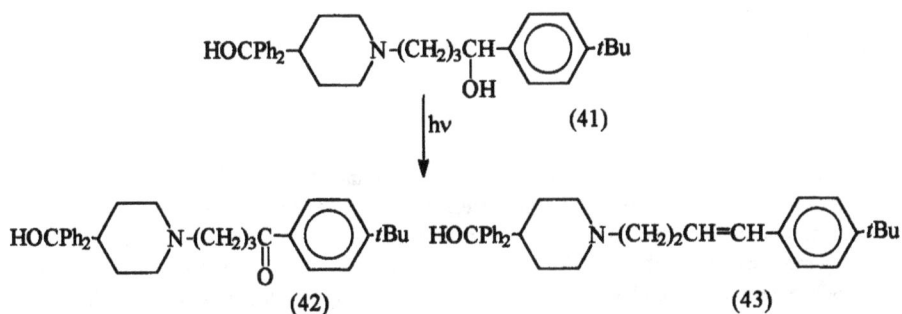

Scheme 17

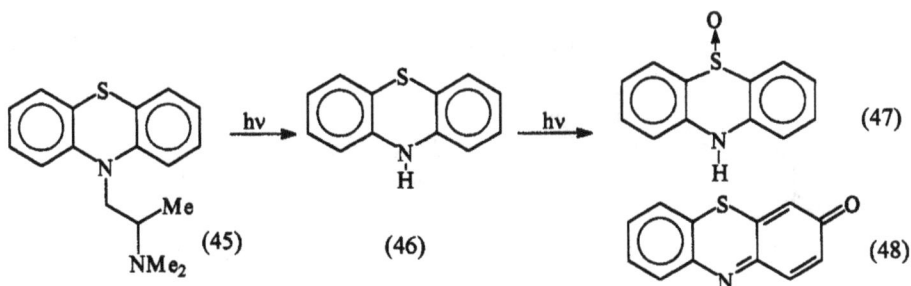

Scheme 18

2.1.4 Glucocorticosteroids. These products show, among other effects, an important anti-inflammatory activity. It is known that they have to be protected from light, and their photoreactivity has been explored both in solution and in the solid state. Hydrocortisone (49), cortisone (50) and their acetates (51, 52) undergo photo-oxidation in the solid state; the main process involves loss of the side-chain at C(17) to give androstendione and trione derivatives respectively (Scheme 19).[58] Molecular packing has an important role in determining the photostability in the solid. As an example, irradiation of crystalline hydrocortisone *tert*-butylacetate leads to photo-oxidation (to give the corresponding cortisone) for two out of five of the polymorphs investigated, while the other ones are photostable. This fact has been correlated with the possibility of oxygen to penetrate in the crystal in such structures.[59]

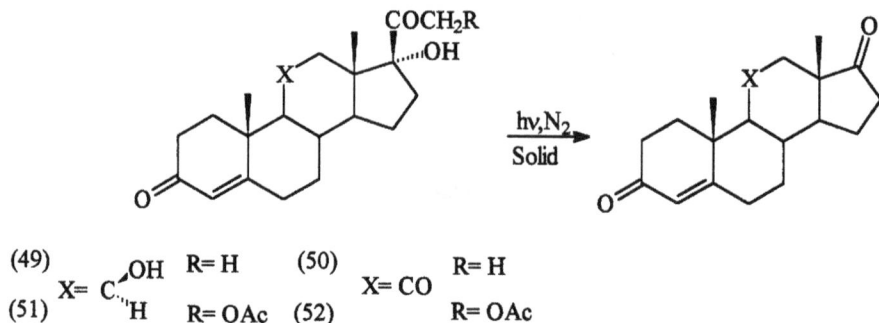

(49)   X= C$\overset{OH}{\underset{H}{\diagup}}$ R= H   (50)   X= CO   R= H

(51)   R= OAc   (52)   R= OAc

Scheme 19

Cross-conjugated glucocorticosteroids such as prednisolone, prednisone, bethmetasone, triamcinolone and others are quite photoreactive, as one may expect since the efficient photorearrangement of cyclohexadienones to bicyclo[3.1.0]hexenones is well known.[60] This rearrangement has been observed for prednisone (53),[61] prednisolone (54),[62] dexamethasone (55),[63, 64] betamethasone (56) and some of their acetates (see Scheme 20).[65, 66] The primary photoproducts may undergo further transformation with cleavage of the three-membered ring, resulting in rearomatisation or cleavage of ring A or in expansion of ring B according to conditions.

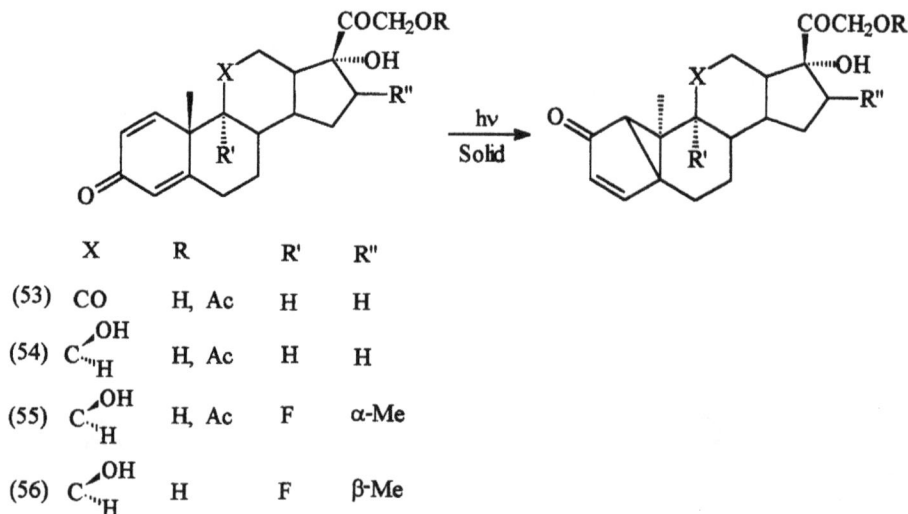

	X	R	R'	R"
(53)	CO	H, Ac	H	H
(54)	C$\overset{OH}{\underset{H}{\diagup}}$	H, Ac	H	H
(55)	C$\overset{OH}{\underset{H}{\diagup}}$	H, Ac	F	α-Me
(56)	C$\overset{OH}{\underset{H}{\diagup}}$	H	F	β-Me

Scheme 20

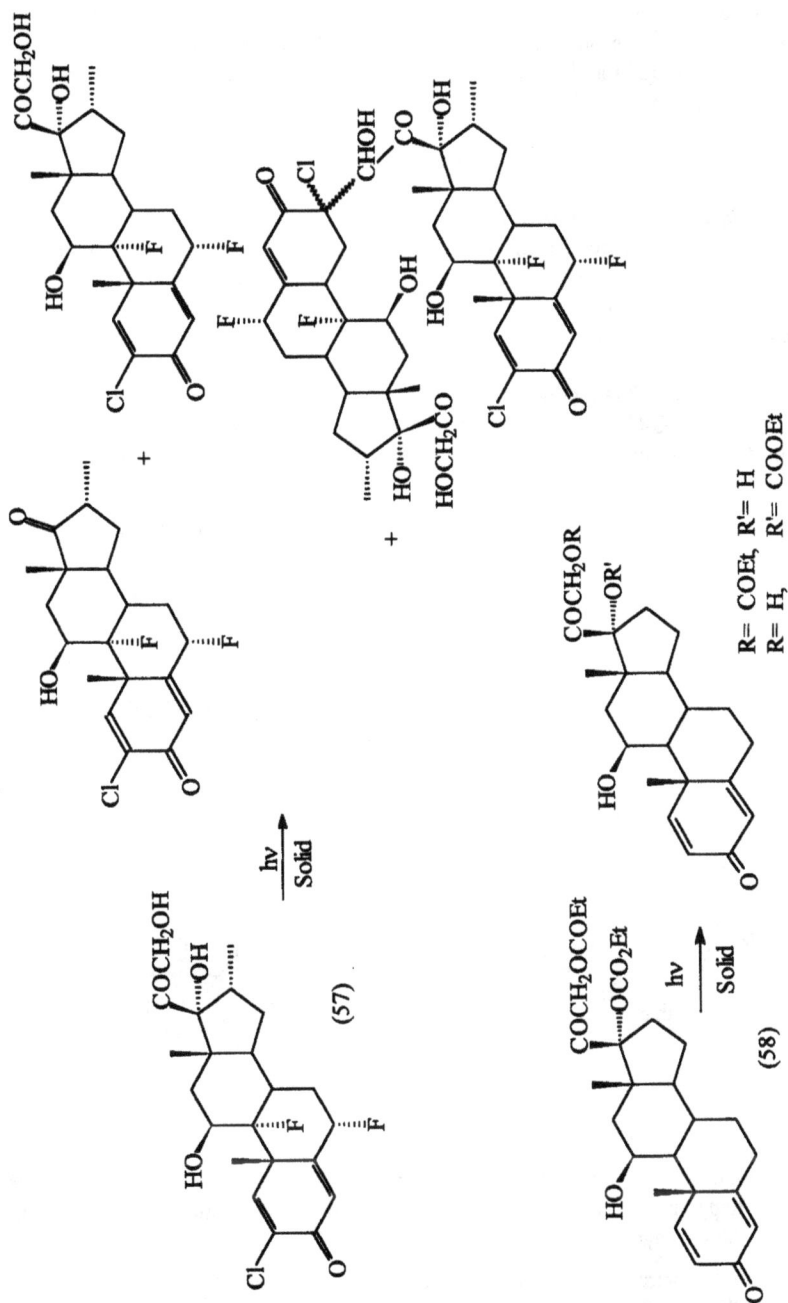

Scheme 21

   The above derivatives and some further ones have been found to be photoreactive also in the solid state.[63, 67-70] In this case the reaction may take a different course, however. Thus, with halomethasone (57) and prednicarbate (58) the observed processes involve the C(17) side chain (see Scheme 21).[70]

2.2 Drugs Acting on the Central Nervous System

   2.2.1 Barbituric acid derivatives.   5,5-Dialkyl derivatives of barbituric acid (59) are used as hypnotics and tranquillisers. These compounds undergo two types of photochemical processes (Scheme 22). The first one involves initial cleavage of the C(4)-C(5) bond and leads to an isocyanate, positive evidence for which has been obtained by irradiation in the solid state (path a).[71] This intermediate in turn adds nucleophilic solvents to give an amide (60) in water and a urethane in ethanol (61). This reaction is observed for barbital (59a, R, R'=Et, R"=H) and its *N*(3)-methyl derivative (59b, R, R'=Et, R"=Me).[72] In a variation of this process, a second C-C bond is cleaved and CO is eliminated (path b). This leads to a hydantoin (62), as it occurs with mephobarbital (59c, R=Et, R'=Me, R"=H).[73, 74]

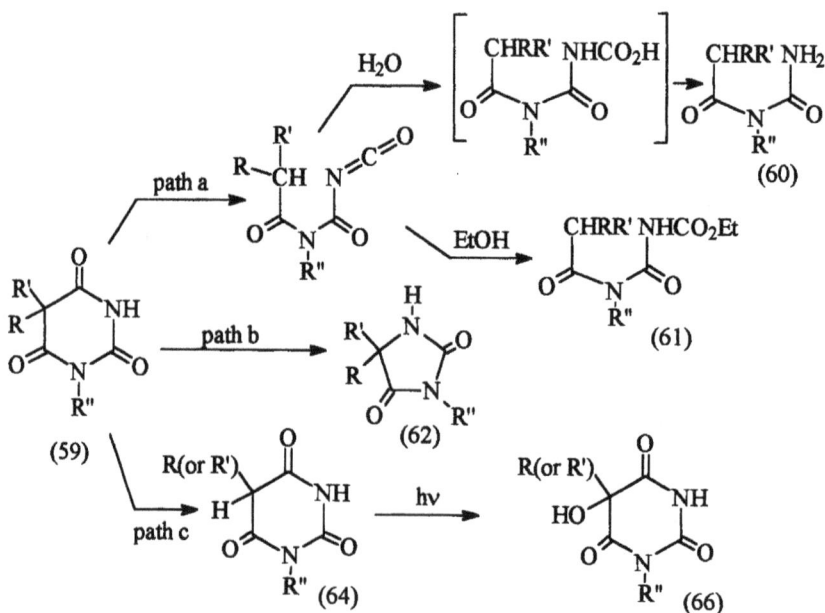

Scheme 22

   A nucleophilic group in the main side chain intervenes in the process via intramolecular addition, as happens with the tranquilliser proxibarbital (59d, R=allyl, R'=2-hydroxypropyl) which gives the tetrahydrofuranone (63) (Scheme 23).[75]

   The second general reaction is dealkylation to give products (64) (path c), and this is the main path followed when one of the substituents is a stabilised alkyl group (secondary or allylic). This is the case with pentobarbital (59e, R=Et, R'=2-pentyl, R"=H)[76] and secobarbital (59f, R=allyl, R'=2-pentyl, R"=H) (Scheme 22).[77]

This general scheme holds for the monoanion (the predominant species at pH 10), and the acidity of the medium affects to some extent the product distribution.[75, 76, 78, 79] A different process occurs for the acidic form of cyclobarbital (59g, R=Et, R'=1-cyclohexyl, R"=H),[80] which is photo-oxidised to the ketone (65) rather than cleaved (Scheme 23). Monoalkylbarbiturates (64) undergo hydroxylation at position 5 to give products (66) (see Scheme 22).[76] The 2-thio analogue of phenobarbital (67) gives (68) by selective reduction of the thiocarbonyl function by irradiation in alcohols (Scheme 24).[81]

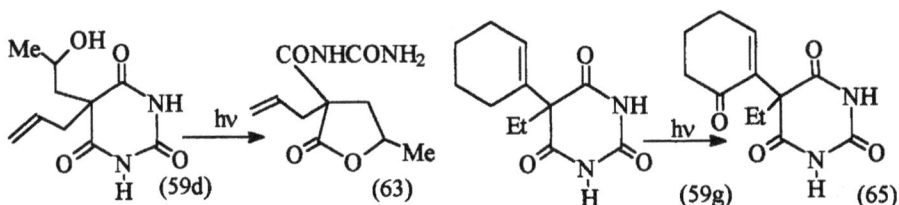

Scheme 23

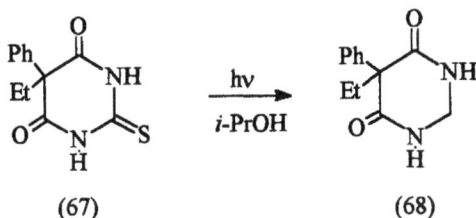

Scheme 24

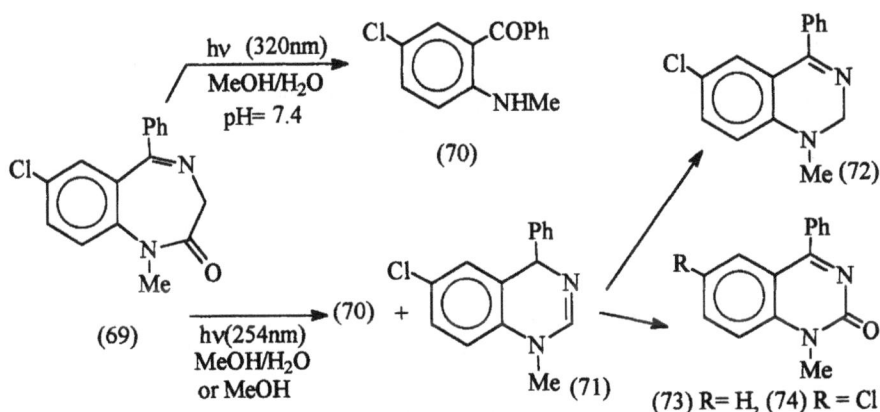

Scheme 25

2.2.2   Benzodiazepines.   Benzodiazepines are generally photolabile, but the path followed in the degradation depends on the structure of each derivative and on conditions. Thus, diazepam (69) undergoes cleavage of the heterocyclic ring, the main products being the benzophenone (70) by irradiation at 320 mn in MeOH-H$_2$O, and the dihydroquinazoline (71) by irradiation at 254 nm in MeOH (Scheme 25). The latter compound then suffers slow isomerisation to (72) as well as dechlorination and oxidation to (73) and (74).[82]   The intravenous anaesthesic midazolam (75) undergoes ring restriction to a quinazoline (76) (main path with solar light) as well as oxidation of the 5-fluorophenyl moiety to give (77) and then (78) (main path with artificial light, Scheme 26).[83, 84]

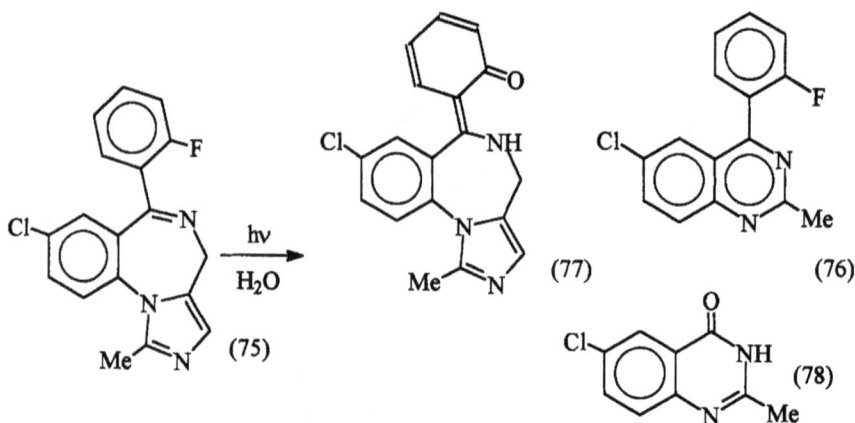

Scheme 26

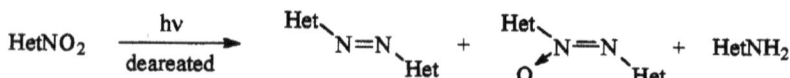

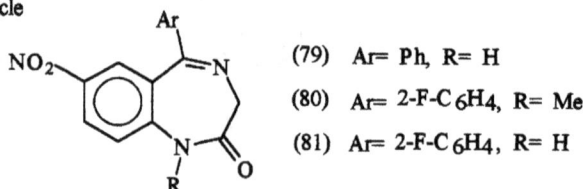

(79)   Ar= Ph, R= H

(80)   Ar= 2-F-C$_6$H$_4$, R= Me

(81)   Ar= 2-F-C$_6$H$_4$, R= H

Scheme 27

Insertion of a nitro substituent, as in the anticonvulsant nitrazepam (79), changes the photochemistry, that is now dominated by reaction at that group (compare Scheme 1, b). As expected, this abstracts hydrogen and is reduced to the azoxy, azo, and amino functions by irradiation in organic solvents under nitrogen (Scheme 27).[85] In the presence of oxygen the molecule is rather photostable, but produces singlet oxygen.[86] Analogously, the hypnotic flunitrazepam (80) undergoes a multistep reduction finally leading to the 7-amino derivative under anaerobic conditions,[87, 88] while it is *N*-demethylated to give (81) in the presence of oxygen (Scheme 27).[87]

When a *N*-oxide function is present, as in the sedative chlordiazepoxide (82), this is the photochemically active moiety. As it is usual with nitrones,[89] rearrangement to the oxaziridine (83) takes place (compare Scheme 1, c), and this compound further reacts to give compounds (84) and (85) through ring contraction and ring expansion respectively (Scheme 28).[90, 91] In the presence of a reducing agent such as glutathione (GSH) the main process is *N*-deoxygenation to (86), occurring in part from the oxaziridine, in part directly from the excited state of the *N*-oxide.[92] Irradiation in the solid state leads again to the quinazoline (84), but a ring-opened imine (87) and a diazepinone (88) are also obtained, independent of whether the free base or the hydrochloride are irradiated (Scheme 28).[93]

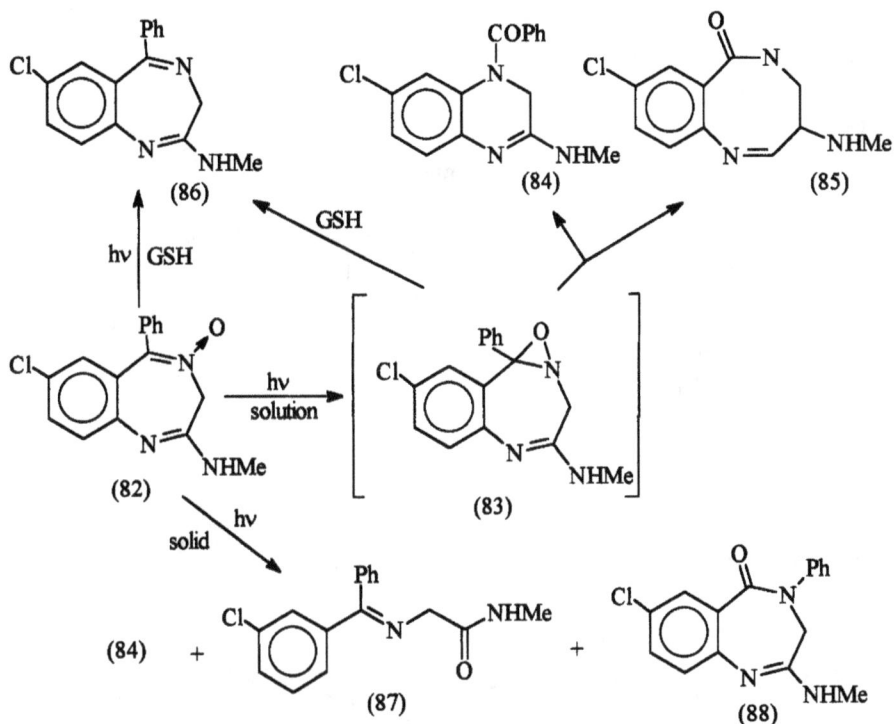

Scheme 28

2.2.3 Thioxanthine and Phenothiazine Psychotherapeutic Agents.

Many derivatives of these two heterocycles are largely used in therapy, mainly as antipsychotic agents. The alkylenethioxanthines flupenthixol (89), clopenthixol (90), thiotixene (91) and chlorprothixene (92) are used as *E/Z* isomeric mixtures. The *E/Z* ratio changes upon irradiation (>300 nm, see

Scheme 29) and, as it has been pointed out, this may affect the potency of such drugs.[94] Furthermore, a slow oxidation to thioxanthone derivatives (93) occurs.[94]

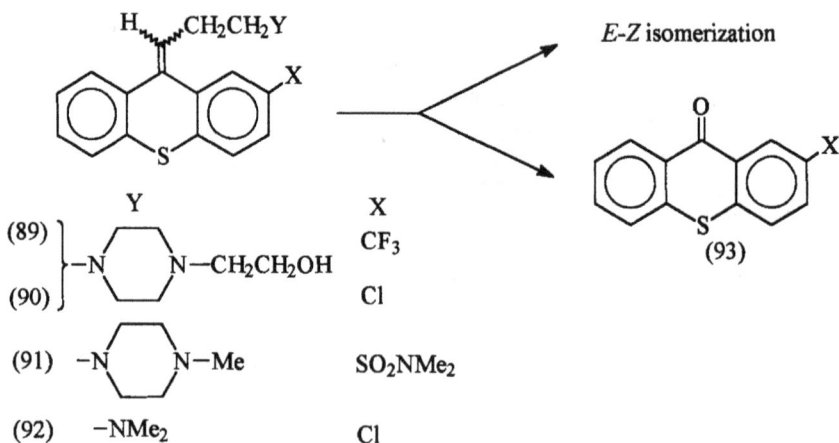

E-Z isomerization

	Y	X
(89)		CF$_3$
(90)	$-$N$\diagup$N$-$CH$_2$CH$_2$OH	Cl
(91)	$-$N$\diagup$N$-$Me	SO$_2$NMe$_2$
(92)	$-$NMe$_2$	Cl

Scheme 29

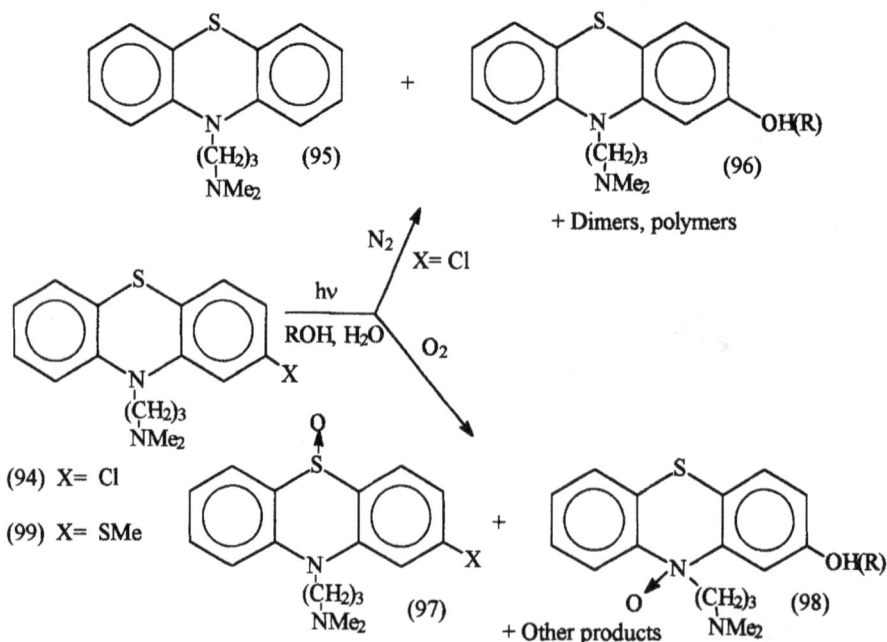

+ Dimers, polymers

(94) X= Cl

(99) X= SMe

+ Other products

Scheme 30

Among phenothiazines, chlorpromazine (94) has been extensively investigated (see Scheme 30). The main process under anaerobic conditions is dechlorination.[95-99] This process occurs through a radical pathway,[100-101] and causes reduction to (95), substitution to give (96) as well as dimerisation and/or polymerisation. In the presence of oxygen, oxidation to the sulfoxide (97) and the *N*-oxide (98) takes place.[102-104] It has been proved that singlet oxygen is formed under these conditions.[105] Thioridazine (99) is likewise *S* and *N*-oxidised (as well as *N*-dealkylated) upon irradiation in aqueous solution.[106] Several other phenothiazines of this type have been investigated.[107] The relative photoreactivity has been found to depend on the ring substituent X and its liability to substitution.[108] B. D. Glass has carried out an in depth study of various phenothiazines, and has established a photoreactivity versus activity relationship (p. 134).

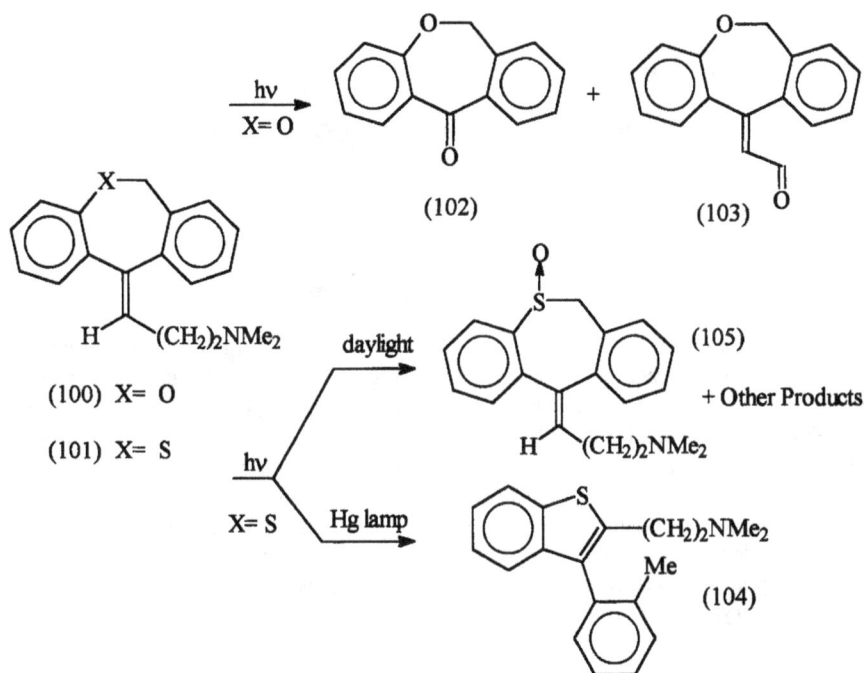

Scheme 31

2.2.4   Other Psychotropic Agents.   The tricyclic antidepressants doxepin (100) and dothiepin (101) are photodegraded upon exposure to natural, and more rapidly, to artificial light. Besides *E/Z* isomerisation, the processes occurring for the former drug are oxidation of the side-chain to give (102) and (103).[109] For the latter compound either cleavage and rearrangement to give (104) or *S*-oxidation to (105) are observed according to conditions (see Scheme 31).[110] The related dibenzocycloheptene protriptyline, also an antidepressant, reacts at the double bond C_{10}-C_{11}: the epoxide is formed by irradiation of the hydrochloride in water in the presence of oxygen,[111] and a [2+2] dimer by irradiation in water or ethanol under nitrogen.[112]

The solid-state photodegradation of carbamazepine (106), an anticonvulsant drug, has been found to depend on the polymorf considered.[113] Another anticonvulsant, phenyltoin (107), has been found to undergo cleavage of the heterocyclic ring upon irradiation in methanol, and to give benzophenone (108) and benzyl (109) (Scheme 32).[114]

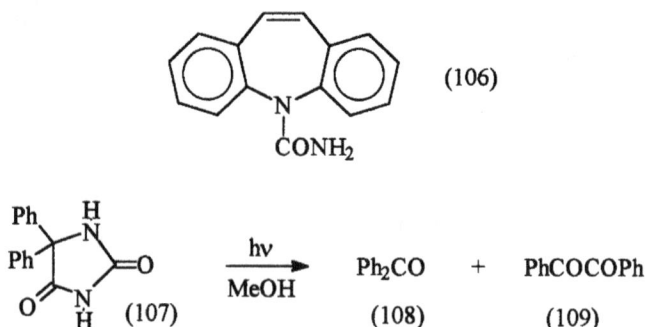

(106)

$$\text{(107)} \xrightarrow[\text{MeOH}]{h\nu} \text{Ph}_2\text{CO} + \text{PhCOCOPh}$$

(107)   (108)   (109)

Scheme 32

2.3 Cardiovascular, Diuretic and Hemotherapeutic Drugs

2.3.1 Cardiac Agents. The cardiotonic digitoxin (110) has been found to give (111) through intramolecular addition of the 14-hydroxy group to the conjugated double bond by irradiation in the solid state (Scheme 33).[115] Ubidecarenone (112) has been found to be photolabile in the solid-state.[116] (Z) 2-amino-5-chlorobenzophenoneamidinohydrazone acetate (113), an antiarrhythmically active compound, undergoes geometrical isomerisation and, more slowly, decomposition when exposed to light in aqueous solution.[117] Antianginal and antiarrhythmic amiodarone (114) (Scheme 34) was found to deiodinate sequentially upon irradiation in deaerated ethanol to give (115) and finally the iodine-free ketone (116); in turn, the last compound underwent α-cleavage to give (117) when irradiated for a longer time.[118] Formation or aryl radicals during the deiodination process was evidenced by a spin trapping study.[119]

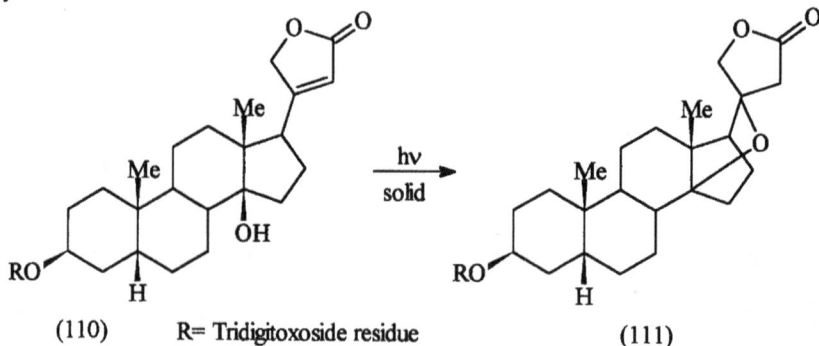

(110)   R= Tridigitoxoside residue   (111)

Scheme 33

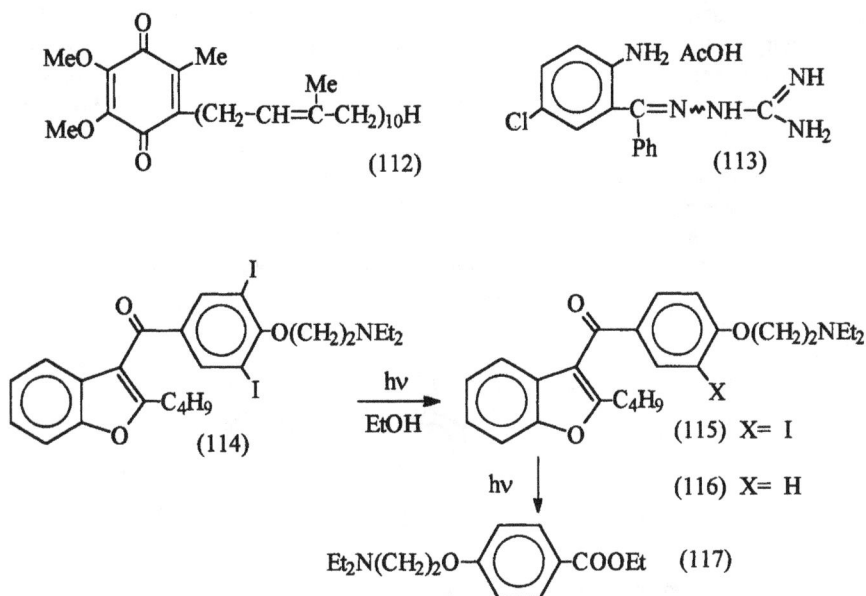

(112)

(113)

Scheme 34

2.3.2 Blood pressure regulating drugs. The largely used 4-nitrophenyldihydropyridine vasodilators are quite photosensitive, as one may expect due to the copresence of the hydrogen abstracting nitro group and an easily abstractable benzylic hydrogen. Aromatisation of the heterocyclic ring and reduction of the nitro to the nitroso group result. In the presence of oxygen the nitroso function is in turn re-oxidised (see Scheme 35). Thus nifedipine (118a, R, R'= Me) gives the corresponding nitrosophenylpyridine (119a), in part oxidised to the nitro derivative (120).[120-122] Irradiation under oxygen gives directly the latter compound.[123] The reaction occurs also in the solid state,[124, 125] and in this case clean formation of the nitroso derivative takes place (> 95%).[126, 127]

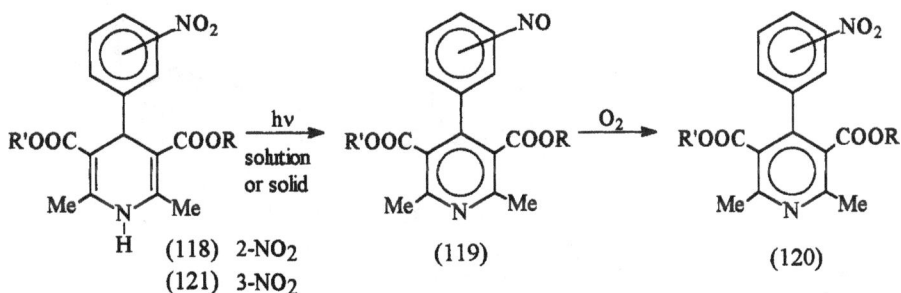

Scheme 35

Other 4-(2-nitrophenyl)dihydropyridines such as furnidipine (118b, R, R'= Et, 2-tetrahydrofurylmethyl)[128] react similarly, and the same holds for 4-(3-nitrophenyl) derivatives such as nitrendipine (121a, R, R'= Me, Et),[129] nimodipine (121b, R, R'= *iso*-propyl, 2-methoxyethyltetrahydrofurylmethyl)[130] and nicardipine [121c, R=Me, R'=(*N*-methyl-*N*-benzyl)-2-aminoethyl].[131]

Several other vasodilators have been found to be photolabile. These include diltiazem (122), which is deacetylated by UV irradiation in aqueous solution both at pH 2 and at pH 7, but rather stable in the solid state[132] and the pyrimidine *N*-oxide minoxidil (123), photoreactive in ethanol-water solution, but again stable in the solid state.[133]

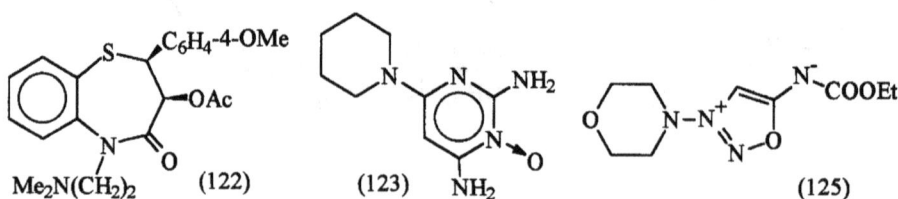

Me$_2$N(CH$_2$)$_2$ (122)   (123)   NH$_2$ (125)

Reserpine (124) photoreacts both in aqueous solution and in chloroform, and the processes occurring are epimerisation at C-3 and stepwise dehydrogenation of the tetrahydro β-carboline skeleton (Scheme 36).[134, 135] The coronary vasodilator molsidomine (*N*-ethoxycarbonyl-3-morpholinosydnonimine, 125) is photochemically fragmented, as is the rule with sydnone derivatives.[136, 137]

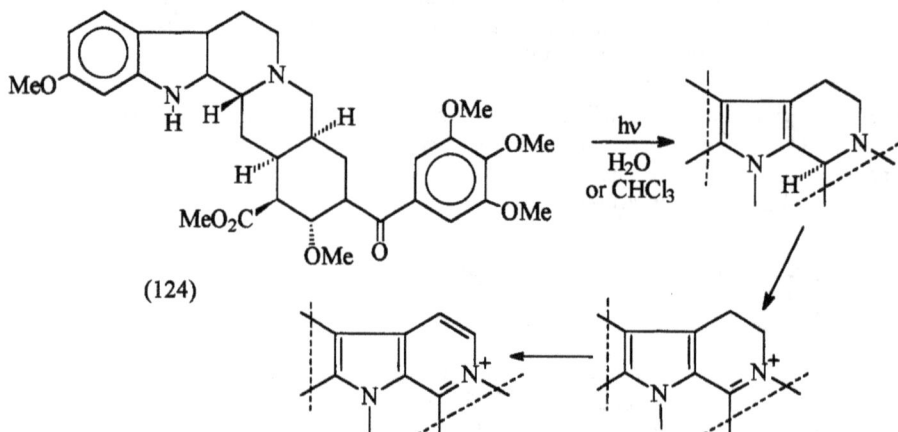

(124)

Scheme 36

Sodium nitroprusside, also used as an antihypertensive, is degraded when exposed to ambient light in solution. The process involves aquation of the iron complex with liberation of nitrite anion.[138, 139]

$$[Fe(NO)(CN)_5]^{2-} + 2\,H_2O \quad \rightarrow \quad [Fe(H_2O)(CN)_5]^{3-} + NO_2^- + 2\,H^+$$

Ergotamine (126), used for its vasoconstrictor action, undergoes hydration of the 9,10 double bond when the tartrate is UV irradiated in aqueous solution under nitrogen (Scheme 37).[140] 9,10-Dihydroergotamine (127), on the other hand, is oxidised at position 2 when an aqueous solution of the methanesulfonate is exposed to sunlight (Scheme 37).[141]

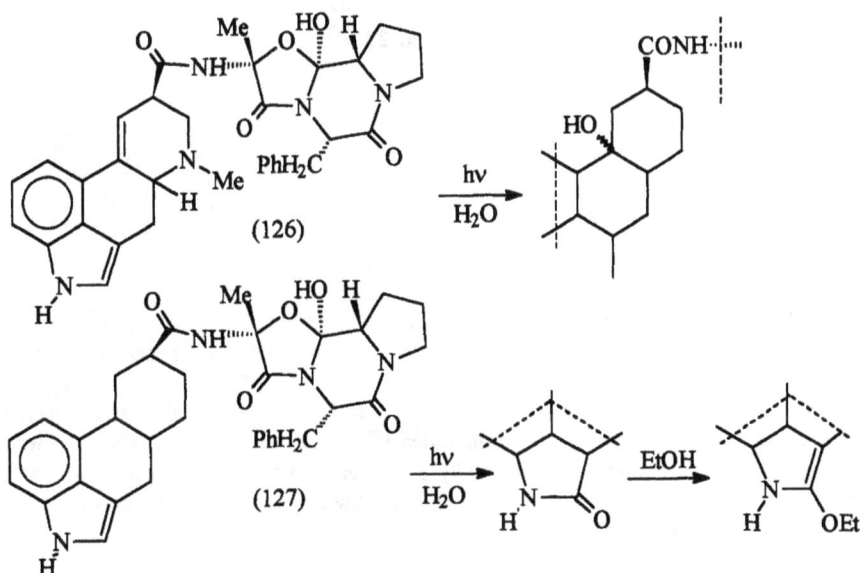

Scheme 37

2.3.3 Adrenergics. Adrenaline (128) and isoprenaline (129) are cyclised to the corresponding (highly coloured) aminochromes (130) upon irradiation in aqueous solution (Scheme 38).[142] The role of superoxide, singlet oxygen,[143, 144] and semiquinone radicals[145] in this reaction has been discussed. Ephedrine (131) is cleaved to benzaldehyde (132); that reacts with the starting material to give oxazolidine (133) (Scheme 39).[146]

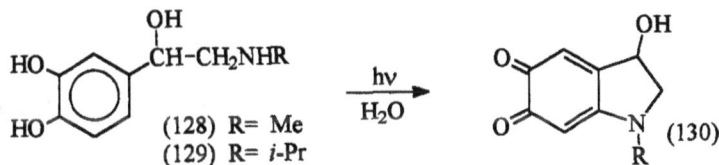

Scheme 38

2.3.4 Diuretic Drugs. Commonly used diuretics are aromatic or heteroaromatic compounds, in many cases chlorinated and bearing a sulphonamide function. Reductive dechlorination is often a major path, as shown in Scheme 40 for trichlormethiazide (134) which gives (135).[147] A detailed discussion of the photochemistry of diuretic drugs by D. E. Moore is presented in this book (p. 100).

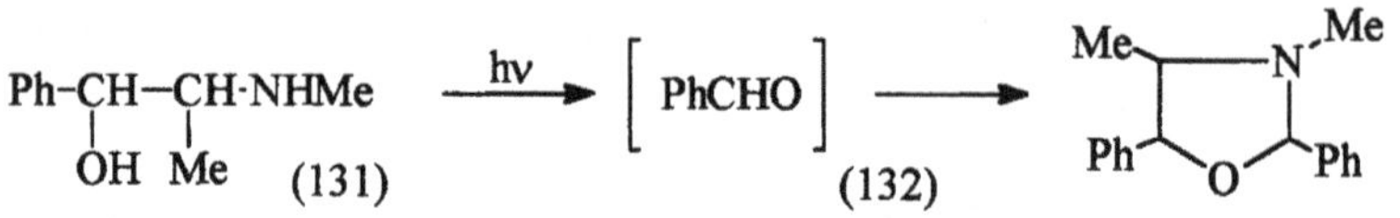

(131) (132) (133)

Scheme 39

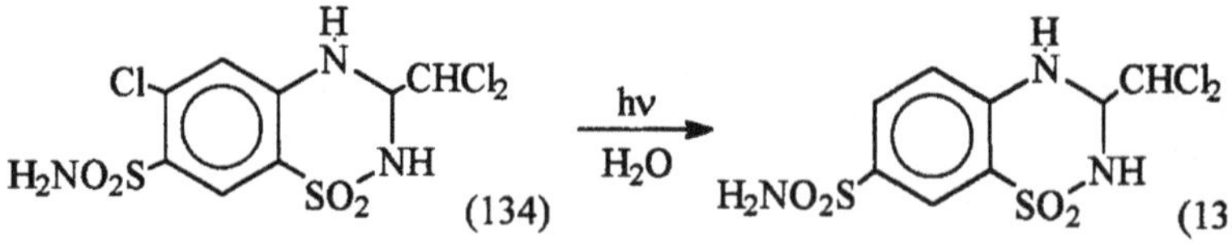

(134) (135)

Scheme 40

2.3.5 Serum Lipid Reducing Agents and Antithrombotic Agents. The antihyperlipoproteinemic drug fenofibrate (136) is photolabile under aerobic conditions, yielding peracid (137) among other compounds (scheme 41).[148] Several other drugs of this group [gemfibrozil (138), bezafibrate (139), clofibric acid (140) and clofibrate (141)] have also been found to be oxidatively photodegraded.[149] Thyroxine (142), the D form of which is used as antihyperlipoproteinemic, is rapidly photodeiodinated. Initial reaction involves the iodine atom *ortho* to the phenoxy function, and then the reaction proceeds stepwise to give 3-iodothyronine, the fourth iodine being more resistant to elimination.[150]

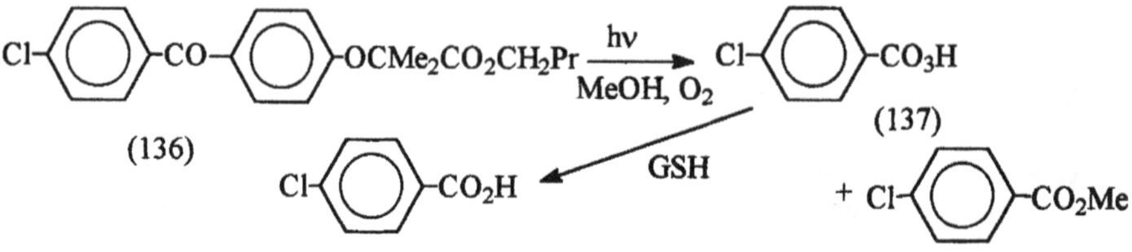

(136) (137)

Scheme 41

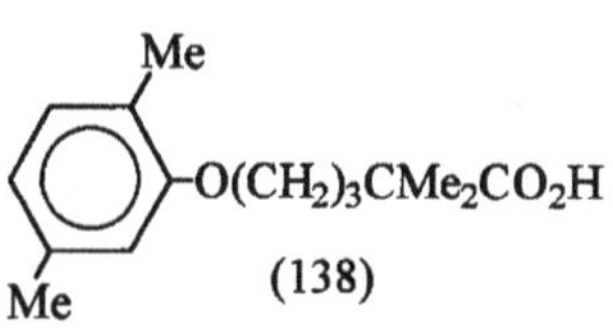

(138)

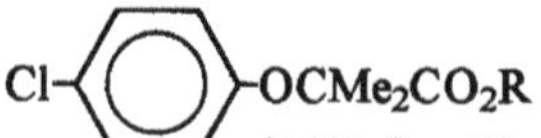

(140) R= H
(141) R= Et

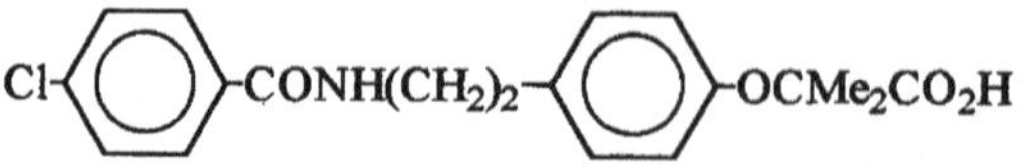

(139)

The anticoagulant dipyridamole (143), a pteridine derivative, undergoes oxidation of one of the piperidine side chains giving (144) upon irradiation in solution.[151] Another anticoagulant, warfarin sodium (145), has been found to be photostable in the solid state.[152] The photochemistry of vitamins K is discussed in Sec. 2.7.1.

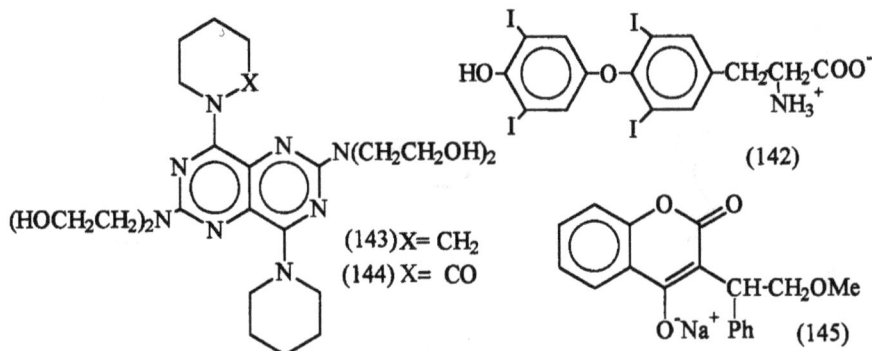

(143) X= CH$_2$
(144) X= CO

(142)

(145)

2.4 Gonadotropic Steroids and Synthetic Estrogens

When irradiated in buffered (pH 7.4) aqueous solution the progestinic norethisterone (146a, R=Me, R'=H) (Scheme 42) is photo-oxidised to give the 4,5-epoxide (147a) (β,β diastereoisomer predominating) and the 5-hydroxy (148a) (α epimer predominating) derivatives as well as photo-dimerised.[153-155] The phenolic ring present in estrogens makes these substrates quite labile to photo-oxidation. The reaction can be conveniently carried out by photosensitisation, under conditions where singlet oxygen is produced avoiding direct irradiation of the substrate. Thus, the photosensitised reaction of ethinyl estradiol (149) in solution involves addition of singlet oxygen to the electron rich phenolic ring to yield a the ketohydroperoxide (150) (Scheme 43).[156] Estrone (151) reacts similarly (Scheme 43).[157] Δ$^{9(11)}$-dehydroestrone (152) gives complex mixture of products upon photosensitised oxygenation, while the corresponding methyl ether (153) undergoes clean cleavage of the C ring to give the ketoaldehyde (154) (Scheme 44).[158, 159]

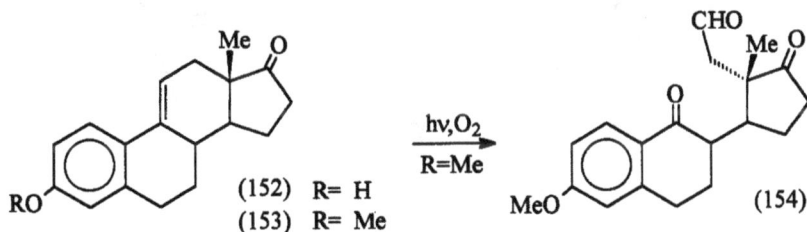

$$\xrightarrow[R=Me]{hv, O_2}$$

(152) R= H
(153) R= Me

(154)

Scheme 44

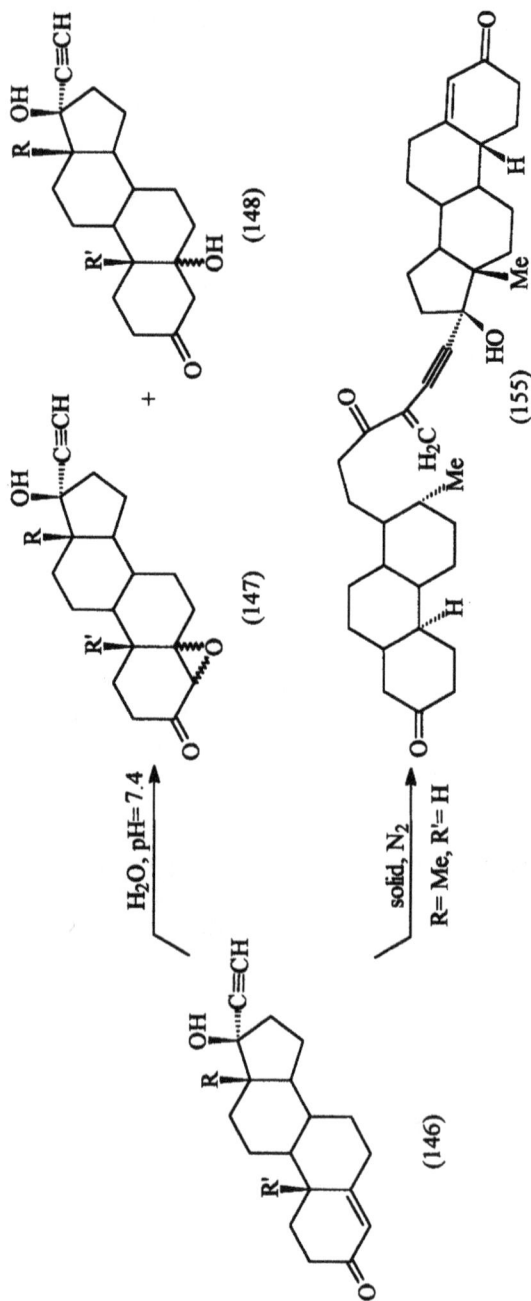

Scheme 42

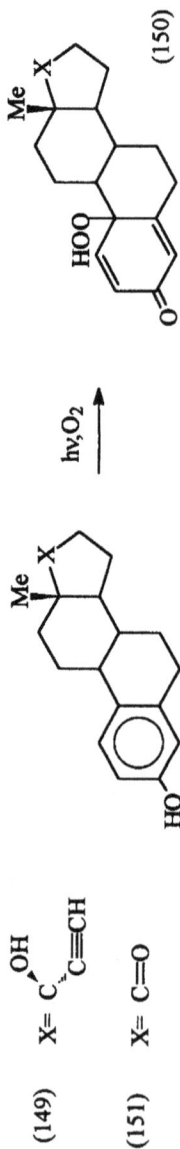

Scheme 43

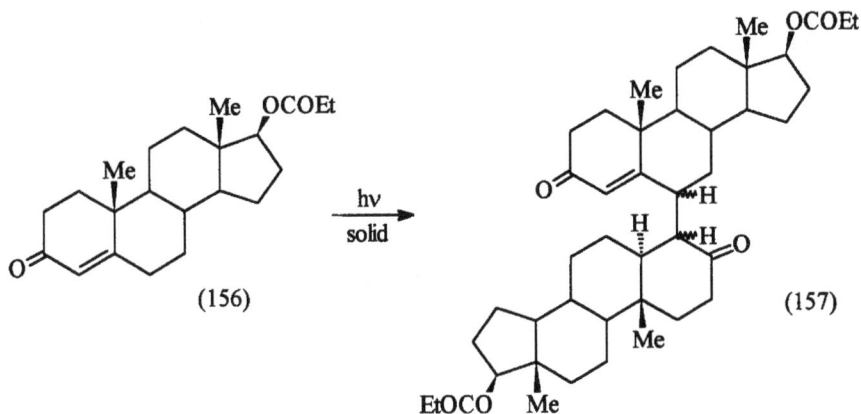

Scheme 45

In the solid state different paths are observed, and can be rationalised with reference to the crystal packing. Thus, levonorgestrel[160] (146b, R=Et, R'=H) as well as ethisterone (146c, R, R'=Me) and norethisterone (146a)[161] undergo photo-dimerisation in the solid state under nitrogen; compound (155) is formed from (146a) (Scheme 42). Crystalline lynestrenol is also photoreactive, while mestranol and progesterone have shown good photostability.[162] Testosterone propionate (156) gives dimer (157) when irradiated in the solid state (Scheme 45).[163] Testosterone (158) disporportionates to androstenedione (159) and androstanedione (160) and 17-methyltestosterone (161) give the seco derivative (162) under that condition (Scheme 46).[164]

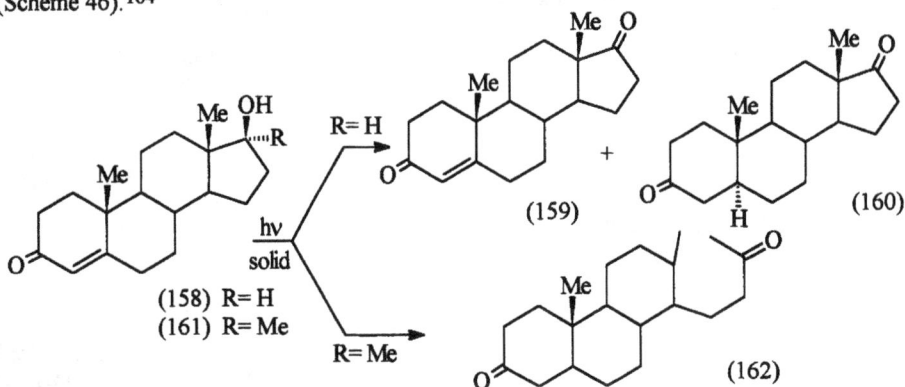

Scheme 46

A number of stilbene derivatives are used as synthetic estrogens. The photochemistry of stilbenes has been investigated in depth. *E-Z* photo-isomerisation is always a fast process and the *Z* isomer undergoes conrotatory electrocyclic ring closure to give a *trans* fused dihydrophenanthrene; in the presence of various oxidants (often atmospheric oxygen is sufficient) this intermediate is then aromatised.[165] As expected, clomiphene (163a, R=Cl) isomerises when irradiated in chloroform, and each isomer cyclises to the corresponding phenanthrene, (164) and (165) respectively (Scheme 47).[166]

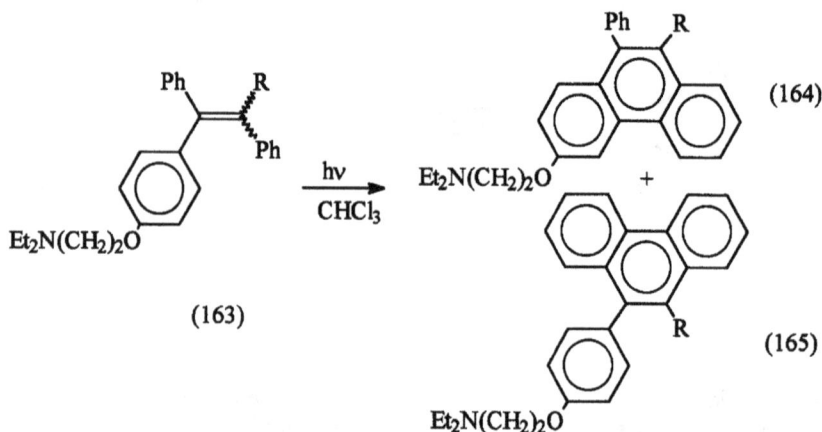

Scheme 47

Cyclisation is observed also with tomoxifen (163b, R=Et).[167] Dihydroxysubstituted stilbenes such as stilboestrol (166) and dienoestrol (167) are likewise cyclised (Schemes 48 and 49). From the former drug either a tetrahydrophenanthrenedione (168) or an aromatised dihydroxyphenanthrene (169) have been obtained under different conditions. From the latter one, hydrogen sigmatropic shift, isomerisation and cyclisation finally give dione (170).[168-170]

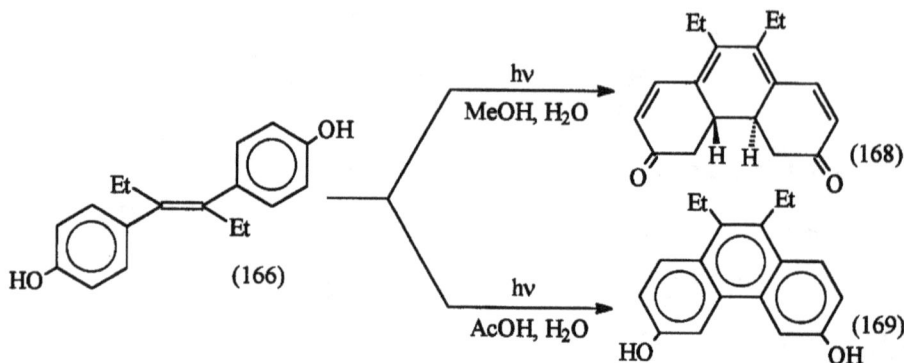

Scheme 48

2.5 Dermatologicals

Photostability is a particularly significant characteristic for drugs for external use. Some indications about the photostability of drugs in ointments and eye-drops are given in Sec. 3.

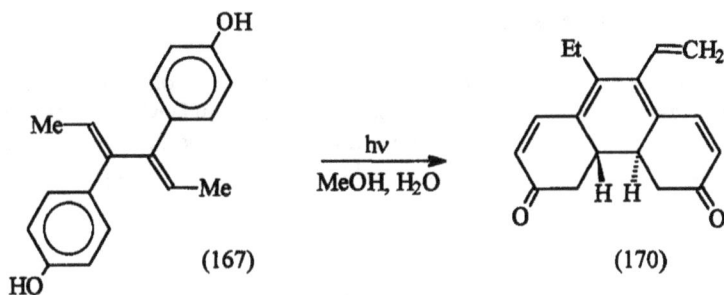

Scheme 49

As for drugs for skin pathologies, it is known that the antipsoriatic dihydroxyanthrone dithranol (171) undergoes self-sensitised oxidation to the corresponding anthraquinone (172) and dianthrone (173) derivatives as well as polymerisation (Scheme 50).[171] The action of psoralenes (e.g. methoxsalen, 174), used in phototherapy for controlling the skin disorder vitiligo, is based on their capacity of binding covalently with DNA through a photochemical reaction and/or of acting as photosensitisers and causing a specific damage to the cell via singlet oxygen.[172, 173] Photostability is obviously important with these drugs which are used under irradiation. Their structure makes them liable to photodimerisation, however.

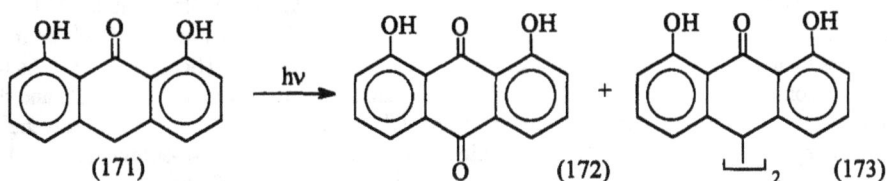

Scheme 50

Another important group of chemicals pertaining to this category is that of sunscreens. These are supposed to absorb UV light and degrade it to heat avoiding skin photosensitisation. The most largely used derivative is *p*-aminobenzoic acid (PABA, 175).[174] However, most of these compounds do show some photoreactivity and this requires careful evaluation before their use as active ingredients in sunscreens is accepted. The subject is discussed by J. M. Allen in this book (p. 171)

2.6 Chemotherapeutic Agents

2.6.1 Antibacterials: Sulfa Drugs. The photochemistry of sulfa drugs has been extensively investigated. The main process with N-substituted sulfanylamide derivatives (176) is cleavage of the S-N and C-S bonds, with extrusion of SO_2 and formation of aniline (177) and of the appropriate amine (178) (Scheme 51), although the yield of the photoproducts changes greatly with the structure.[175]

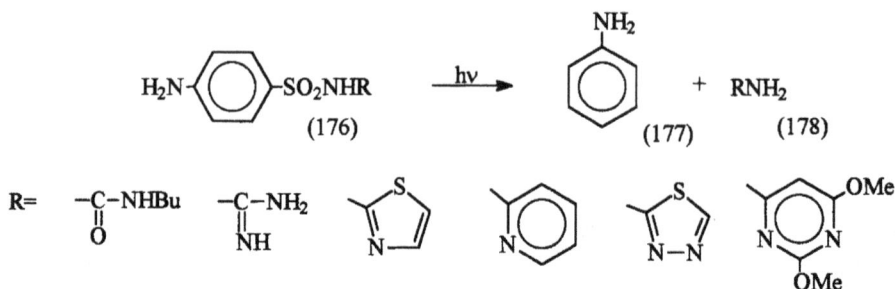

Scheme 51

Radicals are produced during the photolysis.[176] With N-[2-(5-methylisoxazolyl)] sulfanylamide, sulfamethoxazole (179), rearrangement to the corresponding 2-(5-methyloxazolyl) derivative (180) also takes place (Scheme 52).[177] Sulfacetamide (181) has been the subject of a photophysical study, and the triplet has been characterised;[178] the main process by irradiation in water is deacetylation to sulfanilamide (182) (Scheme 53).[179] As for the last compound, this undergoes oxidation of the amino group to give the azo (183) and the nitro (184) derivative when irradiated in water;[180] in ethanol oxidation of the solvent to acetaldehyde (revealed by the formation of 2-methyl-quinoline-6-sulfonamide, 185) accompanies the above process (Scheme 53).[181] Irradiation of the secondary amine sulfadimetoxine (176, R=2,6-dimethoxy-4-pyrimidinyl) in methanol causes methylation of the amino group.[182]

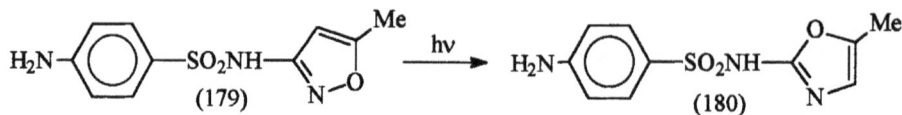

Scheme 52

2.6.2 Antibacterials and Antivirals: Aromatic Derivatives. Irradiation of aqueous solutions of tetracycline (186) under oxygen causes homolytic deamination followed by oxygen addition of the resulting radical to finally yield quinone (187) (Scheme 54).[183, 184] It has been shown that both singlet oxygen and superoxide anion are formed during irradiation of this and related compounds.[185] On the other hand, different photoprocesses occur with some substituted tetracyclines. Thus, the 7-chloro derivative of (186), chlortetracycline, undergoes homolytic dechlorination when irradiated in aqueous buffer at pH 7.4, while no such reaction occurs in demeclocycline, the 7-chloro derivative lacking the 6-methyl group.[186]

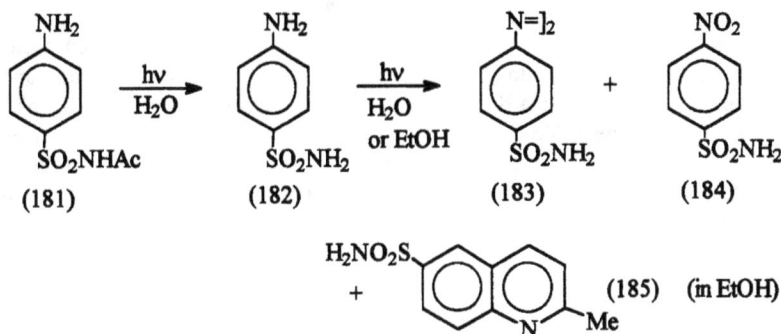

Scheme 53

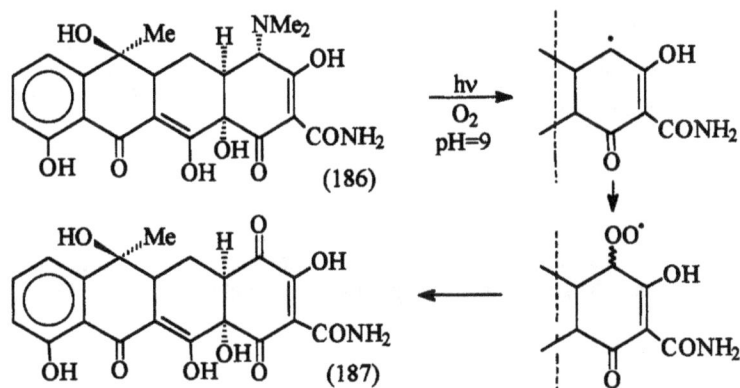

Scheme 54

Chloramphenicol (188) is well known for its instability under irradiation (also under solar light). The initial process in water is homolytic cleavage of the C-C bond and further evolution of the two radicals to give glycolic aldehyde (189), dichloroacetamide (190) and 4-nitrobenzaldehyde (191) (Scheme 55).[187]

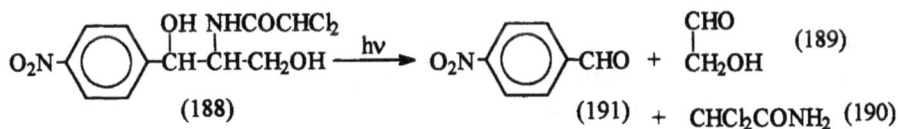

Scheme 55

Product (191) in turn undergoes secondary photoreactions involving the known redox chemistry of this compound, and leading to amino and nitrosobenzoic acid, the corresponding azoxy derivative, as well as to aminobenzaldehyde.[188-190]

Further aromatic derivatives for which photolability has been reported include the antihepatitis drug catechin (cyanidol, 192). This is photoreactive both in solution and in the solid state, in the latter case to a different degree for the different crystalline forms.[191, 192] The topic antibacterial hexachlorophene (193) is dechlorinated when irradiated in degassed ethanol, the faster reaction involving the chlorine atoms *ortho* or *para* to the hydroxy groups.[193] Likewise, tetrachlorosalicylanilide (194) is photodechlorinated preferentially from the position *ortho* to the phenolic function.[194]

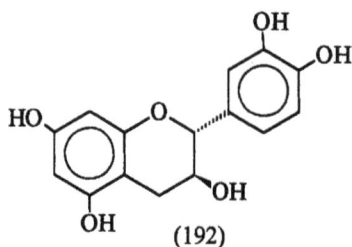

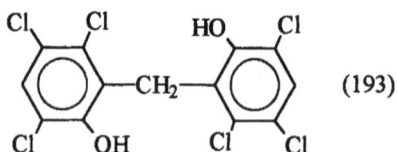

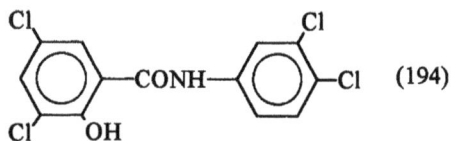

2.6.3 β-Lactam Antibiotics. In compounds containing an oxime function such as aztreonam (195)[195] and cefotaxime (196),[196] *syn-anti* isomerisation occurs rapidly upon irradiation (thus diminishing the pharmacological activity), and is accompanied by degradation processes.

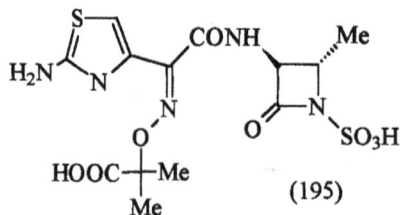

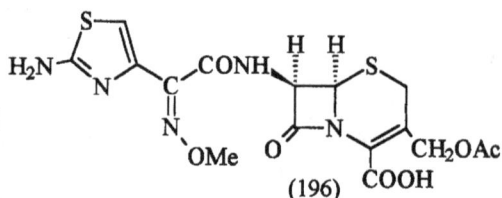

Penems are quite labile. One derivative of this group (197) has been shown to undergo retrocyclisation to give the thiazole (198) and 3-hydroxybutyric acid (detected as the methyl ester, 199) by irradiation in solution. In the solid state a different reaction takes place, viz intramolecular hydrogen abstraction followed by addition of oxygen to a radicalic centre to finally yield compound (200) (see Scheme 56).[197]

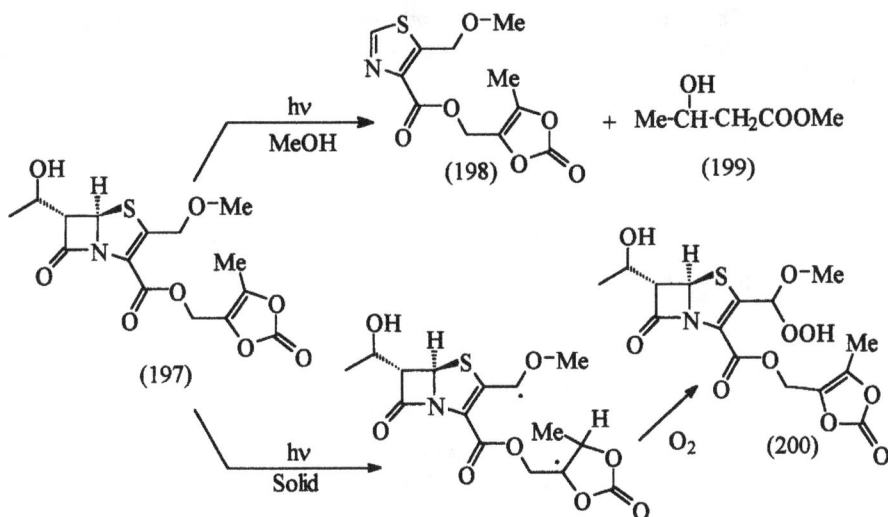

Scheme 56

2.6.4 Antibacterials: Heterocyclic Drugs. Among five-membered heterocyclics, one should mention metronidazole (201, R=CH$_2$CH$_2$OH) and related antibacterials,[198-200] which show a typical nitro group photoreaction. This is initiated by nitro - nitrite rearrangement (see Scheme 1, b) to give (202) followed by shift of the NO group to the vicinal position. Hydrolytic ring opening – ring closure from oxime (203) finally leads to 1,2,4-oxadiazol-3-carboxamide (204) (Scheme 57). These in turn undergo photoaccelerated hydrolysis to oxaldiamides (205). This rearragement is also one of the processes occurring upon radiolysis and does not involve the radical anion of the substrate.[200, 201] The photochemistry of metronidazole is changed in the presence of citrate.[202]

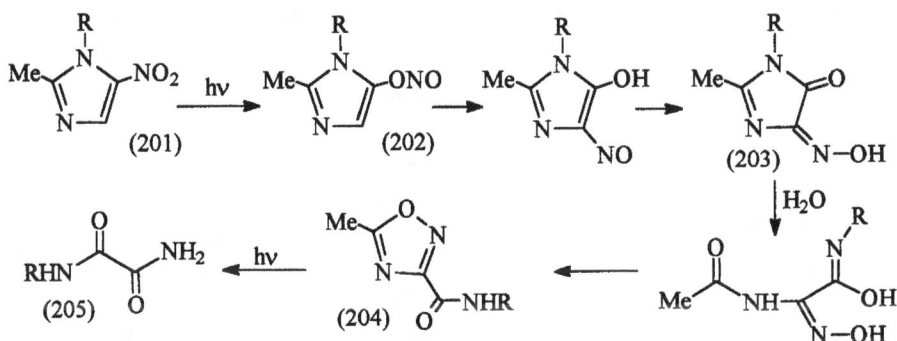

Scheme 57

The urinary antibacterial nitrofurantoin (206) is cleaved to nitrofurancarboxyaldehyde (207) upon UV irradiation (Scheme 58).[203] Among nitrofuran antiseptics for topical use, nitrofurazone (208) undergoes rapid *syn-anti* isomerisation to give (209) and fragmentation of the N-N bond leading to the azine (210).[204] Furazolidone (211) is cleaved and hydrolysed to aldehyde (207) (Scheme 58).[203]

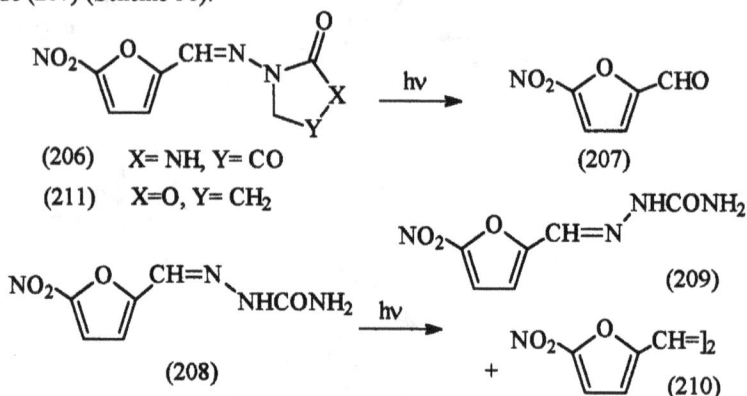

(206)   X= NH, Y= CO
(211)   X=O, Y= CH$_2$

(207)

(208)

(209)

(210)

Scheme 58

As for six-membered heterocycles, it has been reported that irradiation (254 nm) of the antitubercular drug isoniazid (212) in ethanol caused oxidation of the solvent to acetaldehyde, which is trapped by the substrate as the hydrazone (213). The last compound then adds a further molecule of alcohol giving (214).[205] Using a high pressure mercury lamp, C-N and N-N bond fragmentation also occurred, leading to the dipyridylhydrazone (215) and bis-pyridylhydrazide (216) (Scheme 59).[206]

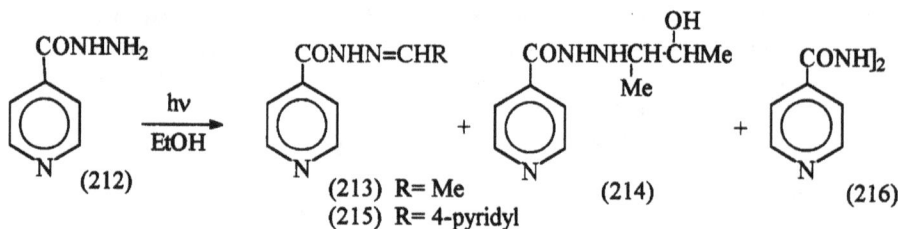

(212)

(213) R= Me
(215) R= 4-pyridyl

(214)

(216)

Scheme 59

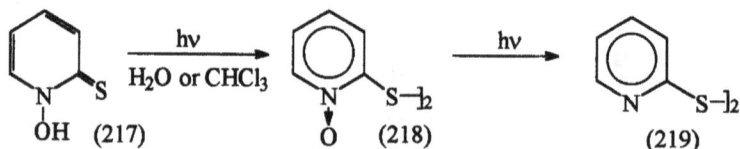

OH   (217)

(218)

(219)

Scheme 60

1-Hydroxy-2-thiopyridine, the antibacterial pyrithione (217), dimerises to the disulfide di *N*-oxide (218) upon irradiation in chloroform or aqueous buffer; further irradiation causes *N*-

deoxygenation to (219) (Scheme 60).[207] The antibacteric trimethoprim (220) undergoes oxidation at the benzyl position (e. g. 221) as well as hydrolysis of the amino groups in the pyrimidine ring (e.g. 222) and of the 4'-methoxy function (some of the products are shown in Scheme 61).[208]

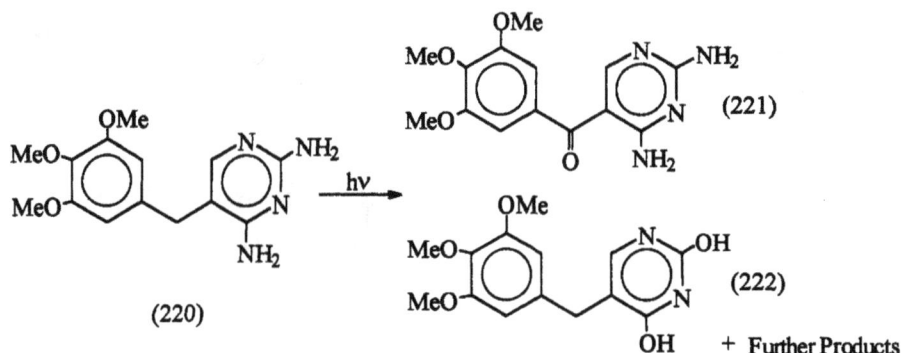

Scheme 61

The antibacterial nalidixic acid (223, Scheme 62) is decarboxylated by irradiation in 0.1 M sodium hydroxide solution and yields the corresponding naphthyridinone (224) and naphthyridindione (225).[209] Both (223) and related oxolinic acid (226) sensitise oxygen.[210]

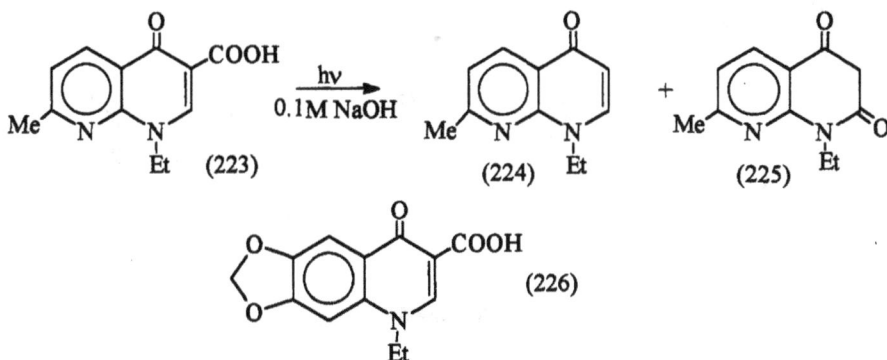

Scheme 62

Nalidixic acid has been the precursor of the largely used fluoroquinolone antibiotics. These are weak oxygen sensitisers[211] and are photolabile to a degree which changes greatly with the structure. Two different paths have been observed. The first one involves oxidative degradation of the amine side chain, as it is the case with ofloxacin (227) which gives products (228) (Scheme 63)[212] and ciprofloxacin.[213] The latter path involves defluoruration,[214] and is the dominating process when an 8-fluoro substituent is present, as is the case for the highly reactive lomefloxacin (229, Scheme 64) from which products (230)-(233) have been isolated.[215]

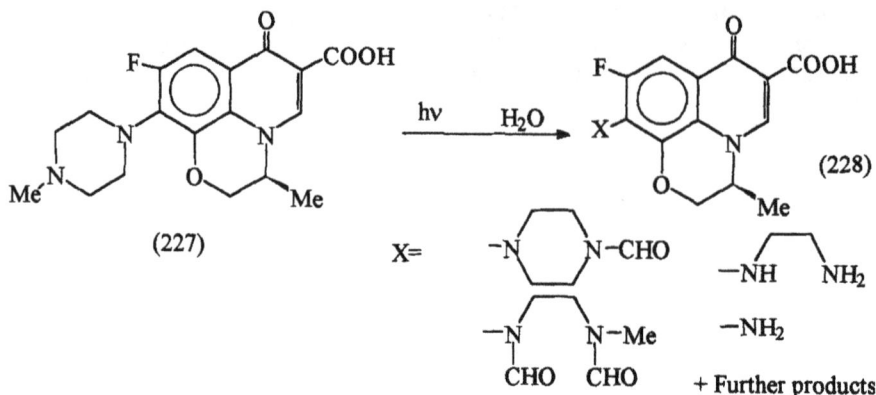

X=   −N⟨   ⟩N−CHO   −NH   NH₂

     −N⟨   ⟩N−Me   −NH₂
      | |
     CHO   CHO   + Further products

Scheme 63

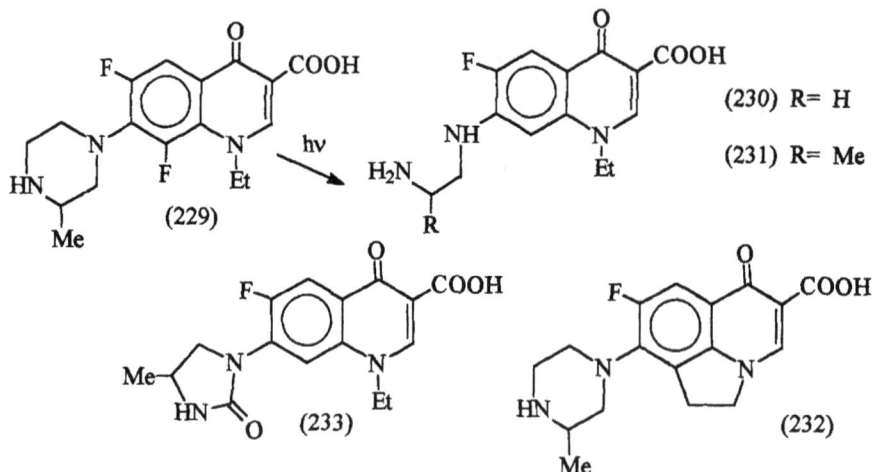

(230) R= H

(231) R= Me

Scheme 64

2.6.5 Antiprotozoal, Antiamebic and Antimicotic Drugs.   Quinine (234) has long been known to be photolabile,[216] and the photoreactions occurring in citric acid solutions have been extensively investigated.[217-219] Synthetic quinoline antimalarials are also photoreactive. Mefloquine (235) undergoes degradation of the alkyl side-chain by irradiation in methanol. The main products are (236) and (237) (Scheme 65).[220, 221] 4-Alkylaminoquinolines as chloroquine (238),[222-224] hydroxychloroquine (239)[225] and amodiaquine (240)[226] as well as 6-alkylamino derivatives as primaquine (241)[227, 228] all undergo photochemical oxidative degradation of the alkylamino chain. Some of the products are shown in Schemes 66-68. A further study of the photochemistry of some of these compounds is presented in this book (H. H. Tønnesen, p. 87).

(234)

(240)

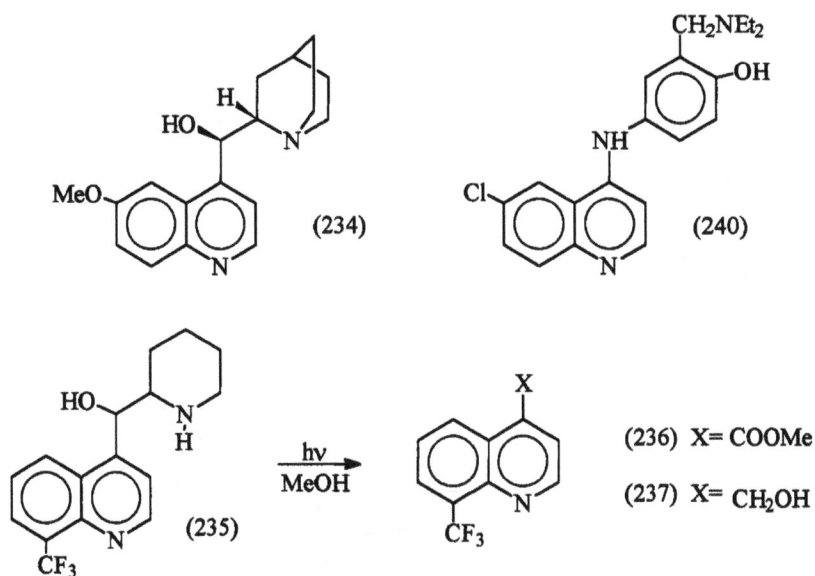

(236) X= COOMe

(237) X= CH$_2$OH

(235)

Scheme 65

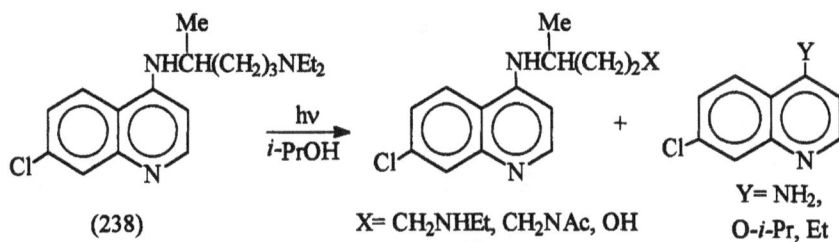

(238)

X= CH$_2$NHEt, CH$_2$NAc, OH

Y= NH$_2$, O-i-Pr, Et

Scheme 66

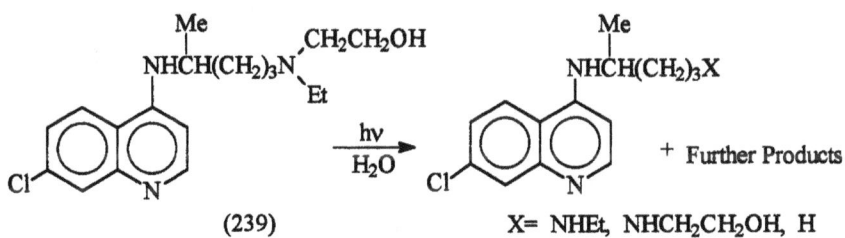

(239)

+ Further Products

X= NHEt, NHCH$_2$CH$_2$OH, H

Scheme 67

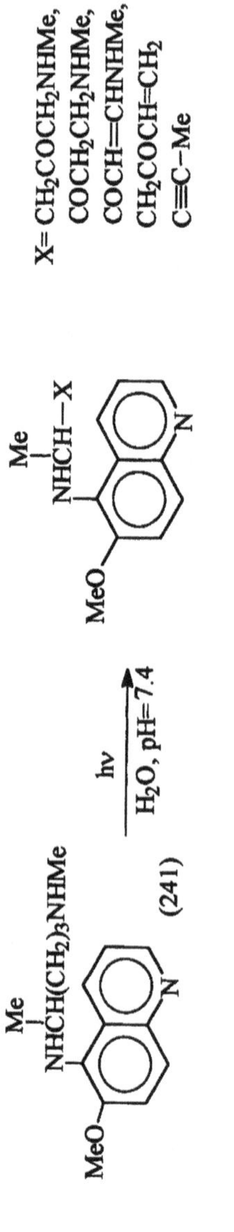

X= CH₂COCH₂NHMe,
COCH₂CH₂NHMe,
COCH=CHNHMe,
CH₂COCH=CH₂
C≡C–Me

Scheme 68

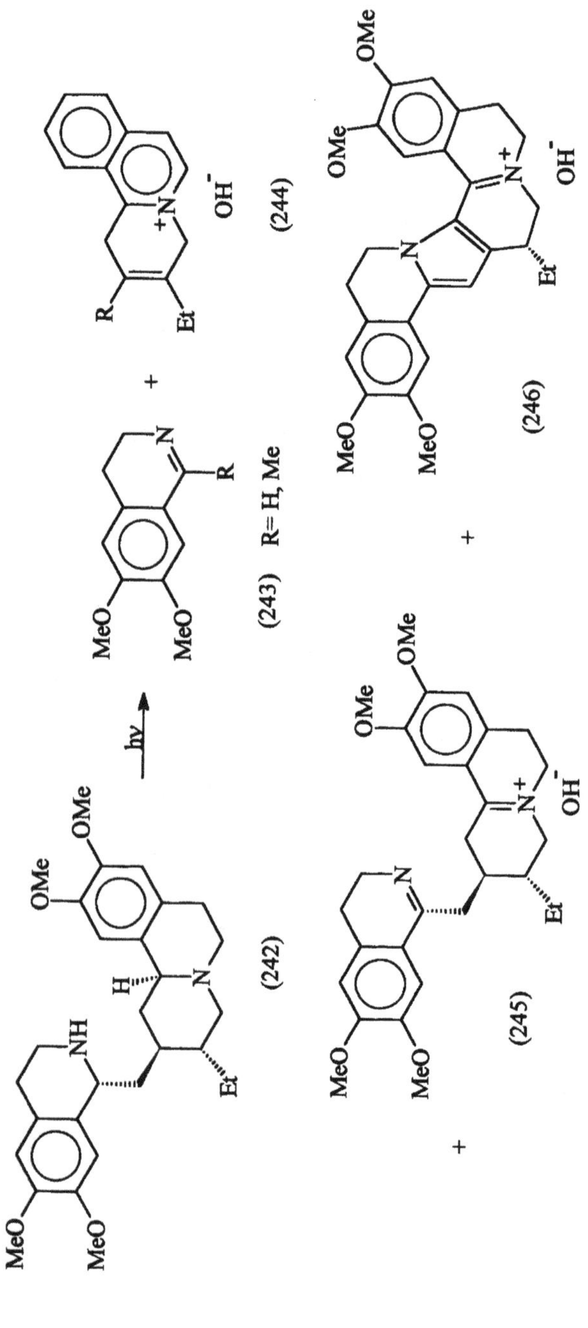

Scheme 69

(241)

(242)

(243) R= H, Me

(244)

(245)

(246)

The ipecac alkaloid emetine (242), used as antiamebic, is degraded when the dihydrochloride is irradiated in aqueous solution. The observed paths include oxidative cleavage of the methylene bridge between the two heterocyclic rings to give products (243) and (244) as well as oxidation to immonium salt (245) and oxidative cyclisation to (246) (Scheme 69).[229]

Antimicotics are another class of drugs where many photoreactive compounds are found, among both drugs for systemic and for topical use. A review of recent results in this field is presented by K. Thoma in this book (p. 116).

2.6.6 Antineoplastic Drugs. The yellow antineoplastic cisplatin (247) is solvated by irradiation in aqueous solution and the rate of the process depends on the pH.[230, 231] The products characterised have been the *cis*-diamineaquochloro complex (248) in the absence,[232] and the aminetrichloro complex (249) in the presence of sodium chloride (Scheme 70).[231, 233]

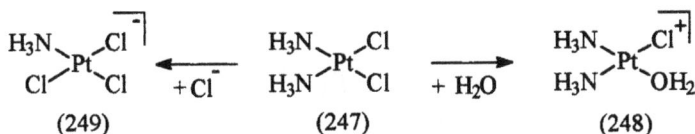

Scheme 70

The nitroso ureides carmustine (250)[234] and tauromustine (251)[235] are light sensitive, as it is the case for mitonaftide (252).[236] The antitumor alkaloid vinblastine sulfate (253) is both thermally and photochemically unstable.[237]

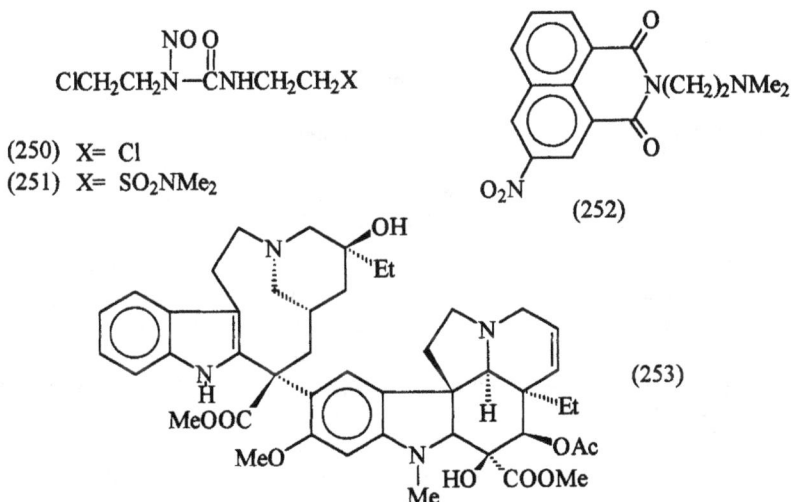

(250) X= Cl
(251) X= SO$_2$NMe$_2$

(252)

(253)

6-Mercaptopurine (38) is oxidised when irradiated in oxygen-equilibrated aqueous solution by near-UV light, giving as the primary product the sulfinate (254), which is then further oxidised to the sulfonate (255) (Scheme 71).[238]

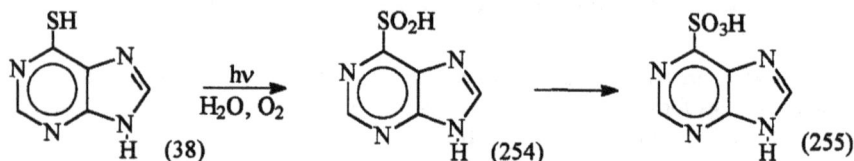

Scheme 71

Irradiation of dacarbazine (256) in solution causes elimination of dimethylamine to give the diazo derivative (257). This is then hydrolysed to give the hydroxyimidazole (258), in turn coupling with the above diazo to give the azo derivative (259) or alternatively cyclise to 2-azahypoxanthine (260) (Scheme 72).[239]

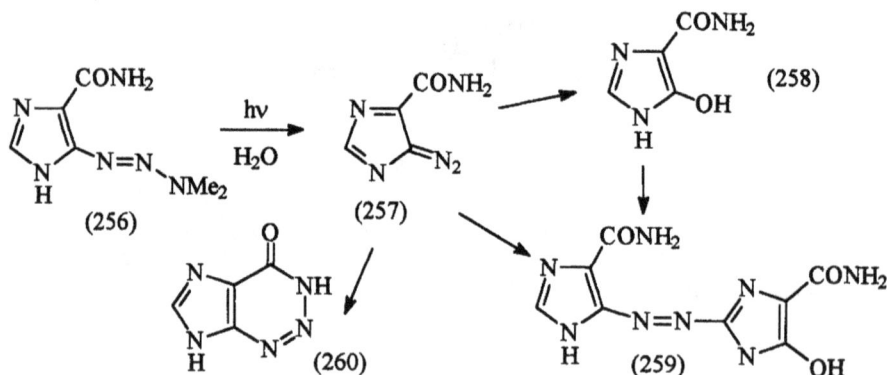

Scheme 72

UV irradiation (as well as hematoporphyrin photosensitisation) of methotrexate (261) leads to benzylic cleavage. The products include 2,4-diaminopteridincaboxyladehyde (262) and the corresponding acid (263) as well as 4-aminobenzoylglutamic acid (264) (in turn oxidised to the azo derivative by sensitisation by the pteridinaldehyde, see Scheme 73).[240-242]

Doxorubicin (265) and daunorubicin (266) are photolabile.[243, 244] In particular, the latter drug undergoes homolytic deacylation.[245] The antitumor antibiotic hedamycin (267) reacts photochemically via initial intramolecular hydrogen abstraction from the tetrahydropyranyl ring in position 8; in the presence of oxygen the final product is the corresponding dihydropyranyl derivative (268), while in nitrogen–flushed solution this process is accompanied by reductive cleavage of an epoxide ring of the chain at C-2 to give products (269) and (270) (see Scheme 74); the related derivative kidamycin reacts similarly.[246] The glycopeptides bleomycins[247-249] and the protein zinostatin,[250] both used as antineoplastics, are photoreactive.

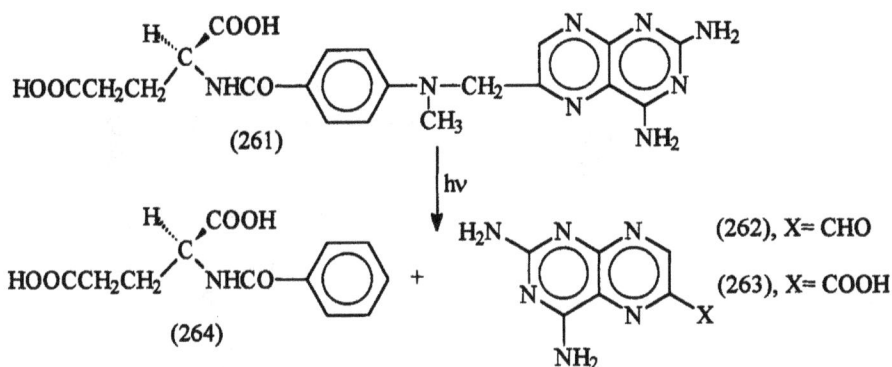

Scheme 73

(261)

hv ↓

(264)

(262), X= CHO

(263), X= COOH

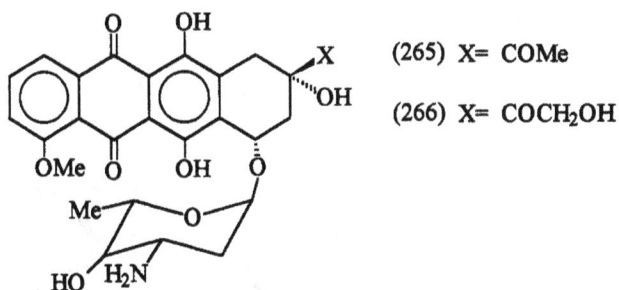

(265) X= COMe

(266) X= COCH₂OH

2.7 Vitamins

2.7.1 Liposoluble Vitamins. Retinol, vitamin A (271), is liable to photo-induced geometrical isomerisation and oxidation,[251] as it is generally the case for polyenes. The composition of the photostationary mixture of geometric isomers has been determined for a related compound, retinoic acid (272) (used as keratolytic).[252]

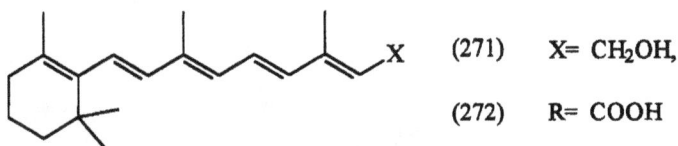

(271) X= CH₂OH,

(272) R= COOH

Vitamins D have offered to organic photochemistry one of the most exciting fields of research, and the unravelling of the geometric isomerisation and electrocyclic cyclisation reactions occurring from various vitamins and provitamins of this group has been of great value for the advancement of this science.[253]

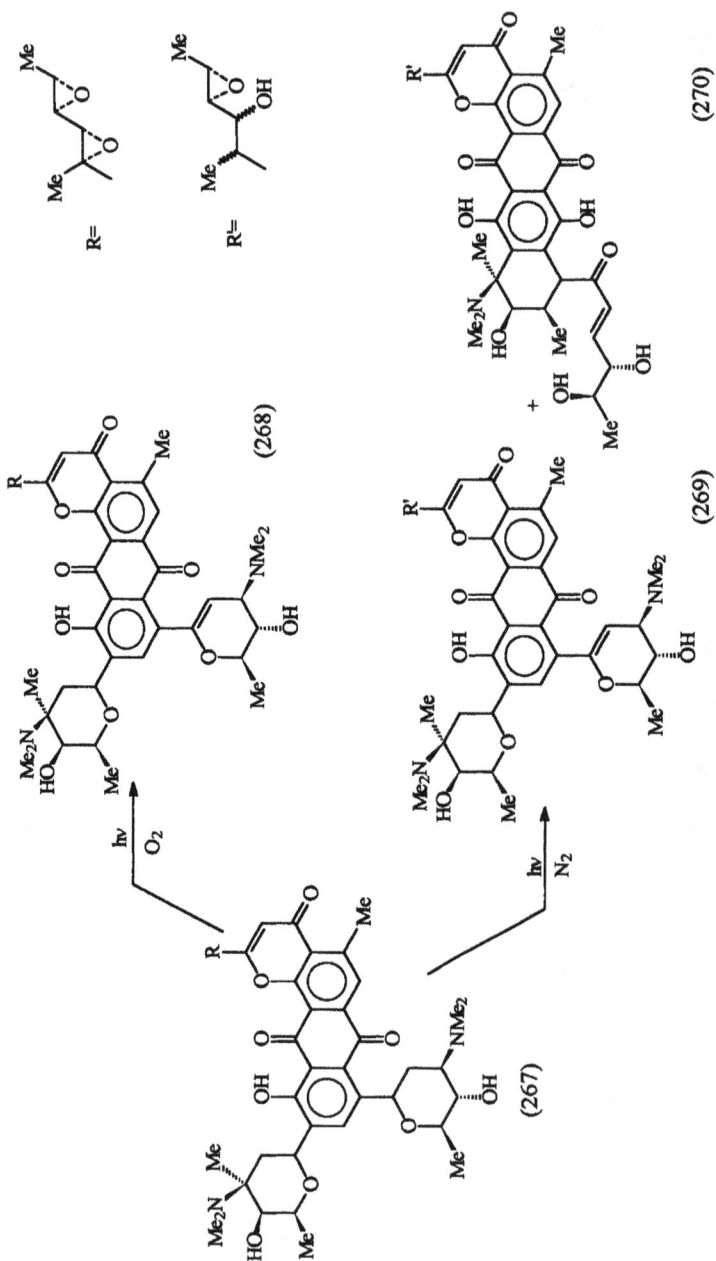

Scheme 74

This work cannot be detailed here; a typical example is the photorearrangement of calciferol (vitamin D_2) (273) which undergoes both sigmatropic hydrogen shift to give (274) and electrocyclic ring-closure to give (275) (Scheme 75).[254, 255] Obviously vitamins D must be protected from light.[256]

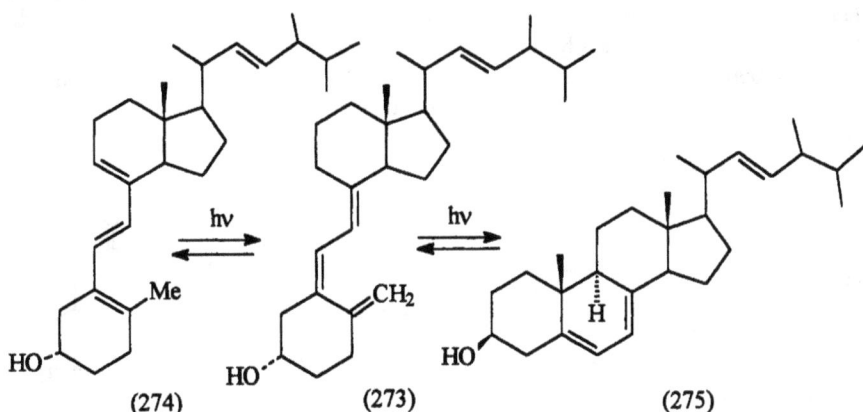

Scheme 75

α-Tocopherol (276) (vitamin E) is well known as an antioxidant. The sensitised photo-oxygenation involves singlet oxygen addition to the phenol moiety giving hydroperoxycyclohexadienone (277). This in turn slowly decomposes to give benzoquinone derivatives (278) and (279) (Scheme 76).[257]

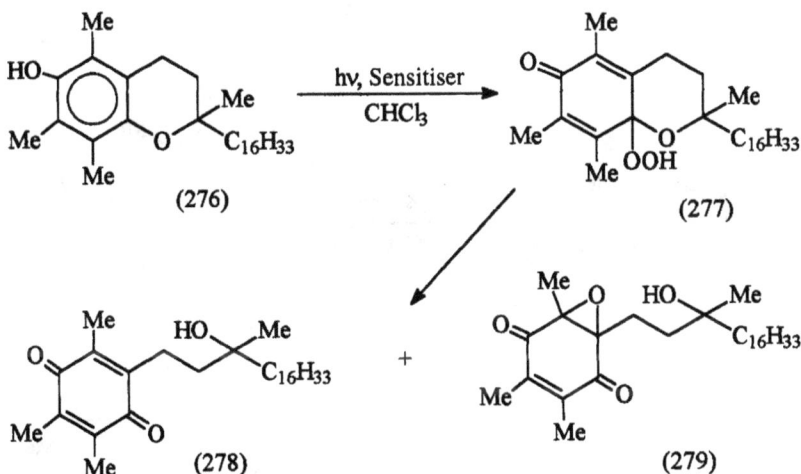

Scheme 76

Irradiation of menadione (280) (vitamin K_3 or vitamin $K_{2(0)}$) leads to epoxidation in aerated ethanol (product 281) and to hydroxylation and disproportionation (products 282 and 283) in micellar solution (Scheme 77).[258, 259] This molecule also undergoes photodimerisation: the *syn* cyclobutane dimers (284) and (285) were obtained by irradiation in the solid state (Scheme 77),[260] and the *anti* dimers by irradiation in acetone solution.[261] Other vitamins K bearing an allylic chain undergo *E-Z* isomerisation[262] as well as ene addition with singlet oxygen to give allyl hydroperoxides or related derivatives[263-267] [see as an example the peroxides (286) and (287) formed by singlet oxygen addition to menaquinone-1, vitamin $K_{2(1)}$, (288) in Scheme 78].[268] Menadiol derivatives are also photoreactive, as it has been shown for solid menadiol acetate (vitamin K_4).[260]

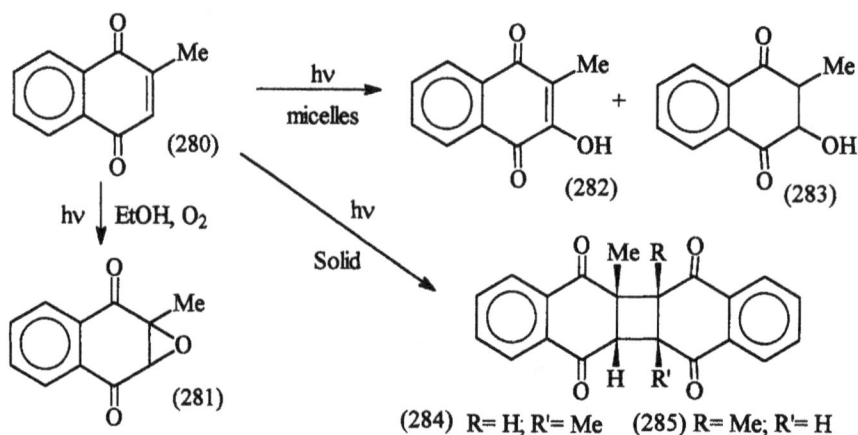

Scheme 77

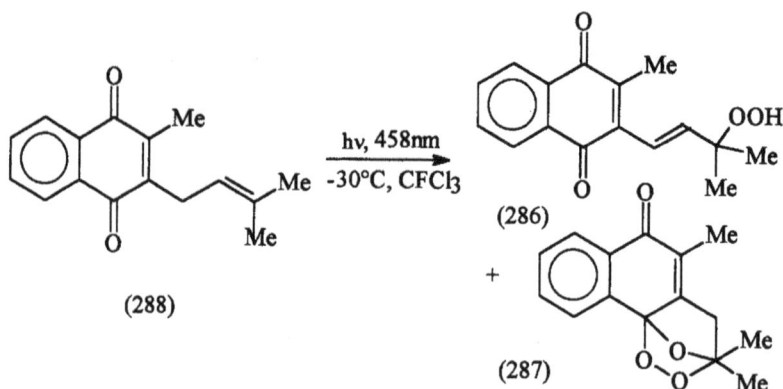

Scheme 78

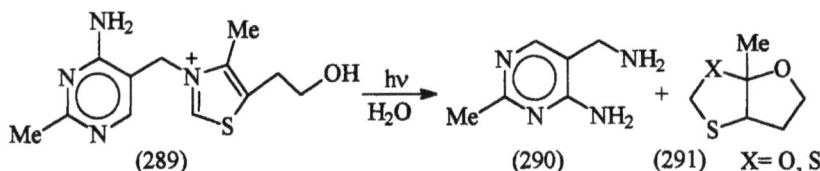

Scheme 79

2.7.2 *Hydrosoluble Vitamins.* Thiamine (289) undergoes cleavage of the thiazolyium ring when irradiated, giving an aminopyrimidylmethylamine (290)[269] and sulfurated (and odoriferous) aliphatic fragments (291) (Scheme 79).[270, 271] Riboflavin (292) is known for its high sensitivity to solar light. Irradiation under anaerobic conditions causes reversible reduction,[272] while in the presence of oxygen the N_{10}-ribityl side chain is oxidised; the main product is lumichrome (293), with elimination of the chain (see Scheme 80), but various products of intermediate oxidation have been characterised.[273-278]

Both niacinamide (294) (vitamin PP) and the related vitamin factor nicotinic acid (295) are hydroxylated at position 6 upon irradiation under acidic conditions to give (296) and (297) respectively; compound (295) is decarboxylated and gives the dipyridyl derivative (298) in neutral solution (Scheme 81).[279]

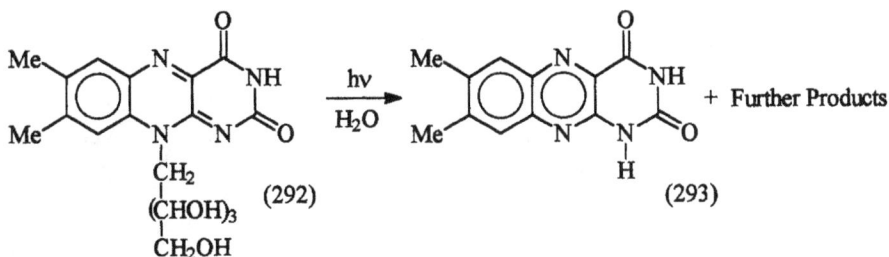

Scheme 80

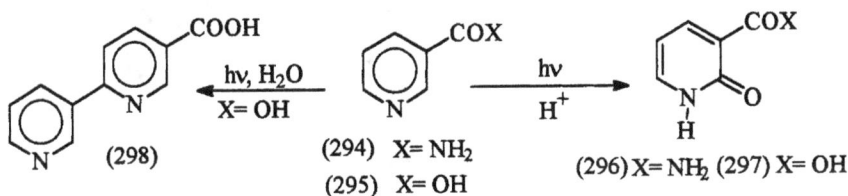

Scheme 81

In the case of pyridoxine (299), both hydroxymethyl groups are sequentially oxidised to carboxy acids under oxygen to finally give (300), while under nitrogen dimeric pyridinium-3-olate zwitterion (301) is obtained (Scheme 82).[280]

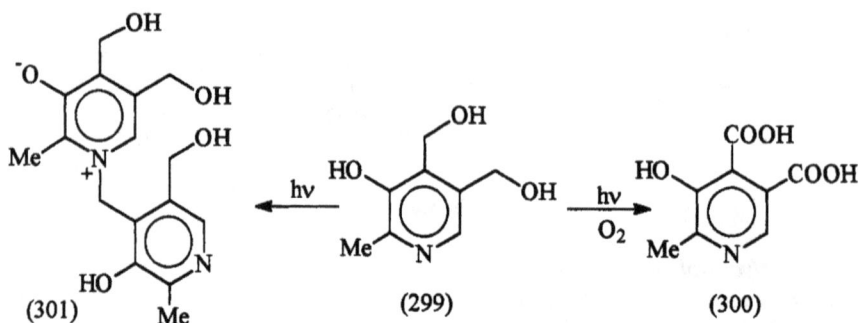

Scheme 82

With pyridoxal-5-phosphate (302) the reaction involves the aldehyde function, giving the acyloin (303) under anaerobic conditions and various oxidised products, such as the carboxylic acid (304) and the dipyridyl α-diketone (305) in the presence of oxygen (Scheme 83).[281] Vitamin B_{12} (306) is liable to photo-oxygenation. The related dicyanocobirinate (307) undergoes two consecutive oxidative cleavages (via singlet oxygen, compare Scheme 1, f) to give the corrin macrocycle (308) (Scheme 84).[282]

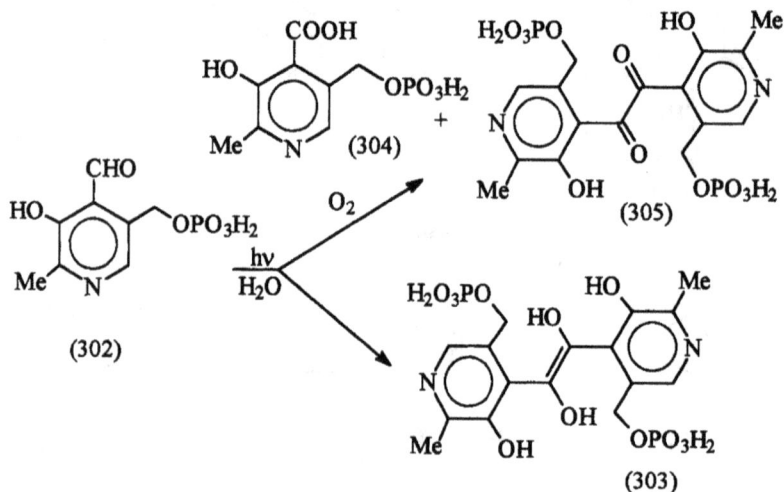

Scheme 83

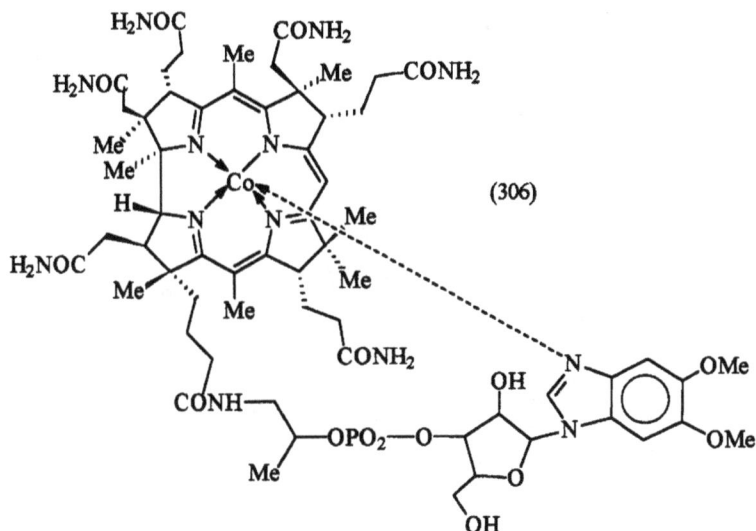

(306)

3 PHOTODECOMPOSITION OF DRUGS: A PRACTICAL APPROACH

3.1 Dependence on Physical Conditions

An overview of the photoreactivity of pharmaceutically active compounds has been presented in Sec. 2. It is certainly desirable that the photoreactivity of every molecule used as a drug is investigated, in particular since this should alert about the possibility that the drug is phototoxic. However, this does not imply that every compound found to be reactive under some condition must be considered photolabile during normal storage or use conditions.

First of all, many drug substances do not significantly absorb ambient light (solar or artificial) and are sensitive only to irradiation at a shorter wavelength. In such a case it should at any rate be checked that no other component of the drug preparation, such as an associated drug or an eccipient, sensitises a reaction of an otherwise photostable substrate. As an example, an added ketone may abstract a hydrogen from such a substrate and initiate radicalic decomposition or another additive may sensitise oxygen and thus promote a photo-oxidation (many dyes are good oxygen sensitisers).[283] On the other hand, an additive present in the formulation may also affect positively the photostability of the active principle (see Sec. 4).

Even when the drug is not transparent to ambient light, absorption (and consequently photoreaction) may not involve a significant fraction of the substrate. In a dilute solution, absorption takes place across the bulk of the solution, as it is apparent by inspection of Beer-Lambert's equation (see p. 82). This is, as it has been mentioned, the best condition for studying the photochemical reaction occurring. However, this does not apply to concentrated solutions, where absorption occurs only at the thin layer at the interface, for suspensions,

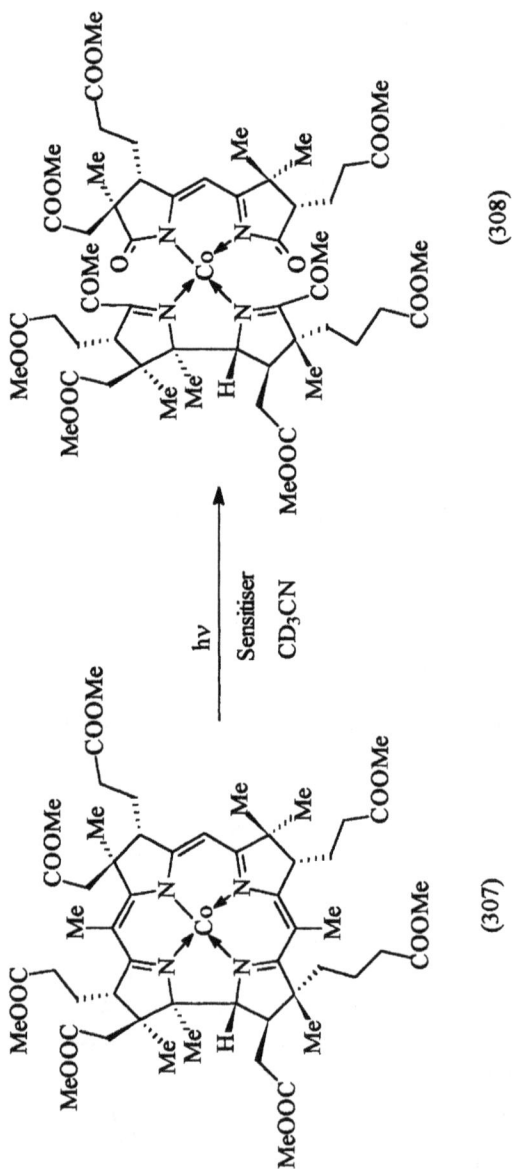

Scheme 84

where part of the light is lost by reflection, or for solids, where again light is completely absorbed by the first thin layers of molecules.

Therefore, the photochemical study of a new drug should proceed along two lines. On one hand an investigation should be carried out under the best conditions for characterising the photochemical reactions, viz in dilute solutions (see e.g. the protocol proposed by Beijersbergen, p. 85) with the main aim of recognising the photoproducts and of evidencing a possible phototoxicity in an early stage of the development. On the other hand, in order to establish the stability of the dosage form the investigation must be extended to the actual pharmaceutical form and be carried out under conditions representative of those under which the drug is used and the illumination conditions to which it will be exposed. This demands that one takes care of photostability, as a part of stability studies in general, at the preformulation and formulation stage.[284]

In the present section two key points concerning the dependence of photochemical reaction on the conditions under which drugs are prepared or used are briefly addressed. These are solid state vs solution and anaerobic vs aerobic conditions.

3.1.1   Photoreactions in the Solid State.   The different chemistry occurring by irradiating a drug in the solid state vs the solution for most drugs (and in general most molecules) has been often mentioned in Sec. 1 (see e.g. Schemes 13, 15, 28, 42). According to the different light penetration in the crystal, two different situations may arise.

1. The first case is characterised by the fact that penetration does not increase with reaction. This may be either because the photoproducts absorb themselves, thus screening the bulk of the solid from light or because loss or change of the crystalline form with reaction leads to reflection of all the light. In this case the photoreaction is limited to the first molecular layers at the surface. This may lead to a change in the appearance of the preparation, e. g. a conspicuous coloration or discoloration (usually evaluated by tristimulus colorimetry),[285, 286] but may involve no serious loss of the active principle as evidenced by the appropriate assay (of course the change of colour may be unacceptable per se and force to suitably protect the preparation). Indeed, several drugs are quite sensitive in dilute solution but rather stable in the solid [e.g. indomethacin (13),[19, 32] some steroids, e.g. (146),[162] dilthiazem (122),[132] minoxidil (123)].[133] In some of these reactions the conversion proceeds up to some percent conversion and new product(s) are formed in a detectable concentration, although the reaction stops when only a fraction of the substrate is consumed. As an example, photoinduced hydrogen abstraction followed by oxygen addition to give a hydroperoxide proceeds up to ca 10% conversion by irradiating the crystalline penem derivative (197) as shown in Scheme 56. Likewise, in most reactions of steroids in the solid state (see e. g. Schemes 20 and 21) the product is formed only in a few percent yield.

2. In the latter case the photoproduct is transparent at some wavelength absorbed by the substrate. Thus, light penetrates through successive layers while the reaction proceeds and a significant part of the crystal, or the whole of it, may be transformed. This is the case for π cycloaddition processes, as in the cases of cynnamic acid[287] and of thymine,[288] where the adducts absorb at a much shorter wavelength than the reagents, but applies to every reaction satisfying the above criterion. Menadione offers another case of photodimerisation, where the stereoisomers obtained in the solid state differ from those obtained in solution (Scheme 77).

Schemes 20, 21 and 56 quoted above show that reactions in the solid state and in solution often involve a different moiety in the molecule, since restrictions imposed by the crystal lattice to molecular motions preclude some otherwise viable paths and vice versa introduce new paths involving interaction between functions that are close one to another in the solid state (topochemical control).[289] There are several other examples, such as those of estrogens (Scheme 42) and of some benzodiazepines (Scheme 28). In several cases intra- or intermolecular (between two closely stacked molecules) hydrogen abstraction predominates in the crystals, whereas different reactions are observed in solution (e.g. aminopyrine, Scheme 13, as well as the penem in Scheme 56 above). Azapropazone (Scheme 15) shows sigmatropic shift occurring only in the solid state.

However, even when the photochemical process remains the same, the lattice constraint may still change the end result. A typical example is that of metyrapone (309), a diagnostic aid for pituitary function determination. As one may expect from the structure this ketone undergoes a very fast (τ_T 12 ns) α cleavage from the $n\pi^*$ triplet state to give a pair of radicals. In solution, the two fragments migrate and give different reactions, the main one being recombination in the para position to give conjugated alkene (310) and a polymer from it. This reaction is little affected by the nature of the solvent and the presence of oxygen.[290]

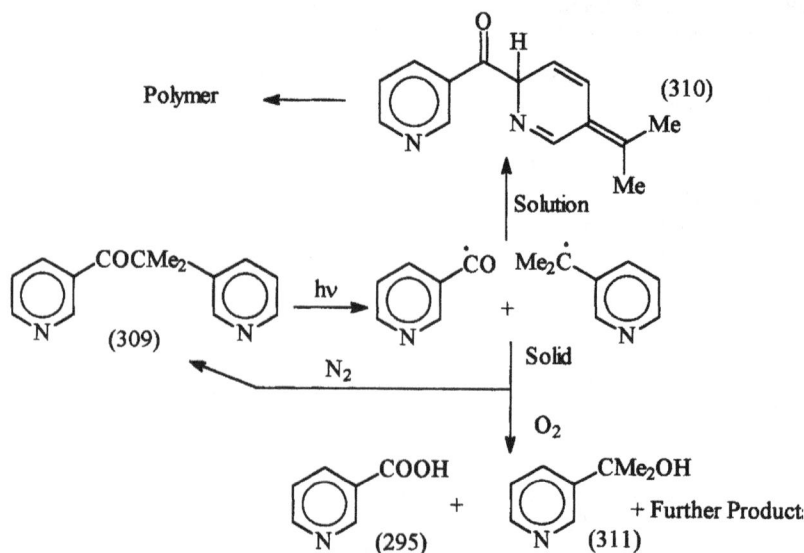

Scheme 85

There is no reason to expect that the primary process is different in the solid, but there the radicals cannot migrate, and thus recombine to the starting material. Therefore, the solid is rather photostable under anaerobic conditions. On the other hand, when oxygen is present in the atmosphere, it adds to the radicals formed at the surface to give peroxy radicals thus making cleavage irreversible and finally leading to various oxidised products such as (295) and (311) (Scheme 85).[291]

It has also been noted with several drugs that different polymorphs give different results. Again, this is not surprising, since the crystal lattice is one of the factors determining the photoreactivity[292] as it has been found in the case of carbamazepine (106),[113] cyanidol (192),[191, 192] and furosemide.[284] The dimension of the crystals and micronisation can also greatly affect the sensitivity.[293]

3.1.2. *Photoreactions in the Presence of Oxygen.* The effect of oxygen is also complex. This molecule is an efficient quencher of excited states, viz causes decay to the ground state with no chemical reaction. Thus, it may have a photo-stabilising effect on reactions involving relatively long-lived excited states such as triplets (short-lived singlet states cannot be quenched by oxygen since the concentration of the latter is too low to make a bimolecular reaction such as quenching significant in comparison with monomolecular reactions). However, in the quenching process the strong electrophile singlet oxygen is generated, and this may lead to reactions of its own, if activated alkenes, polyenes or aromatics are present (typical singlet oxygen processes are observed with estrogens, Scheme 43 and 44, vitamin E, Scheme 76, vitamin K, Scheme 78 and Vitamin B_{12} analogues, Scheme 84).

Another reactive species, the superoxide anion is formed during photoinduced electron transfer. Finally, ground state oxygen itself may trap radicalic intermediates, often formed during the course of a photochemical reaction, and thus change the product formed, even when it has no influence on the *primary* process from the excited state (see as example Scheme 85 above, or aminopyrine in Scheme 13). Quite complex situations result in some cases, as in the example of hydrocortisone *tert*-butylacetate,[59] where the photoreactivity of one of the polymorphs has been attributed to the existence in this lattice of a cavity allowing molecular oxygen to penetrate within the crystal. *S* and *N*-heterocycles are often liable to *S* and *N*-oxidation (see e.g. Scheme 29 and 31), although it is not clear whether ground state or activated oxygen is involved.

3.1.3 *Wavelength Effect.* It is a general postulate in organic photochemistry that internal conversion between higher lying singlet and triplet states (S_n, T_n) to the lowest singlet (S_1) and triplet (T_1) respectively is always so fast that precludes any reaction of such states. As a consequence, the photochemical reaction of a molecule remains the same, independently from the wavelength of irradiation. Thus, while using different lamps (or solar light) may lead to a different rate of reaction (see Sec. 3.4), no change in the product distribution is expected. However, when the products are themselves photoreactive, the equilibrium reached certainly depends on the wavelength used (typical examples, the rearrangements of Vitamin D, Scheme 75, as well as all geometric isomerisations). Other cases of documented difference in the products by using different light sources, e. g. midazolam (Scheme 26) or indomethacin (Scheme 6) probably likewise depend on the further reactivity of the photoproducts. Alternatively it may be that the species absorbing is the substrate at one wavelength, a sensitiser (possibly an impurity) at another one. This demands that particular attention to impurities is given when a photochemical process is studied. On the other hand, a sensitiser may be formed during the photoproces, as shown in the case of methotrexate, where one of the products, a pteridinaldehyde (see Scheme 73), sensitises further degradation.

3.2 Photoreactivity of Drugs under Conditions of Practical Use

The photostability of a drug preparation has to be assayed under conditions of practical use. Thus, the commercially available form (tablets, capsules, suspension, solution, etc.) is examined both within the package and outside it. The problem here is significant when the drug preparation is exposed to indoor or solar light for a significant time, as is the case for eye drops[294] or dermatological ointments, as well as infusion or parentheral fluids or solution in disposable plastic syringes. The degradation occurring under such conditions has been assessed in several cases. Representative examples concern lipid-[295, 296] and water-soluble[297] vitamins in parentheral solutions or emulsions[298] (particularly if they are administered under phototherapy light),[299] ranitidine[300] in parentheral emulsions, molsidomine,[301] methotrexate (261)[302] or sodium nitroprusside in infusion fluids,[303, 304] cathecolamines[305] or amphotericin B[306] in dextrose injections, terbutaline sulfate[307] or sodium nitroprusside[308] in plastic syringes.

3.3 Photoreactivity of Drugs and Analytical Procedures

Formation of small amounts of photoproducts may seriously undermine the efficiency of a drug. Thus, photoreactions should be followed by appropriate instruments ensuring the required sensitivity.[309, 310] Preliminary tests may be carried out simply by illuminating the drug deposed on a tlc plate and then revealing the products by developing the plate with the suitable eluent, although data obtained in this way need not   reproduce the photolability in solution or in the solid state.[287]

The high photolability of a drug may lead to artefacts during assays. Ordinary fluorescent laboratory light causes a significant degradation of dilute solutions of many drugs [e.g. anti-inflammatory steroids,[311] many antineoplastic agents,[312] thioridazine (99) and other neuroleptics].[313] Therefore, these must be protected during preparation and storage for the appropriate assay. On the other hand, one may take advantage of the photoreactivity of a drug to transform the analyte in a more sensitively determined derivative. As an example, tamoxifen (163, R=Et), a triphenylethylene derivative used as anti-estrogen and anti-neoplastic, can be photocyclised to a phenanthrene (see Scheme 47) a derivative that can be more conveniently determined by densitometry on a TLC plate[314] and by fluorescence in solution.[167]

3.4 Assessment of Drug Photostability

In order to be significant for industry and regulatory agencies, photostability tests must be carried out under controlled and internationally accepted conditions. The tests must be such as that an accelerated degradation is obtained, yet that this is representative of the degradation under actual use conditions. Thus, the conditions of the experiments must be specified. Different experimental set ups have been used, which have led to problems in confronting the conclusions from different laboratories.[315-317] ICH guidelines for drug photostability have been formulated and after a careful discussion have been recently published (implemented from January 1998). These are found as an appendix to the present chapter, and their application is discussed in several respects by Drew and Helboe (p. 227 and 243, respectively) in this book.

Detailed studies have been carried out for the comparison of the photodegradation of drugs (both absorbing in the UV and in the visible) with different lamps in various laboratories.[285, 286, 318-322] In general, it turns out that the qualitative course of the reaction does not change with the lamp, as one may expect, unless the primarily formed photoproducts are thermally or photochemically unstable. The rate of degradation changes greatly, but the results become comparable when the actually absorbed light dose, not the overall lamp flux, is taken into account. The choice of an appropriate actinometer has also been discussed.[323, 324]

Both these key aspects, the choice of the lamp (p. 247, 272 and 288, contributions by Piechocki, Boxhammer et al, Forbes) and that of the actinometer (p. 295 and 305, contributions by Favaro and Bovina et al,) are discussed in this book.

4   PROTECTION OF PHOTOLABILE DRUGS

4.1   External Protection

Photolabile drugs must be protected from light under every phase, from the preparation to the final use. In particular, the commercialised dosage form must be such that no significant decomposition of the active principle takes place. This can be done either by avoiding that the light reaches the preparation, viz using an opaque container or, in the case of tablets and capsules, by means of an opaque coating (external protection), or by incorporating in the preparation some additive that either absorbs the active light competitively with the drug principle or quenches the photoreaction of the latter (internal protection).

A sensible choice of the package is obviously important. ICH guidelines (see Appendix) demand that the photostability is assessed both with and without the original package.[325, 326] Incorporation of an UV adsorber,[327-329] a dye[330-331] or an opaciser (e. g. titanium dioxide)[331-333] in the coating of tablets or capsules is often used. The effectiveness of such protection is usually determined either directly on the coated samples[327] or in model studies, where the drug is covered by a coloured film.[331]

Special recommendations must be added for extemporaneously prepared solutions or suspensions. As an example, syringes that contain sodium nitroprusside must be wrapped in aluminium foil before adding the solutions to avoid photoinitiated decomposition.[308] When this is impractical, e.g. because the level of the liquid must be visible in order to allow monitoring of the infusion rate, containers opaque to the light absorbed by the drug are used. Brown glass (e. g. Amberset) bottles are mostly employed,[138, 334-338] but their effectiveness should be checked in every case, since such glasses often do not completely absorb blue light. Coloured plastic containers are also used.[339]

4.2 Internal Protection

An alternative procedure is incorporating in the pharmaceutical preparation an additive able to quench the photochemical reaction. This can be made on the basis of two different principles, viz. either by competitive absorption by adding a compound the spectrum of which is superimposed to that of the labile drug, or by adding a physical quencher, a compound which causes the decay of the excited state to the ground state before it has the time to undergo a chemical reaction.

The first idea has been developed systematically by Thoma, who has shown that by choosing a food dye with the absorption spectrum matching that of the drug a much improved stability of uncoated tablets is obtained. Thus, yellow food dyes photostabilise nifedipine, red dyes daunorubicin, blue dyes methylene blue.[340] Colorants may be added both to solid formulations and to dry emulsion.[341]

Synthetic iron oxides have also been successfully used for the protection of yellow drugs.[342] UV absorbers of the type used as sunscreens have been used for drugs absorbing in that region. As an example *p*-aminobenzoic acid (PABA) and uric acid solubilised by lithium carbonate enhance the photostability of colchicine[343, 344] and PABA with urocanic acid and sodium urate stabilise adriamicyn.[345] Vanillin is useful for protecting molsidomine[346] and has been positively tested, together with various UV absorbers, for protecting glucocorticoids, both in solution and in gels for topical application.[328]

As for quenching of excited states, this is a largely used principle in many fields of applied photochemistry, e.g. for the stabilisation of plastic, rubber, dyes.[347] Obviously, addition of a further chemical component to a drug must be considered with care, taking into account possible toxic effects. This is probably why this path has received little attention up to now.

Examples of moves in this direction are the use of cyclodextrins, which have been shown to affect (in some case positively, in other ones adversely) the photostability of drugs such as emetine,[348] cephaeline,[348] nicardipine,[349, 350] hydrochlorothiazide,[351] clofibrate,[351] the antiulcer agent 2'-carboxymethoxy-4,4'-bis(3-methyl-2-butenyloxy)chalcone,[352] and some vitamins.[351] This effect is due to the limitation to the molecular motions and the change in the microenvironment introduced by cyclodextrin complexation (the latter fact may introduce new photochemical paths, e.g. irreversible binding to the carbohydrate, however).

Other additives act not on the excited state per se, but rather on the primary product, be it a ground state molecule or a reaction intermediate. These are converted back to the starting material. Reducing agents have been used for this purpose. As an example, glutathione is photoprotective for doxorubicin,[353] dacarbazine,[354] menadione sodium bisulfite,[355] demeclocycline[356] and tetracycline;[357] uric acid has been found be effective with sulfathiazole[358] and tocopherol and phytonadione for ubidecarenone.[359] It may be that PABA acts not only as an internal filter (see above) but also interferes with the photoreaction in a similar way. Another mechanism of action is formation of a complex between the additive and the ground state of the drug, when such complex is less photolabile than the original substrate. This appears to be the case with sodium urate and metronidazole.[360]

Acknowledgement⸝ Financial support to our work on the photochemistry of drugs by 'Istituto Superiore di Sanita', Rome, in the frame of the programme 'Proprietà chimico-fisiche dei medicamenti e loro sicurezza d'uso', is gratefully acknowledged.

References

1. W. H. Horspool and P. S. Song, 'CRC Handbook of Organic Photochemistry and Photobiology', CRC Press, Boca Raton, 1995.
2. J. D. Coyle, R. R. Hill and D. R. Roberts, 'Light, Chemical Change and Life: a Source Book in Photochemistry', The Open University Press, Milton Keynes, 1982.
3. J. Kopecky, 'Organic Photochemistry: a Visual Approach', VCH Publ., New York, 1992.
4. W. H. Horspool and D. Armesto, 'Organic Photochemistry: a Comprehensive Treatment', Ellis Horwood, New york, 192.
5. N. J. Turro, 'Modern Molecular Photochemistry', Benjamin Cummings, Menlo Park, 1978.
6. H. H. Tønnesen (Ed.), 'Photostability of Drugs and Drug Formulations', Taylor & Francis, London, 1996.
7. J. V. Greenhill, in ref. 1, p. 83.
8. J. V. Greenhill and M. A. McLelland, *Prog. Med. Chem.*, 1990, **27**, 51.
9. G. M. J. Beijersbergen van Henegouwen, *Adv. Drug. Res.*, 1997, **29**, 79.
10. D. E. Moore, *J. Pharm. Biomed. Anal.*, 1987, **5**, 441.
11. D. E. Moore, *TrAC, Trends Anal. Chem. (Pers. Ed.)*, 1987, **6**, 234.
12. G. M. J. Beijersbergen van Henegouwen, *J. Photochem. Photobiol. B: Biol.*, 1991, **10**, 183.
13. K. Thoma and N. Kuebler, *Pharmazie*, 1996, **51**, 660.
14. J. V. Greenhill in J. Swarbrick and J. C. Boyland (Eds.), 'Encyclopedia of Pharmaceutical Technology, Dekker, New York, 1995, Vol. 12, p. 105.
15. T. Oppenlaender, *Chimia*, 1988, **42**, 331.
16. H. H. Tønnesen and J. Carlsen, *Pharmeuropa*, 1995, **7**, 137.
17. H. H. Tønnesen, in ref 6, p.371.
18. F. Vargas, C. Rivas, M. A. Miranda and F. Bosca, *Pharmazie*, 1991, **46**, 767.
19. D.E. Moore and P. P. Chappuis, *Photochem. Photobiol.*, 1988, **47**, 173.
20. G. Condorelli, G. De Guidi, G. Giuffrida and L. L. Costanzo, *Coord. Chem. Rev.* 1993, **125**, 115.
21. D. Budac and P. Wan, *J. Photochem. Photobiol.*, 1992, **67**, 135.
22. K. Reszka and C. F. Chignell, *Photochem. Photobiol.*, 1983, **38**, 281.
23. F. Bosca, M. A. Miranda, L. Vano and F. Vargas, *J. Photochem. Photobiol., A: Chem.*, 1990, **54**, 131.
24. J. V. Castell, M.J. Gomez-Lechon, M. A. Miranda and I. M. Morrea, *Photochem. Photobiol.*, 1987, **46**, 991.
25. J. V. Castell, M. J. Gomez-Lechon, M. A. Miranda and I. M. Morera, *J. Photochem. Photobiol. B: Biol.*, 1992, **13**, 71.
26. S. Monti, S. Sortino, G. De Guidi and G. Marconi, *J. Chem. Soc., Faraday Trans.*, 1997, **93**, 2269.
27. F. Bosca, M. A. Miranda, G. Carganico and D. Mauleon, *Photochem. Photobiol.*, 1994, **60**, 96

28. J. V. Castell, M. J. Gomez-Lechon, C. Grassa, L. A. Martinez, M. A. Miranda and P. Tarrega, *Photochem. Photobiol.*, 1994, **59**, 35.

29. S. Navaratnam, J. L. Hughes, B. J. Parsons and G. O. Phillips, *Photochem. Photobiol.*, 1985, **41**, 375.

30. F. Bosca, M. A. Miranda and F. Vargas, *J. Pharm. Sci.*, 1992, **81**, 181.

31. L. Gu, H. S. Chiang and D. Johnson, *Int. J. Pharm.*, 1988, **41**, 105.

32. N. Ekiz-Guecer and J. Reisch, *Pharm. Acta Helv.*, 1991, **66**, 66.

33. A. C. Weedon and D. F. Wang, *J. Photochem. Photobiol., A: Chem.*, 1991, **61**, 27.

34. R. Dabestani, R. H. Sik, D. G. Davis, G. Dubay and C. F. Chignell, *Photochem. Photobiol.*, 1993, **58**, 367.

35. A. B. Wu, H. W. Cheng, C. M. Hu, F. A. Chen, T. C. Chou and C. Y. Chen, *Tetrahedron Lett.*, 1997, **38**, 621.

36. D. E. Moore, S. Roberts-Thomson, D. Zhen and C. C. Duke, *Photochem. Photobiol.*, 1990, **52**, 685.

37. J. Philip and D. H. Szulczewski, *J. Pharm. Sci.*, 1973, **62**, 1479.

38. F. Vargas, C. Rivas, R. Machado and Z. Sarabia, *J. Pharm. Sci.*, 1993, 82, 371.

39. J. V. Castell, M. J. Gomez-Lechon, V. Mirabet, M. A. Miranda and I. M. Morea, *J. Pharm. Sci.*, 1987, **76**, 374.

40. J. Jensen, C. Cornett, C. E. Olsen, J. Tjoernelund and S. H. Hansen, *Int. J. Pharm.*, 1992, **88**, 177.

41. J. Reisch and R. Schildgen, *Arch. Pharm. (Weinheim)*, 1972, **305**, 40.

42. J. Reisch and R. Schildgen, *Arch. Pharm. (Weinheim)*, 1972, **305**, 49.

43. D. D. Denson, J. C. Crews, K. W. Grummich, E. J. Stirm and C. A. Sue, *Am. J. Hosp. Pharm.*, 1991, **48**, 515.

44. F. Gimenez, E. Postaire, P. Prognon, M. D. Le Hoang, J. M. Lecomte, D. Pradeau and G. Hazebroucq, *Int. J. Pharm.*, 1988, **43**, 23.

45. J. Reisch and A. Fitzek, *Tetrahedron Lett.*, 1967, 4513.

46. S. N. Ege, *Chem. Commun.*, 1967, 488.

47. B. Marciniec, *Pharmazie*, 1984, **39**, 103.

48. H. Fabre, N. Hussan-Eddine, D. Lerner and B. Mandrou, *J. Pharm. Sci.*, 1984, **73**, 1709.

49. J. Reisch, N. Ekiz and T. Guneri, *Pharmazie*, 1986, **41**, 287.

50. B. Marciniec, *Acta Pol. Pharm.*, 1985, **42**, 448.

51. B. Marciniec, *Pharmazie*, 1983, **38**, 848.

52. J. Reisch and M. Abdel-Khalek, *Pharmazie*, 1979, **34**, 408.

53. J. Reisch, N. Ekiz-Gucer, M. Takacs, G. M. Gunaherath and B. Kamal, *Arch. Pharm. (Weinheim)*, 1989, **322**, 295.

54. V. J. Hemmens and D. E. Moore, *Photochem. Photobiol.*, 1986, **43**, 257.

55. T. M. Chen, A. D. Sill and C. L. Housmyer, *J. Pharm. Biomed. Anal.*, 1986, **4**, 533.

56. G. M. J. Beijersbergen van Henegouwen, H. J. van de Zijde, J. van de Griend and H. de Vries, *Int. J. Pharm.*, 1987, **35**, 259

57. S. Hashiba, M. Tatsuzawa and A. Eijma, *Eisei Shikensho Hokoku*, 1979, 97, 73 through *Chem. Abstr.*, 1980, **93**, 16854n.

58. J. Reisch, L. Iranshani and N. Ekiz-Guecer, *Liebigs Ann. Chem.*, 1992, 1199.

59. C. T. Lin, P. Perrier, G. G. Clay, P. A. Sutton and S. R. Byrn, *J. Org. Chem.*, 1982, **47**, 2978.

60. K. Shaffner and M. Demuth, in 'Rearrangements in the Ground and Excited State', P. de Mayo, Ed., Academic Press, New York, 1980, Vol 3, p. 281.

61. J. R. Williams, R. H. Moore, R. Li and J. F. Blount, *J. Am. Chem. Soc.*, 1979, **101**, 5019

62. J. R. Williams, R. H. Moore, R. Li and C. M. Weeks, *J. Org. Chem.*, 1980, **45**, 2324.

63. J. Reisch, Y. Topaloglu and G. Henkel, *Acta Pharm. Technol.*, 1986, **32**, 115.

64. Y. Arakawa, C. Fukaya, K. Yamanouchi and K. Yokoyama, *Yakagaku Zasshi*, 1985, **105**, 1029.

65. T. Hidaka, S. Huruumi, S. Tamaki, M. Shiraishi and H. Minato, *Yakagaku Zasshi*, 1980, **100**, 72.

66. K. Thoma, R. Kerker and C. Weissbach, *Pharm. Ind.*, 1987, **49**, 961.

67. N. Ekiz-Guecer, J. Reisch, and G. Nolte, *Eur. J. Pharm. Biopharm.*, 1991, **37**, 234.

68. M. Takacs, N. Ekiz-Guecer, J. Reisch and A. Gergely-Zobin, *Pharm. Acta Helv.*, 1991, **66**, 137.

69. J. Reisch, J. Zappel, A. Raghu and R. Rao, *Acta Pharm. Turc.*, 1995, **37**, 13.

70. J. Reisch, G. Henkel, N. Ekiz-Guecer and G. Nolte, *Liebigs Ann. Chem.*, 1992, 63.

71. K. Jochym, H. Barton and J. Bojarski, *Pharmazie*, 1988, **43**, 623.

72. H. Barton, J. Bojarski and J. Mokrosz, *Tetrahedron Lett.*, 1982, **23**, 2133.

73. H. Barton, Z. Alexandra, J. Boiarski and W. Welna, *Pharmazie*, 1983, **38**, 268.

74. H. Barton, J. Bojarski and A. Zurowska, *Arch. Pharm. (Weinheim, Ger.)*, 1986, **319**, 457.

75. J. Mokrosz, A. Zurowska and J. Bojarski, *Pharmazie*, 1982, **37**, 832.

76. H. Barton, J. Mokrosz, J. Bojarski and M. Klimczak, *Pharmazie*, 1980, **35**, 155.

77. H. Barton and J. Bojarski, *Pharmazie*, 1983, **38**, 630.

78. J. Mokrosz and J. Bojarski, *Pharmazie*, 1982, **37**, 768.

79. J. Mokrosz, M. Klimczak, H. Barton and J. Bojarski, *Pharmazie*, 1980, **35**, 205.

80. R. Bouche, M. Draguet-Brughmans, J. P. Flandre, C. Moreaux and M. Van Meersche, *J. Pharm. Sci.*, 1978, **67**, 1019.

81. M. H. Paluchowska and H. J. Bojarski, *Arch. Pharm. (Weinheim)*, 1988, **321**, 343.

82. D. E. Moore, *J. Pharm. Sci.*, 1977, **66**, 1282.

83. R. Andersin, J. Ovaskainen and S. Kaltia, *J. Pharm. Biomed. Anal.*, 1994, **12**, 165.

84. R. Andersin and M. Mesilaakso, *J. Pharm. Biomed. Anal.*, 1995, **13**, 667.

85. H. J. Roth and M. Adomeit, *Arch. Pharm. (Weinheim)*, 1973, **306**, 889.

86. P. J. G. Cornelissen and G. M. J. Beijersbergen van Henegouwen, *Photochem. Photobiol.*, 1979, **30**, 337.

87. R. W. Busker, G. M. J. Beijersbergen van Henegouwen, B. M. C. Kwee and J. H. M. Winkens, *Int. J. Pharm.*, 1987, **36**, 113.

88. R. S. Givens, J. Gingrich and S. Mecklemburg, *Int. J. Pharm.*, 1986, **29**, 67.

89. A. Albini and S. Pietra, 'Heterocyclic *N*-Oxides', CRC Press, Boca Raton, FL, 1991.

90. A. Bakri, G. M. J. Beijersbergen van Henegouwen and J. L. Chanal, *Photodermatology*, 1985, **2**, 205.

91. P. J. G. Cornelissen, G. M. J. Beijersbergen van Henegouwen and K. W. Gerritsma, *Int. J. Pharm.*, 1979, **3**, 205.

92. P. J. G. Cornelissen and G. M. J. Beijersbergen van Henegouwen, *Pharm. Weekblad, Sci. Ed.*, 1980, **2**, 39.

93. J. Reisch, N. Ekiz-Guecer and G. Tewes, *Liebigs Ann. Chem.*, 1992, 69.

94. A. L. W. Po and W. J. Irwin, *J. Pharm. Pharmacol.*, 1980, **32**, 25.

95. D. E. Moore and S. R. Tamat, *J. Pharm. Pharmacol.*, 1980, **32**, 172.

96. C. L. Huang and F. L. Sands, *J. Pharm. Sci.*, 1967, **56**, 259.

97. A. S. W. Li and C. F. Chignell, *Photochem. Photobiol.*, 1987, **45**, 695.

98. F. W. Grant and J. Greene, J. Toxicol., *Appl. Pharmacol.*, 1972, **23**, 71.

99. I. Rosenthal, E. Ben-Hur, A. Prager and E. Riklis, *Photochem. Photobiol.*, 1978, **28**, 591.

100. A. K. Davies, S. Navaratnam and G. O. Phillips, *J. Chem. Soc., Perkin Trans. 2*, 1976, 25.

101. A. G. Motten, G. R. Buettner and C. F. Chignell, *Photochem. Photobiol.*, 1985, **42**, 9.

102. C. L. Huang and F. L. Sands, *J. Chromatogr.*, 1964, **13**, 246.

103. A. Felmeister and C. A. Discher, *J. Pharm. Sci.*, 1964, **53**, 756.

104. D. Sharples, *J. Pharm. Pharmacol.*, 1981, **33**, 242.

105. R. D. Hall, G. R. Buettner, A. G. Motten and C. F. Chignell, *Photochem. Photobiol.*, 1987, **46**, 295.

106. E. Pawelczyk and B. Marciniec, *Pharmazie*, 1974, **29**, 585.

107. E. M. Abdel-Moety, K. A. Al-Rashood, A. Rauf and N. A. Khattab, *J. Pharm. Biomed. Anal.*, 1996, **14**, 1639.

108. E. Pawelczyk and B. Marciniec, *Pol. J. Pharmacol. Pharm.*, 1977, **29**, 143.

109. P. S. Song, E. C. Smith and D. E. Metzler, *J. Am. Chem. Soc.*, 1965, **87**, 4181.

110. S. Tammilehto and K. Torniainen, *Int. J. Pharm.*, 1989, **52**, 123.

111. G. E. Jones and D. Sharples, *J. Pharm. Pharmacol.*, 1984, **36**, 46.

112. F. P. Gasparro and I. E. Kochevar, *Photochem. Photobiol.*, 1982, **35**, 351.

113. Y. Matsuda, R. Akazawa, R. Teraoka and M. Otsuka, *J. Pharm. Pharmacol.*, 1994, **46**, 162.

114. H. C. Chiang and S. Y. Li, *J. Taiwan Pharm. Assoc.*, 1977, **29**,70.

115. N. Ekiz-Guecer and J. Reisch, *Liebigs Ann. Chem.*, 1991, 1105.

116. Y. Matsuda and R. Masahara, *J. Pharm. Sci.*, 1983, **72**, 1198.

117. M. Schleuder, P. H. Richter, A. Keckeis and T. Jira, *Pharmazie*, 1993, **48**, 33.

118. N. Paillous and M. Verrier, *Photochem. Photobiol.*, 1988, **47**, 337.

119. A. S. W. Li and C. F. Chignell, *Photochem. Photobiol.*, 1987, **45**, 191.

120. P. Pietta, A. Rava and P. A. Biondi, *J. Chromatogr.*, 1981, **210**, 516.

121. W. A. Al-Turk, I. A. Majeed, W. J. Murray, D. W. Newton and S. Othman, *Int. J. Pharm.*, 1988, **41**, 227.

122. G. S. Sadana and A. B. Ghogare, *Int. J. Pharm.*, 1991, **70**, 195.

123. F. Vargas, C. Rivas and R. Machado, *J. Pharm. Sci.*, 1992, **81**, 399.

124. Y. Matsuda, R. Teraoka and I. Sugimoto, *Int. J. Pharm.*, 1989, **54**, 211.

125. S. Ogawa, Y. Itagaki, N. Hayase, I. Takemoto, N. Kasahara, S. Akutsu and S. Inagaki, *Byoin Yakugaku*, 1990, **16**, 189 through *Chem. Abstr.*, 1991, **114**, 12062g.

126. B. Marciniec and W. Rychcik, *Pharmazie*, 1994, **49**, 894.

127. K. Thoma and R. Kerker, *Pharm. Ind.* 1992, **54**, 465.

128. L. J. Nunez-Vergara, C. Sunkel and J. A. Squella, *J. Pharm. Sci.*, 1994, **83**, 502

129. J. A. Squella, A. Zanocco, S. Perna and L. J. Nunez-Vergara, *J. Pharm. Biomed. Anal.*, 1990, **8**, 43.

130. A. L. Zanocco, L. Diaz, M. Lopez, L. J. Nunez-Vergara and J. A. Squella, *J. Pharm. Sci.*, 1992, **81**, 920.

131. M. C. Bonferoni, G. Mellerio, P. Giunchedi, C. Caramella and U. Conte, *Int. J. Pharm.*, 1992, **87**, 133.

132. M. S. Suleiman, M. E. Abdulhameed, N. M. Najib and H. Y. Muti, *Int. J. Pharm.*, 1989, **50**, 71.

133. N. Ekiz-Guecer and J. Reisch, *Acta Pharm. Turc.*, 1990, **32**, 103.

134. N. Jamil, H. Afrozrizvi, I. Ahmed and A. Jazbeg, *Pharmazie*, 1983, **38**, 467.

135. G. E. Wright and T. Y. Tang, *J. Pharm. Sci.*, 1972, **61**, 299.

136. Y. Asaki, K. Shinzaki and M. Nagaoka, *Chem. Pharm. Bull.*, 1971, **19**, 1079.

137. C. F. Carney, *J. Pharm. Sci*, 1987, **76**, 393

138. S. W. Davidson and D. Lyall, *Pharm. J.*, 1987, 599.

139. O. R. Leuwenkamp, E. J. Van der Merk, W. P. Bennekom and A. Bult, *Int. J. Pharm.*, 1985, **24**, 27.

140. R. Karlicek and V. Klimesova, *Cesk. Farm.*, 1977, **26**, 413 through *Chem. Abstr.*, 1978, **89**, 12047.

141. H. Auterhoff and A. Walter, *Dtsch. Apoth. Ztg.*, 1975, **115**, 1931.

142. De Mol, N. J. and G. M. J. Beijersbergen van Henegouwen, *Photochem. Photobiol.*, 1979, **29**, 479.

143. I. Kruk, *Z. Phys. Chem. (Leipzig)*, 1985, **226**, 1239.

144. L. S. Jahnke and A. W. Frenkel, *Photochem. Photobiol.*, 1978, **28**, 517.

145. M. Yoshioka, Y. Kirino, Z. Tamura and T. Kwan, *Chem. Pharm. Bull.*, 1977, **25**, 75.

146. K. Usmanghani, I. Ahmad and S. M. S. Zoha, *Pakistan J. Sci. Ind. Res.*, 1975, **18**, 229.

147. V. Ulvi, M. Mesilaakso and J. Matikainen, *Pharmazie*, 1996, **51**, 774.

148. F. Vargas, C. Rivas and N. Canudas, *J. Pharm. Sci.*, 1993, **82**, 590.

149. F. Vargas and N. Canudas, *Pharmazie*, 1993, **48**, 900.

150. B. van der Walt and H. J. Cahnmann, *Proc. Nat. Acad. Sci. USA*, 1982, **79**, 1492.

151. K. Kigasawa, H. Shimizu, S. Hayashida and K. Ohkubo, *Yakugaku Zasshi*, 1984, **104**, 1191.

152. J. Reisch and J. Zappel, *Sci. Pharm.*, 1993, **61**, 283.

153. A. G. J. Sedee, G. M. J. Beijersbergen van Henegouwen, H. de Vries, W. Guijt and C. A. G. Haasnoot, *Pharm. Weekbl., Sci. Ed.*, 1985, **7**, 194.

154. A. G. J. Sedee, G. M. J. Beijersbergen van Henegouwen, W. Guijt and C. A. G. Haasnoot, *Pharm. Weekbl., Sci. Ed.*, 1985, **7**, 202.

155. A. G. J. Sedee, G. M. J. Beijersbergen van Henegouwen and H. J. A. Blaauwgeers, *Int. J. Pharm.*, 1983, **15**, 149

156. A. G. J. Sedee and G. M. J. Beijersbergen van Henegouwen, *Arch. Pharm. (Weinheim)*, 1985, **318**, 111.

157. A. G. J. Sedee, G. M. J. Beijersbergen van Henegouwen, N. J. de Mol and G. Lodder, *Chem. Biol. Interact.*, 1984, **51**, 357.

158. P. Lupon, F. Grau and J. J. Bonet, *Helv. Chim. Acta*, 1984, **67**, 332.

159. P. Lupon, J. Gomez and J. J. Bonet, *Angew. Chem., Int. Ed. Engl.*, 1983, **22**, 711.

160. J. Reisch, J. Zappel, A. R. R. Akinepalli and G. Henkel, *Pharm. Acta Helv.*, 1994, **69**, 97.

161. J. Reisch, J. Zappel, G. Henkel and N. Ekiz-Guecer, *Monatsh. Chem.*, 1993, **124**, 1169.

162. N. Ekiz-Guecer, J. Zappel and J. Reisch, *Pharm. Acta Helv.*, 1991, **66**, 2.

163. J. Reisch, N. Ekiz-Guecer, M. Takacs and G. Henkel, *Liebigs Ann. Chem.*, 1989, 595.

164. J. Reisch, N. Ekiz-Guecer and M. Takacs, *Arch. Pharm. (Weinheim)*, 1989, **322**, 173.

165. K. A. Muszkat, *Top. Curr. Chem.*, 1980, **88**, 89.

166. R. G. Frith and G. Phillipou, *J. Chromatogr.*, 1986, **367**, 260.

167. D. W. Mendenhall, H. Kobayashi, F. M. L. Shih, L. A. Sternson, T. Higuchi and C. Fabian, *Clin. Chem.*, 1978, 1518.

168. T. D. Doyle, W. R. Benson and N. Filipescu, *J. Am. Chem. Soc.*, 1976, **98**, 3262.

169. P. Hugelshofer, J. Kalvoda and K. Schaffner, *Helv. Chim. Acta*, 1960, **43**, 1322.

170. T. D. Doyle, W. R. Benson and N. Filipescu, *Photochem. Photobiol.*, 1978, **27**, 3.

171. K. Muller, K. K. Mayer and W. Wiegrebe, *Arch. Pharm. (Weinheim)*, 1987, **320**, 59.

172. F. Dall'Acqua, S. M. Magno, F. Zambon and G. Rodighiero, *Photochem. Photobiol.*, 1979, **29**, 489.

173. N. J. de Mol and G. M. J. Beijersbergen van Henegouwen, *Photochem. Photobiol.*, 1981, **33**, 815.

174. S. A. Langford, J. K. Sugden and R. W. Fitzpatrick, *J. Pharm. Biomed. Anal.*, 1996, **14**, 1615.

175. F. Golpashin, B. Weiss and H. Durr, *Arch. Pharm. (Weinheim)*, 1984, **317**, 906.

176. A. G. Motton and C. F. Chignell, *Photochem. Photobiol.*, 1983, **37**, 17.

177. W. Zhou and D. E. Moore, *Int. J. Pharm.*, 1994, **110**, 55.

178. E. J. Land, S. Navaratnam, B. J. Parsons and G. O. Phillips, *Photochem. Photobiol.*, 1982, **35**, 637.

179. S. Tammilehto, *Acta Pharm. Fenn.*, 1989, **98**, 225.

180. J. Pawlaczyck and W. Turowska, *Acta Pol. Pharm.*, 1976, **33**, 505.

181. J. Reisch and D. H. Niemeyer, *Arch. Pharm. (Weinheim)*, 1972, **305**, 135.

182. H. C. Chiang, S. J. S. Chen and L. C. Yeh, *J. Taiwan Pharm. Assoc.*, 1976, **28**, 36.

183. A. K. Davies, J. F. McKellar, G. O. Phillips and A. G. Reid, *J. Chem. Soc., Perkin Trans. 2*, 1979, 369.

184. D. E. Moore, M. P. Fallon and C. D. Burt, *Int. J. Pharm.*, 1983, **14**, 133.

185. A. S. W. Li, H. P. Roethling, K. B. Cummings and C. F. Chignell, *Biochem. Biophys. Res. Commun.*, 1987, 1191.

186. A. S. W. Li, C. F. Chignell and R. D. Hall, *Photochem. Photobiol.*, 1987, **46**, 379.

187. I. K. Shih, *J. Pharm. Sci.*, 1971, **60**, 1889.

188. J. Reisch and K. G. Weidmann, *Arch. Pharm. (Weinheim)*, 1971, **304**, 911.

189. S. I. M. Mubarak, J. B. Standford and J. K. Sudgen, *Pharm. Acta Helv.*, 1982, **57**, 226.

190. S. I. M. Mubarak, J. B. Standford and J. K. Sudgen, *Pharm. Acta Helv.*, 1983, **58**, 343.

191. K. Akimoto, K. Inoue and I. Sugimoto, *Chem. Pharm. Bull.*, 1985, **33**, 4050.

192. K. Akimoto, H. Nakagawa and I. Sugimoto, *Drug. Devel. Ind. Pharm.*, 1985, **11**, 865. (contr)

193. G. W. Schaffer, E. Nikawitz, M. Manowitz and H. U. Daeniker, *Photochem. Photobiol.*, 1971, **13**, 347.

194. A. K. Davies, N. S. Hilal, J. F. McKellar and G. O. Phillips, *Brit. J. Dermatol.*, 1975, **92**, 143.

195. H. Fabre, H. Ibork and D. A. Lerner, *J. Pharm. Biomed. Anal.*, 1992, **10**, 645.

196. D. A. Lerner, G. Bannefond, H. Fabre, B. Mandrou and M. Simeon de Buochberg, *J. Pharm. Sci.*, 1988, **77**, 699.

197. A. Albini, M. Alpegiani, D. Borghi, S. Del Nero, E. Fasani and E. Perrone, *Tetrahedron Lett.*, 1995, **36**, 4633.

198. B. J. Wilkins, G. J. Gainsford and D. E. Moore, *J. Chem. Soc., Perkin Trans. 1*, 1987, 1817.

199. K. H. Pfoertner and J. J. Daly, *Helv. Chim. Acta*, 1987, **70**, 171.

200. D. E. Moore and B. J. Wilkins, *Radiat. Phys. Chem.*, 1990, **36**, 547.

201. D. E. Moore, C. F. Chignell, R. H. Sik and A. G. Motten, *Int. J. Radiat. Biol. Relat. Stud. Phys., Chem.. Med.*, 1986, **50**, 885.

202. R. Godfrey and R. Edwards, *J. Pharm. Sci.*, 1991, **80**, 212.

203. A. B. Narbutt-Mering and W. Weglowska, *Acta Pol. Pharm.*, 1980, **37**, 301.

204. M. A. Quillian, B. E. McCarry, K. H. Hoo, D. R. McCalla and S. Vaitekunas, *Can. J. Chem.*, 1987, **65**, 1128.

205. I. Ninomiya and O. Yamamoto, *Heterocycles*, 1976, **4**, 475.

206. H. C. Chiang and L. J. Lin, *J. Chin. Chem. Soc.*, 1978, **25**, 125.

207. P. G. E. Evans, J. K. Sugden and N. J. Van Abbe, *Pharm. Acta Helv.*, 1975, **50**, 94.

208. J. J. Bergh, J. C. Breytenbach and P. L. Wessels, *J. Pharm. Sci.*, 1989, **78**, 348.

209. N. Detzer and B. Huber, *Tetrahedron*, 1975, **31**, 1937.

210. D. E. Moore, V. J. Hemmens and H. Yip, *Photochem. Photobiol.*, 1984, **39**, 57.

211. P. Bliski, L. J. Martinez, E. B. Koker and C. F. Chignell, *Photochem. Photobiol.*, 1996, **64**, 496.

212. Y. Yoshida, E. Sato and R. Moroi, *Arzneim. Forsch.*, 1993, **43**, 601.

213 K. Torniainen, J. Mattinen, C. P. Askolin and S. Tammiletho, *J. Pharm. Biomed. Anal.*, 1997, **15**, 887.

214. L. J. Martinez, G. Li and C. F. Chignell, *Photochem. Photobiol.*, 1997, **65**, 599.

215. E. Fasani, M. Mella, D. Caccia, S. Tassi, M. Fagnoni and A. Albini, *J. Chem. Soc., Chem. Commun.*, 1997, 1329.

216. L. Pasteur, *Compt. Rend.*, 1853, **37**, 110.

217. W. A. Laurie, D. McHale and K. Saag, *Tetrahedron*, 1986, **42**, 3711.

218. W. A. Laurie, D. McHale, K. Saag and J. B. Sheridan, *Tetrahedron*, 1988, **44**, 5905.

219. D. McHale, W. A. Laurie, K. Saag and J. B. Sheridan, *Tetrahedron*, 1989, **45**, 2127.

220. G. A. Epling and U. C. Yoon, *Chem. Lett.*, 1982, 211.

221. H. H. Tønnesen amd A. L. Grislingaas, *Int. J. Pharm.*, 1990, **60**, 157.

222. K. Nord, J. Karlsen and H. H. Toennesen, *Int. J. Pharm.*, 1991, **72**, 11.

223. J. A. Owoyale, *Int. J. Pharm.*, 1989, **52**, 179.

224. J. J. Aaron and J. Fidanza, *Talanta*, 1982, **29**, 383.

225. H. H. Tønnesen, A. L. Grislingaas, S. O. Woo and J. Karlsen, *Int. J. Pharm.*, 1988, **43**, 215.

226. J. A. Owoyale, *Int. J. Pharm.*, 1989, **56**, 213.

227. A. Brossi, W. P. Gessner, C. D. Hufford, J. K. Baker, F. Homo, P. Millet and I. Landau, *FEBS Lett.*, 1987, **223**, 77.

228. S. Kristensen, A. L. Grislingaas, J. V. Greenhill, T. Skjetne, J. Karlsen and H. H. Toennesen, *Int. J. Pharm.*, 1993, **100**, 15.

229. C. Schuijt, G. M. Beijersbergen van Henegouwen and K. W. Gerritsma, *Pharm. Weekblad, Sci. Ed.*, 1979, **1**, 186.

230. A. D. Villarejo, P. Revilla, P. Alonso and A. Doadrio, *An. R. Acad. Farm.*, 1986, **52**, 265.

231. P. A. Zieske, M. Koberda, J. L. Hines, C. C. Knight, R. Sriram, N. V. Raghavan and B. E. Rabinow, *Am. J. Hosp. Pharm.*, 1991, **48**, 1500.

232. R. F. Greene, D. C. Chatterji, P. K. Hiranaka and J. F. Gallelli, *Am. J. Hosp. Pharm.*, 1979, **36**, 38.

233. M. Macka, J. Borak, L. Semenkova and F. Kiss, *J. Pharm. Sci.*, 1994, **83**, 815.

234. K. Fredriksson, P. Lundgren and L. Landersjoe, *Acta Pharm. Suec.*, 1986, **23**, 115.

235. R. F. Betteridge, A. L. Culverwell and A. G. Bosanquet, *Int. J. Pharm.*, 1989, **56**, 31.

236. A. I. Torres Suarez and M. A. Camacho, *Arzneim. Forsch.*, 1994, **44**, 81.

237. J. Black, D. D. Buetcher, J. W. Chinn, J. Gard and D. E. Thurston, *J. Pharm. Sci.*, 1988, **77**, 630.

238. V. J. Hemmens and D. E. Moore, *J. Chem. Soc., Perkin Trans. 2*, 1984, 209.

239. J. K. Horton, M. F. G. Stevens, *J. Pharm. Pharmacol.*, 1981, **33**, 808.

240. D. C. Chatterji and J. F. Gallelli, *J. Pharm. Sci.*, 1978, **67**, 526.

241. C. Chahidi, M. Giraud, M. Aubailly, A. Valla and R. Santus, *Photochem. Photobiol.*, 1986, **44**, 231.

242. C. Chahidi, P. Morliere, M. Aubailly, L. Dubertret and R. Santus, *Photochem. Photobiol.*, 1983, **38**, 317.

243. M. J. Wood, W. J. Irwin and D. K. Scott, *J. Clin. Pharm. Ther.*, 1990, **15**, 291.

244. D. M. Maniez-Devos, R. Baurain, M. Lesne and A. Trouet, *J. Pharm. Biomed. Anal.*, 1986, **4**, 353.

245. A. S. Li and C. F. Chignell, *Photochem. Photobiol.*, 1987, **45**, 191.

246. A. Fredenhagen and U. Séquin, *Helv. Chim. Acta*, 1985, **68**, 391.

247. K. T. Douglas, H. A. M. Ratwatte and N. Thakrar, *Bull. Cancer*, 1983, **70**, 372.

248. N. Thakrar and K. T. Douglas, *Cancer Lett. (Shannon, Irel.)*, 1981, **13**, 265.

249. K. T. Douglas, *Biomed. Pharmacother.*, 1983, **37**, 191.

250. R. M. Burger, J. Peisach and S. B. Horwitz, *J. Biol. Chem.*, 1978, **253**, 4830.

251. M. C. Allwood and J. H. Plane, *Int. J. Pharm.*, 1986, **31**, 1.

252. R. W. Curley and J. W. Fowble, *Photochem. Photobiol.*, 1988, **47**, 831.

253. H.C.J. Jacobs and E. Havinga, *Adv. Photochem.*, 1979, **11**, 305.

254. R. Mermet-Bouvier, *Bull. Soc. Chim. Fr.*, 1973, 3023.

255. A. E. C. Snoeren, M. R. Daha, J. Lugtenburg and E. Havinga, *Rec. Trav. Chim. Pays-Bas*, 1970, **89**, 261.

256. R. Huettenrauch, S. Fricke and K. Matthey, *Pharmazie*, 1985, **41**, 742.

257. R. L. Clough, B. G. Yee and C. S. Foote, *J. Am. Chem. Soc.*, 1979, **101**, 683.

258. J. M. L. Mee, C. C. Brooks and K. H. Yanagihara, *J. Chromatogr.*, 1975, **110**, 178.

259. K. Nakagawa, E. Kouno, K. Tabuchi, M. Kidawara, A. Osuka and K. Maruyama, *Niihama Kogyo Koto Semmon Gakko Kiyo, Rikogaku Hen*, 1988, **24**, 96.
260. B. Marciniec and D. Witkowska, *Acta Pol. Pharm.*, 1988, **45**, 528.
261. H. Werbin and E. T. Strom, *J. Am. Chem. Soc.*, 1968, **90**, 7296.
262. Y. Yamano, S. Ikenoya, M. Ohmae and K. Kawabe, *Yakagaku Zasshi*, 1979, **99**, 1102.
263. C. Snyder and H. Rapoport, *J. Am. Chem. Soc.*, 1969, **91**, 731.
264. M. Ohmae and G. Katsui, *Vitamins*, 1969, **39**, 190.
265. M. Ohmae and G. Katsui, *Vitamins*, 1969, **39**, 181.
266. Y.Yamano, S. Ikenoya, M. Anze, M. Ohmae and K. Kawabe, *Yakagaku Zasshi*, 1978, **98**, 774.
267. R. Teraoka and Y. Matsuda, *Int. J. Pharm.*, 1993, **93**, 85.
268. R. M. Wilson, T. F. Walsh and S. K. Gee, *Tetrahedron Lett.*, 1980, **21**, 3459.
269. N. Hameed and H. D. C. Rapson, *J. Pharm. Pharmacol.*, 1972, **24** Suppl., 139 P.
270. P. Gygax, *J. Agric. Food Chem.*, 1981, **29**, 172.
271. H. M. van Dort, L. M. van der Linde and D. de Rijke, *J. Agric. Food Chem.*, 1984, **32**, 454.
272. G. Strauss and W. J. Nickerson, *J. Am. Chem. Soc.*, 1961, **83**, 3187.
273. W. C. Dunlap and M. Susic, *Mar. Chem.*, 1986, **19**, 99.
274. G. Oster, J. S. Bellin and B. Holmstrom, *Experientia*, 1962, **18**, 249.
275. P. S. Song, E. C. Smith and D. E. Metzler, *J. Am. Chem. Soc.*, 1965, **87**, 4181.
276. W. Cairns and D. E. Metzler, *J. Am. Chem. Soc.*, 1971, **93**, 2772.
277. M. S. Jorns, G. Schollnhammer and P. Hemmerich, *Eur. J. Biochem.*, 1975, **57**, 35.
278. I. Ahmad and H. D. C. Rapson, *J. Pharm. Biomed. Anal.*, 1990, **8**, 217.
279. F. Takeuchi, T. Sugiyama, T. Fujimori, K. Seki, Y. Harada and A. Sugimori, *Bull. Chem. Soc. Jpn.*, 1974, **47**, 1245.
280. S. Ikeda, T. Oka, N. Ohishi and S. Fukui, *Vitamins*, 1968, **38**, 109.
281. N. P. Bazhulina, M. P. Kirpichnikov, Y. V. Morozov, F. A. Savin, L. B. Sinyavina and V. L. Florentiev, *Mol. Photochem.*, 1974, **6**, 367.
282. B. Kraeutler and R. Stepanek, *Helv. Chim. Acta*, 1985, **68**, 1079.
283. K. Azakane, A. Ryu, K. Takarada, T. Masunaga, K. Shirimoto, R. Kobayashi, S. Mashiko, T. Nagano, and M. Hirobe, *Chem. Pharm. Bull.*, 1996, **44**, 1
284. M. M. de Villers, J. G. van der Watt and A. P. Loetter, *Int. J. Pharm.*, 1992, **88**, 275.
285. Y. Matsuda and R. Masahara, *Yakugaku Zasshi*, 1980, **100**, 953.
286. H. Nyqvist, *Acta Pharm. Suec.*, 1984, **21**, 245.
287. J. Reisch and Y. Topaloglu, *Pharm. Acta Helv.*, 1986, **61**, 142.
288. R. Lisewski and K. L. Wierzchowski, *Photochem. Photobiol.*, 1970, **11**, 327.
289. J. Reisch, *Dtsch. Apoth. Ztg.*, 1979, **119**, 1.
290. E. Fasani, M. Mella, S. Monti, S. Sortino and A. Albini, *J. Chem. Soc., Perkin Trans. 2*, 1996, 1889.
291. E. Fasani, M. Mella and A. Albini, *Chem. Pharm. Bull.*, 1997, **45**, 394.
292. H. Nyqvist and T. Wadsten, *Acta Pharm. Technol.*, 1986, **32**, 130.
293. D. R. Merrifield, P. L. Carter, D. Clapham and F. D. Sanderson, in ref. 1, p. 141.
294. K. Thoma, M. Marschall and O. E. Schubert, *Dtsch. Apoth. Ztg.*, 1991, **34**, 1739.

295. D. P. Bluhm, R. S. Summers, M. M. J. Lowes and H. H. Duerrheim, *Int. J. Pharm.*, 1991, **68**, 277.

296. D. P. Bluhm, R. S. Summers, M. M. J. Lowes and H. H. Durrheim, *Int. J. Pharm.*, 1991, **68**, 281.

297. M. F. Chen, H. W. Boyce and L. Triplett, *J. Parenter. Enteral. Nutr.*, 1983, **7**, 462.

298. D. W. Mendenhall, *Drug Dev. Ind. Pharm.*, 1984, **10**, 1297.

299. J. L. Smith, J. E. Canham and P. A. Wells, *J. Parenter. Enteral. Nutr.*, 1988, **12**, 394.

300. M. F. Williams, L. J. Hak and G. Dukes, *Am. J. Hosp, Pharm.*, 1990, **47**, 1574.

301. G. M. R. Vandenbossche, C. De Muynck, F. Colardyn and J. P. Remon, *J. Pharm. Pharmacol.*, 1993, **45**, 486.

302. O. Dyvik, A. L. Grislingaas, H. H. Toennesen and J. Karlsen, *J. Clin. Hosp. Pharm.*, 1986, **11**, 343.

303. C. Mahony, J. E. Brown, W. W. Stargel, C. P. Verghese and T. D. Bjornsson, *J. Pharm. Sci.*, 1984, **73**, 838.

304. C. J. Vesey and G. A. Batistoni, *J. Clin. Pharm.*, 1977, **2**, 105.

305. D. W. Newton, E. Y. Y. Fung and D. A. Williams, *Am. J. Hosp. Pharm.*, 1981, **38**, 1314.

306. E. R. Block and J. E. Bennet, *Antimicrob. Agents Chemother.*, 1973, **4**, 648.

307. J. C. Glascok, J. DiPiro, D. E. Cadwallader and M. Perri, *Am. J. Hosp. Pharm.*, 1987, **44**, 2291.

308. Y. Pramar, V. Das Gupta, S. N. Gardner and B. Yau, *J. Clin. Pharm. Ther.*, 1991, **16**, 203.

309. K. Thoma and N. Kuebler, *Pharmazie*, 1996, **51**, 919.

310. K. Thoma and N. Kuebler, *Pharmazie*, 1996, **51**, 940.

311. W. E. Hamlin, T. Chulski, R. H. Johnson and J. G. Wagner, *J. Am. Pharm. Assoc., Sci, Ed.*, 1960, **49**, 234.

312. A. G. Bosanquet, *Cancer. Chemother. Pharmacol.*, 1985, **14**, 83.

313. C. B. Eap, L. Koeb and P. Baumann, *J. Pharm. Biomed. Anal.*, 1993, **11**, 451.

314. H. K. Adam, M. A. Gay and R. H. Moore, *J. Endocrinol.*, 1980, **84**, 35.

315. N. H. Anderson, M. A. McLelland and P. Munden, *J. Pharm. Biomed. Anal.*, 1991, **9**, 443.

316. N. H. Anderson, D. Johnston, M. A. McLelland and M. A. Munden, *Manuf. Chem.*, 1991, **6**, 25.

317. S. Nema, R. J. Washkuhn and D. R. Beussink, *Pharm. Technol.*, 1995, **19**, 170.

318. M. Matsuo, Y. Machida, H. Furuichi, K. Nakamura and Y. Takeda, *Drug Stability*, 1996, **1**, 179.

319. K. Thoma and R. Kerker, *Pharm. Ind.*, 1992, **54**, 169.

320. K. Thoma and R. Kerker, *Pharm. Ind.*, 1992, **54**, 287.

321. K. Thoma and N. Kuebler, *Pharmazie*, 1996, **51**, 660.

322. Y. Matsuda, R. Teraoka and I. Sugimoto, *Int. J. Pharm.*, 1989, **54**, 211.

323. H. Sekine, Y. Ohta and T. Nakagawa, *Drug Stability*, 1996, **1**, 135. S. Yoshioka, Y. Ishihara, T. Terazono, N. Tsunakawa, M. Murai, T. Yasuda, K. Kitamura, Y. Kunihiro, K. Sakai, Y. Hirose, K. Tonooka, K. Takayama, F. Imai, M. Godo, M. Matsuo, K. Nakamura, Y. Aso, S. Kojima, Y. Takeda and T. Terao, *Drug Dev. Ind. Pharm.*, 1994; **20**, 2049.

324. J. T. Piechocki and R. J. Wolters, *Pharm. Technol.*, 1993, (6), 46.

325. K. Thoma and R. Klimek, *Int. J. Pharm.*, 1991, **67**, 169.

326. K. Thoma and R. Kerker, *Pharm. Ind.*, 1992, **54**, 359.

327. Y. Matsuda, H. Inouye and R. Nakanishi, *J. Pharm. Sci.*, 1978, **67**, 196.

328. K. Thoma and R. Kerker, *Pharm. Ind.*, 1992, **54**, 551.

329. M. Shahjahan, *East. Pharm.*, 1989, **32**, 37.

330. I. Sugimoto, K. Tohgo, K. Sasaki, H. Nakagawa, Y. Matsuda and R. Masahara, *Yakugaku Zasshi*, 1981, **101**, 1149.

331. R. Teraoka, Y. Matsuda and I. Sugimoto, *J. Pharm. Pharmacol.*, 1989, **41**, 293.

332. S. R. Bechard, O. Quaraishi and E. Kwong, *Int. J. Pharm.*, 1992, **87**, 133.

333. Y. Matsuda, T. Itooka and Y. Mitsuhashi, *Chem. Pharm. Bull.*, 1980, **28**, 2665.

334. B. Ameer, R. J. Callahan and S. C. Dragotakes, *J. Pharm. Technol.*, 1989, **5**, 202.

335. A. M. Yahya, J. C. McElnay and P. F. D'Arcy, *Int. J. Pharm.*, 1986, **31**, 65.

336. D. Milanovic and J. G. Nairn, *Am. J. Hosp. Pharm.*, 1980, **37**, 164.

337. H. Wollmann and R. Gruenert, *Pharmazie*, 1984, **39**, 161.

338. B. Bhadresa and J. K. Sugden, *Pharm. Acta Helv.*, 1981, **56**, 122.

339. R. Gruenert and H. Wollmann, *Pharmazie*, 1982, **37**, 798.

340. K. Toma and R. Klimek, *Pharm. Ind.*, 1991, **53**, 504.

341. H. Takeuchi, H. Sasaki, T. Niwa, T. Hino, Y. Kawashima, K. Uesughi and H. Ozawa, *J. Pharm. Sci.*, 1992, **86**, 25.

342. D. S. Desai, M. A. Abdelnasser, B. A. Rubitski and S. A. Varia, *Int. J. Pharm.*, 1994, **103**, 69.

343. M. J. Habib and A. F. Asker, *Drug Dev. Ind. Pharm.*, 1989, **15**, 845.

344. M. J. Habib and A. F. Asker, *Drug Dev. Ind. Pharm.*, 1989, **15**, 1905.

345. M. J. Habib and A. F. Asker, *J. Parenter. Sci. Technol.*, 1989, **43**, 259.

346. K. Thoma and R. Kerker, *Pharm. Ind.*, 1992, **54**, 630.

347. N. S. Allen and J. F. McKellar, 'Photochemistry of Dyed and Pigmented Polymers', Applied Science Pub., London, 1980.

348. D. Teshima, K. Otsubo, S. Higuchi, F. Hirayama, K. Uekama and T. Aoyama, *Chem. Pharm. Bull.*, 1989, **37**, 1591.

349. J. Mielcarek, *Pharmazie*, 1996, **51**, 477.

350. J. Mielcarek, *J. Pharm. Biomed. Anal.*, 1997, **15**, 681.

351. K. Tomono, H. Gotoh, M. Okamura, H. Ueda, T. Saitoh and T. Nagai, *Yakuzaigaku*, 1988, **48**, 322.

352. T. Utsuki, K. Imamura, F. Hirayama and K. Uekama, *Eur. J. Pharm. Sci.*, 1993, **1**, 81.

353. A. F. Asker and M. J. Habib, *J. Parenter. Sci. Technol.*, 1988, **42**, 153.

354. M. S. Islam and A. F. Asker, *J. Pharm. Sci. Technol.*, 1994, **48**, 38.

355. A. F. Asker and M. J. Habib, *J. Parenter. Sci. Technol.*, 1989, **43**, 204.

356. A. J. Ferdous and A. F. Asker, *Drug Dev. Ind. Pharm.*, 1996, **22**, 119.

357. A. F. Asker and M. J. Habib, *J. Parenter. Sci. Technol.*, 1991, **45**, 113.

358. A. F. Asker and M. Larose, *Drug Dev. Ind. Pharm.*, 1987, **13**, 2239.

359. Y. Matsuda and R. Teraoka, *Int. J. Pharm.*, 1985, **26**, 289.

360. M. J. Habib and A. F. Asker, *Pharm. Res.*, 1989, **6**, 58.

APPENDIX

**Stability Testing: Photostability Testing of New Drug Substances and Products
ICH Harmonised Tripartite Guideline** [a]

1 GENERAL
Having reached *Step 4* of the ICH Process at the ICH Steering Committee meeting on 6
November 1996, this guideline is recommended for adoption to the three regulatory parties to
ICH.

The ICH Harmonised Tripartite Guideline covering the Stability Testing of New Drug
Substances and Products (hereafter referred to as the Parent Guideline) notes that light
testing should be an integral part of stress testing. This document is an annex to the Parent
Guideline and addresses the recommendations for photostability testing.

1.1 Preamble

The intrinsic photostability characteristics of new drug substances and products should be
evaluated to demonstrate that, as appropriate, light exposure does not result in unacceptable
change. Normally, photostability testing is carried out on a single batch of material selected
as described under Selection of Batches in the Parent Guideline. Under some circumstances
these studies should be repeated if certain variations and changes are made to the product
(e.g., formulation, packaging). Whether these studies should be repeated depends on the
photostability characteristics determined at the time of initial filing and the type of variation
and/or change made.

The guideline primarily addresses the generation of photostability information for
submission in Registration Applications for new molecular entities and associated drug
products. The guideline does not cover the photostability of drugs after administration (i.e.
under conditions of use) and those applications not covered by the Parent Guideline.
Alternative approaches may be used if they are scientifically sound and justification is
provided.

A systematic approach to photostability testing is recommended covering, as appropriate,
studies such as:

i)   Tests on the drug substance;
ii)   Tests on the exposed drug product outside of the immediate pack;
and if necessary ;
iii) Tests on the drug product in the immediate pack;
and if necessary ;
iv)   Tests on the drug product in the marketing pack.

---

a. By kind permission of ICH (ICH Secretariat c/o IFPMA, 30, r . de Saint-Jean, P. O. Box 9,
1211 Genève 18, Switzerland.

DECISION FLOW CHART FOR
PHOTOSTABILITY TESTING
OF DRUG PRODUCTS

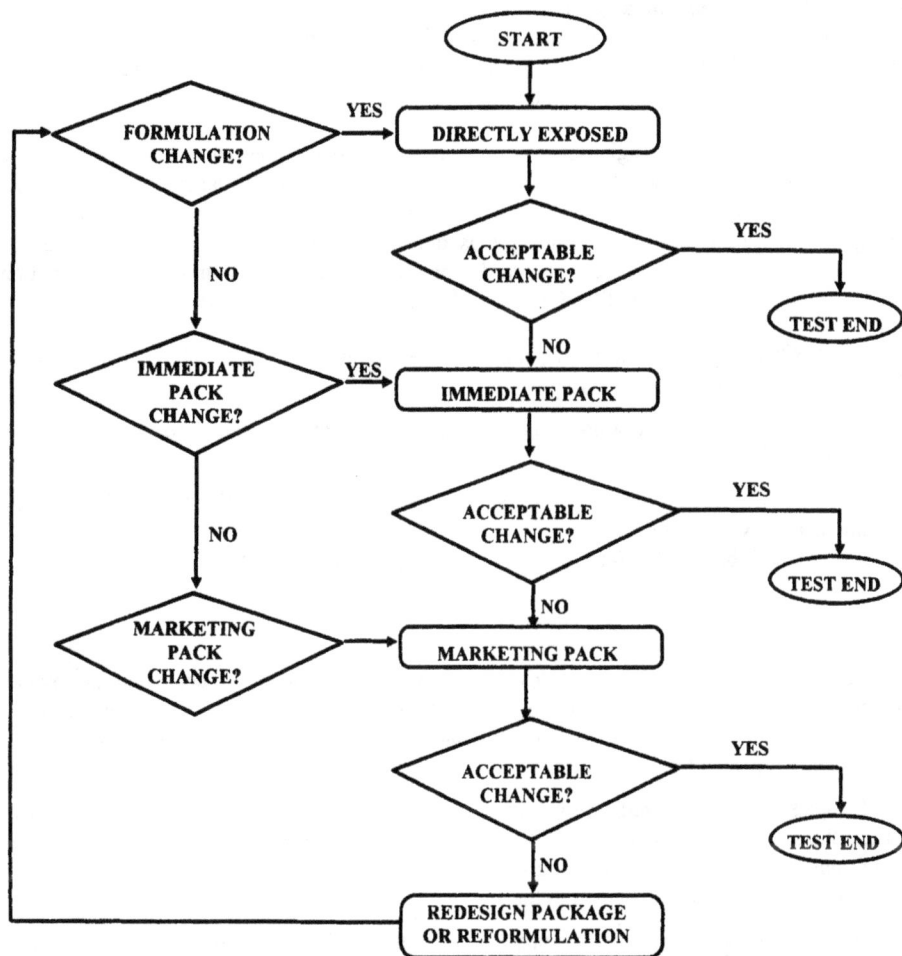

Figure 1. *Decision flow chart for photostability testing of drugs products.*

The extent of drug product testing should be established by assessing whether or not acceptable change has occurred at the end of the light exposure testing as described in the Decision Flow Chart for Photostability Testing of Drug Products (Fig. 1). Acceptable change is change within limits justified by the applicant.

The formal labelling requirements for photolabile drug substances and drug products are established by national/regional requirements.

1.2 Light Sources

The light sources described below may be used for photostability testing. The applicant should either maintain an appropriate control of temperature to minimise the effect of localised temperature changes or include a dark control in the same environment unless otherwise justified. For both options 1 and 2, a pharmaceutical manufacturer/applicant may rely on the spectral distribution specification of the light source manufacturer.

1.2.1 Option 1. Any light source that is designed to produce an output similar to the D65/ID65 emission standard such as an artificial daylight fluorescent lamp combining visible and ultraviolet (UV) outputs, xenon, or metal halide lamp. D65 is the internationally recognised standard for outdoor daylight as defined in ISO 10977 (1993). ID65 is the equivalent indoor indirect daylight standard. For a light source emitting significant radiation below 320 nm, an appropriate filter(s) may be fitted to eliminate such radiation.

1.2.2 Option 2. For option 2 the same sample should be exposed to both the cool white fluorescent and near ultraviolet lamp.

1. A cool white fluorescent lamp designed to produce an output similar to that specified in ISO 10977(1993) ; and

2. A near-UV fluorescent lamp having a spectral distribution from 320 nm to 400 nm with a maximum energy emission between 350 nm and 370 nm; a significant proportion of UV should be in both bands of 320 to 360 nm and 360 to 400 nm.

1.3 Procedure

For confirmatory studies, samples should be exposed to light providing an overall illumination of not less than 1.2 million lux hours and an integrated near-ultraviolet energy of not less than 200 watt hours/square meter to allow direct comparisons to be made between the drug substance and drug product.

Samples may be exposed side-by-side with a validated chemical actinometric system to ensure the specified light exposure is obtained, or for the appropriate duration of time when conditions have been monitored using calibrated radiometers/lux meters. An example of an actinometric procedure is provided in the Annex.

If protected samples (e.g., wrapped in aluminum foil) are used as dark controls to evaluate the contribution of thermally induced change to the total observed change, these should be placed alongside the authentic sample.

2 DRUG SUBSTANCE

For drug substances, photostability testing should consist of two parts: forced degradation testing and confirmatory testing.

The purpose of forced degradation testing studies is to evaluate the overall photosensitivity of the material for method development purposes and/or degradation pathway elucidation. This testing may involve the drug substance alone and/or in simple solutions/suspensions to validate the analytical procedures. In these studies, the samples should be in chemically inert and transparent containers. In these forced degradation studies, a variety of exposure conditions may be used, depending on the photosensitivity of the drug substance involved and the intensity of the light sources used. For development and validation purposes it is appropriate to limit exposure and end the studies if extensive decomposition occurs. For photostable materials, studies may be terminated after an appropriate exposure level has been used. The design of these experiments is left to the applicant's discretion although the exposure levels used should be justified.

Under forcing conditions, decomposition products may be observed that are unlikely to be formed under the conditions used for confirmatory studies. This information may be useful in developing and validating suitable analytical methods. If in practice it has been demonstrated they are not formed in the confirmatory studies, these degradation products need not be further examined.

Confirmatory studies should then be undertaken to provide the information necessary for handling, packaging, and labelling (see Section 1.3, Procedure, and 2.1 Presentation, for information on the design of these studies).

Normally, only one batch of drug substance is tested during the development phase, and then the photostability characteristics should be confirmed on a single batch selected as described in the Parent Guideline if the drug is clearly photostable or photolabile. If the results of the confirmatory study are equivocal, testing of up to two additional batches should be conducted. Samples should be selected as described in the Parent Guideline.

2.1 Presentation of Samples

Care should be taken to ensure that the physical characteristics of the samples under test are taken into account and efforts should be made, such as cooling and/or placing the samples in sealed containers, to ensure that the effects of the changes in physical states such as sublimation, evaporation or melting are minimised. All such precautions should be chosen to provide minimal interference with the exposure of samples under test. Possible interactions between the samples and any material used for containers or for general protection of the sample, should also be considered and eliminated wherever not relevant to the test being carried out.

As a direct challenge for samples of solid drug substances, an appropriate amount of sample should be taken and placed in a suitable glass or plastic dish and protected with a suitable transparent cover if considered necessary. Solid drug substances should be spread across the container to give a thickness of typically not more than 3 millimeters. Drug substances that are liquids should be exposed in chemically inert and transparent containers.

2.2 Analysis of Samples

At the end of the exposure period, the samples should be examined for any changes in physical properties (e.g., appearance, clarity, or color of solution) and for assay and degradants by a method suitably validated for products likely to arise from photochemical degradation processes.

Where solid drug substance samples are involved, sampling should ensure that a representative portion is used in individual tests. Similar sampling considerations, such as homogenisation of the entire sample, apply to other materials that may not be homogeneous after exposure. The analysis of the exposed sample should be performed concomitantly with that of any protected samples used as dark controls if these are used in the test.

2.3 Judgement of Results

The forced degradation studies should be designed to provide suitable information to develop and validate test methods for the confirmatory studies. These test methods should be capable of resolving and detecting photolytic degradants that appear during the confirmatory studies. When evaluating the results of these studies, it is important to recognise that they form part of the stress testing and are not therefore designed to establish qualitative or quantitative limits for change.

The confirmatory studies should identify precautionary measures needed in manufacturing or in formulation of the drug product, and if light resistant packaging is needed. When evaluating the results of confirmatory studies to determine whether change due to exposure to light is acceptable, it is important to consider the results from other formal stability studies in order to assure that the drug will be within justified limits at time of use (see the relevant ICH Stability and Impurity Guidelines).

3 DRUG PRODUCT

Normally, the studies on drug products should be carried out in a sequential manner starting with testing the fully exposed product then progressing as necessary to the product in the immediate pack and then in the marketing pack. Testing should progress until the results demonstrate that the drug product is adequately protected from exposure to light. The drug product should be exposed to the light conditions described under the procedure in Section 1.3.

Normally, only one batch of drug product is tested during the development phase, and then the photostability characteristics should be confirmed on a single batch selected as described in the Parent Guideline if the product is clearly photostable or photolabile. If the results of the confirmatory study are equivocal, testing of up to two additional batches should be conducted.

For some products where it has been demonstrated that the immediate pack is completely impenetrable to light, such as aluminum tubes or cans, testing should normally only be conducted on directly exposed drug product.

It may be appropriate to test certain products such as infusion liquids, dermal creams, etc., to support their photostability in-use. The extent of this testing should depend on and relate to the directions for use, and is left to the applicant's discretion.

The analytical procedures used should be suitably validated.

3.1 Presentation of Samples

Care should be taken to ensure that the physical characteristics of the samples under test are taken into account and efforts, such as cooling and/or placing the samples in sealed containers, should be made to ensure that the effects of the changes in physical states are minimised, such as sublimation, evaporation, or melting. All such precautions should be chosen to provide a minimal interference with the irradiation of samples under test. Possible interactions between the samples and any material used for containers or for general protection of the sample should also be considered and eliminated wherever not relevant to the test being carried out.

Where practicable when testing samples of the drug product outside of the primary pack, these should be presented in a way similar to the conditions mentioned for the drug substance. The samples should be positioned to provide maximum area of exposure to the light source. For example, tablets, capsules, etc., should be spread in a single layer.

If direct exposure is not practical (e.g., due to oxidation of a product), the sample should be placed in a suitable protective inert transparent container (e.g., quartz).

If testing of the drug product in the immediate container or as marketed is needed, the samples should be placed horizontally or transversely with respect to the light source, whichever provides for the most uniform exposure of the samples. Some adjustment of testing conditions may have to be made when testing large volume containers (e.g., dispensing packs).

3.2 Analysis of Samples

At the end of the exposure period, the samples should be examined for any changes in physical properties (e.g., appearance, clarity or color of solution, dissolution/disintegration for dosage forms such as capsules, etc.) and for assay and degradants by a method suitably validated for products likely to arise from photochemical degradation processes.

When powder samples are involved, sampling should ensure that a representative portion is used in individual tests. For solid oral dosage form products, testing should be conducted on an appropriately sized composite of, for example, 20 tablets or capsules. Similar sampling considerations, such as homogenisation or solubilisation of the entire sample, apply to other materials that may not be homogeneous after exposure (e.g., creams, ointments, suspensions, etc.). The analysis of the exposed sample should be performed concomitantly with that of any protected samples used as dark controls if these are used in the test.

3.3 Judgment of Results

Depending on the extent of the change, special labelling or packaging may be needed to mitigate exposure to light. When evaluating the results of photostability studies to determine whether change due to exposure to light is acceptable, it is important to consider the results obtained from other formal stability studies in order to assure that the product will be within proposed specifications during the shelf life (see the relevant ICH Stability and Impurity Guidelines).

4 ANNEX

4.1 Quinine Chemical Actinometry

The following provides details of an actinometric procedure for monitoring exposure to a near-UV fluorescent lamp (based on FDA/National Institute of Standards and Technology study).[1] For other light sources/actinometric systems, the same approach may be used, but each actinometric system should be calibrated for the light source used.

Prepare a sufficient quantity of a 2 per cent weight/volume aqueous solution of quinine monohydrochloride dihydrate (if necessary, dissolve by heating).

4.1.1 Option 1 Put 10 milliliters (ml) of the solution into a 20 ml colorless ampoule, seal it hermetically, and use this as the sample. Separately, put 10 ml of the solution into a 20 ml colourless ampoule (see note 1), seal it hermetically, wrap in aluminum foil to protect completely from light, and use this as the control. Expose the sample and control to the light source for an appropriate number of hours. After exposure determine the absorbances of the sample (AT) and the control (Ao) at 400 nm using a 1 centimeter (cm) path length. Calculate the change in absorbance, $\Delta A = AT - Ao$. The length of exposure should be sufficient to ensure a change in absorbance of at least 0.9.

4.1.2 Option 2 Fill a 1 cm quartz cell and use this as the sample. Separately fill a 1 cm quartz cell, wrap in aluminum foil to protect completely from light, and use this as the control. Expose the sample and control to the light source for an appropriate number of hours. After exposure determine the absorbances of the sample (AT) and the control (Ao) at 400 nm. Calculate the change in absorbance, $\Delta A = AT - Ao$. The length of exposure should be sufficient to ensure a change in absorbance of at least 0.5.

Alternative packaging configurations may be used if appropriately validated. Alternative validated chemical actinometers may be used.

Note 1: Shape and Dimensions (See Japanese Industry Standard (JIS) R3512 (1974) for ampoule specifications)

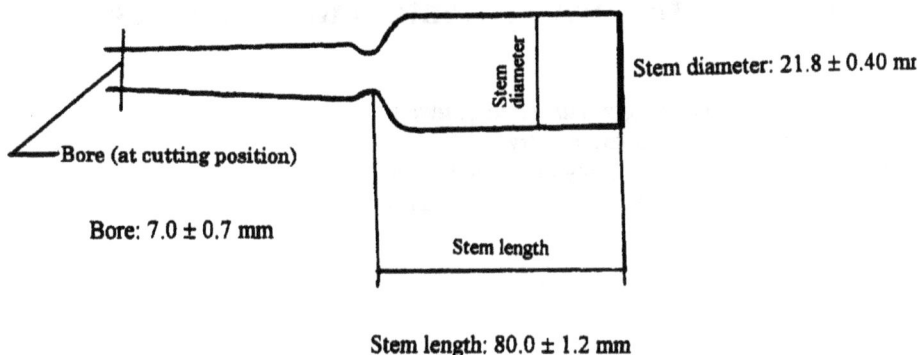

Stem diameter: 21.8 ± 0.40 mr

Bore (at cutting position)

Bore: 7.0 ± 0.7 mm

Stem length

Stem length: 80.0 ± 1.2 mm

5 GLOSSARY

Immediate (primary) pack is that constituent of the packaging that is in direct contact with the drug substance or drug product, and includes any appropriate label.

Marketing pack is the combination of immediate pack and other secondary packaging such as a carton.

Forced degradation testing studies are those undertaken to degrade the sample deliberately. These studies, which may be undertaken in the development phase normally on the drug substances, are used to evaluate the overall photosensitivity of the material for method development purposes and/or degradation pathway elucidation.

Confirmatory studies are those undertaken to establish photostability characteristics under standardised conditions. These studies are used to identify precautionary measures needed in manufacturing or formulation and whether light resistant packaging and/or special labelling is needed to mitigate exposure to light. For the confirmatory studies, the batch(es) should be selected according to batch selection for long-term and accelerated testings which are described in the Parent Guideline.

References.

1. S. Yoshioka, Y. Ishihara, T. Terazono, N. Tsunakawa, T. Murai, T. Yasuda, Kitamura, Y. Kunihiro, K. Sakai, K. Hirose, K. Tonooka, K. Takayama, F. Imai, M. Godo, M. Matsuo, K. Nakamura, Y. Aso, S. Kojima, Y. Takeda and T. Terao, *Drug Dev. Ind. Pharm.*, 1994, **20**, 2049.

Medicinal Photochemistry
(An Introduction with Attention to Kinetic Aspects)

Gerard M. J. Beijersbergen van Henegouwen
Department of Medicinal Photochemistry
Leiden/Amsterdam Center for Drug Research, Leiden University
P.O.Box 9502, NL-2300 RA Leiden, The Netherlands

1 GENERAL REMARKS

More than 25 years ago, I started my research in the photostability of drugs. As a young and enthusiastic PhD in organic photochemistry, I got my job in the faculty of pharmacy and expected much interest in this topic. I based my expectations on the fact that very many monographs in the pharmacopoeias, mention under storage conditions that the drug should be: "protected from light" or "preserved in light-resistant containers".

Soon, however, I learned that these indications for storage not so much resulted from scientific research as from rather superficial observations. Senior faculty members regarded the interaction of light with drugs as complicated and insight necessary for controlling the problem as being hard to obtain. Their reaction to my research plans was: "We avoid this problem by keeping the drug in a box as long as it is not yet administered to the patient". And I almost lost my credits as a young colleague by suggesting that many of these photoinstable drugs could give side-effects when they are exposed to light in the body.

Since then, I experienced that this evasive response to the problem of photoinstability of drugs is found not only with academics but also with pharmacists and people from pharmaceutical industries. The consequence of this is that instability and (enzymatic) reactivity in vitro and in vivo get much attention nowadays, but that photoinstability and phototoxicity of drugs is still treated as a Cinderella. The main cause of the continuance of this unjustified situation is that organic photochemistry and photobiology do not form part of the regular curriculum for undergraduates at the university.

Recently, I have published an extensive review on Medicinal Photochemistry;[1] it examines at length phototoxic and phototherapeutic aspects of drugs. Because of the mentioned frequently occurring lack of background, I start this review with an introduction of photochemical and photobiological facts indispensable for any activity in this field of research. By dealing with these basic concepts in a simplified and descriptive manner, I hope to provide pharmacists and pharmaceutical researchers from industry or university with sufficient theoretical background for effective research of photostability of drugs in view of phototoxic and phototherapeutic effects. In Sec. 2, you find a summary of items, accompanied with a brief explanation or comment, which are dealt with in the introduction of this review.[1] Besides, some points of view are discussed which underlie the main part of this review concerning specific examples of research on phototoxic and phototherapeutic drugs. As an extension of this review,[1] kinetic aspects of Medicinal Photochemistry are dealt with in Sec. 3.

2 ITEMS IN THE REVIEW ON MEDICINAL PHOTOCHEMISTRY[1]

2.1 Concerning Photophysics and Photochemistry

• Difference between thermochemistry and photochemistry and the common misconception that light would be a triggering force or accelerator of thermochemical (= dark) reactions. Although the same kind of reactions are found in photochemistry as in thermochemistry, the reactivity of a given molecule in the dark and in the light mostly differ from each other, not only quantitatively but also qualitatively. Examples are given to illustrate the point.

• The difference in lifetime of the two excited states, singlet and triplet, which are important to organic photochemistry. The photoreactivity of a molecule in its triplet excited state is often over-emphasized and is even considered as a prerequisite for the occurrence of a photochemical reaction. The reason for this idea is based on the fact that the intrinsic lifetime of the triplet excited state is much longer and thus the chance to collide with a reaction partner is far greater than when the molecule is in the singlet excited state. However in vivo, many drugs are extensively complexed to biomacromolecules such as proteins, DNA and to lipids as constituents of membranes. In such complexes, the photoreactive drug does not need to collide at all with its potential target and a molecule in a short-lived singlet excited state can photobind equally efficiently as in the triplet excited state.

• The way excited states are obtained and the processes they can undergo apart from reaction. In connection with this, it is explained that it is a misconception to believe that light of shorter wavelengths is more detrimental to drugs than light of longer wavelengths.

• Energy transfer, which complicates the photochemical activity of drugs. This involves transfer of excitation energy from an electronically excited donor (D) molecule to the ground state of an acceptor (A) molecule resulting in deactivation of D and excitation of A. Excited D concerns either the singlet or the triplet excited state. The longer lifetime of a triplet excited state facilitates energy transfer if D and A have to collide. However, as many drugs are extensively complexed in vivo, a short-lived singlet excited state D can transfer energy equally efficiently as in its triplet excited state; for example, to DNA which acts as an acceptor in that case.

• The electronic spectra (absorption, fluorescence and phosphorescence). Whether energy transfer will occur indeed from an excited D to A, depends on a number of conditions. These concern overlap of spectra, lifetime of excited state and difference in deactivation and excitation energy of D and A respectively. Whether these conditions are met with a given D and A can be concluded from the study of the electronic spectra.

• The concept of action spectrum. Energy transfer from an excited D to a ground state acceptor A mostly proceeds very efficiently. If this is the case and D itself is photostable, it can repeat the transfer of absorbed energy numerous times per second and act as a catalyst. The consequence is that the donor, also called photosensitizer, can be present in a very low concentration. As a result of the latter, the presence of D is easily overlooked by superficial inspection of the UV/Vis spectrum. If a photosensitizer in low concentration is suspected of the photoinstability of a drug, an action spectrum can be helpful. This is a plot of the reaction rate against the wavelength of the light. Provided that at each wavelength the same number of photons per m^2 per second is used, the action spectrum resembles the absorption spectrum of the photosensitizer; they are even identical if no other absorbing compounds are present.

• The concept of quencher. A physical quencher can be considered as the counterpart of a photosensitizer; it is a stable acceptor (A) in the process of energy transfer. It converts the excitation energy into heat and without undergoing any reaction, it can repeat this numerous times per second. Quenchers can be useful as protectors against the harmful effects of phototoxic drugs.

• Distinction between photochemical type I, II and III reactions which take place in biological systems. In type I reactions radicals are formed in which oxygen often plays a role. As a result of this, besides superoxide anion, peroxy radicals, hydrogen peroxide and hydroxyl radicals can be formed. In a type II reaction the triplet excited molecule transfers its energy to oxygen resulting in singlet oxygen.

Photoreactions in which molecular oxygen plays a role (type II reactions and part of the type I reactions) are called photodynamic reactions. All other reactions in which both radicals and oxygen are not involved, are called type III reactions (e.g. photocycloaddition of 8-methoxypsoralen to DNA which plays a role in the PUVA therapy).

2.2 Concerning Human Photobiology in Connection with Drugs

• Besides spectral composition of sunlight, subdivision of UV-radiation and optical properties of the skin and eyes, attention is paid to the question whether there is a difference between phototoxic and phototherapeutic drugs.

As far as the molecular processes underlying the occurrence of photobiological effects are concerned, there is no essential difference between phototherapeutics and phototoxicons; with both groups of drugs, type I, II and III reactions occur. One speaks of a phototherapeutic if the photoreaction is intentional and controlled. The latter concerns not only the site in the body where it occurs but also the dose of the drug and the light (wavelength region, intensity and exposure time) applied.

• The sequence of events which eventually leads to the biological effects. With both phototherapeutics and phototoxicons, this starts with the absorption of a photon of UV-radiation or visible light. As a result of this, the photoexcited drug can undergo a number of primary reactions in the body:

a) Photochemical reaction of the compound as such, a unimolecular reaction, e.g. rearrangement, isomerization or decomposition. The reaction products or their metabolites can display their biological activity by interaction with a receptor. At present there are no reports about phototherapeutic or phototoxic compounds acting in this way. Yet, this can be considered as a rather trivial possibility; certainly if one takes into account that interaction of a photometabolite with a receptor is an essential part of a normal endogenous photobiological process in man. This process concerns the UVB induced rearrangement of 7-dehydrocholesterol into previtamin D_3 and subsequent steps.

Unstable photoproducts can react with endogenous molecules resulting in a biological effect. If the half-life of such a photoproduct is long, it can react in inner organs after having been formed in the skin. Examples are given, such as the photoproducts of the urinary tract disinfectant nitrofurantoin which showed to be irreversibly bound in the spleen, liver and lungs of UVA-exposed rats.

b) Photoreaction with endogenous molecules. An example extensively dealt with is the irreversible binding of photoexcited psoralens to lymphocytes as a result of UVA absorption, what is considered to play an important role in the new phototherapy of auto-immune diseases, called photopheresis. This treatment which aims at specific immunosuppression is also used to prevent rejection of heart transplants.

c) Energy transfer to endogenous compounds. For example to oxygen; the resulting singlet oxygen is assumed to play an important role in the dye/visible light therapy of cancer. Phototoxic compounds of quite different molecular structure can also produce singlet oxygen by energy transfer, e.g. phenylpropionic acid derivatives used as inflammatory drugs and tetracyclines.

• Damage to essential biomacromolecules as a result of primary photoreactions in the body.

Damage to DNA can lead to cell death but also to mutation. If DNA repair enzymes (proteins) are inactivated as well, for instance by a simultaneously occurring primary photoreaction, mutation can be transferred and tumour formation can be the result.

Photoreaction with proteins, or with lipids as constituents of membranes, can lead to cell death, but can also trigger an immune response which eventually can lead to allergy.

Besides biomacromolecules such as DNA, proteins and membrane constituents, small endogenous molecules are target of primary photoreactions, for instance glutathione which is important to the cellular defence.

• The biological effects which can be observed with a given phototherapeutic or phototoxicon. These depend on a variety of factors, such as:

1) The extent to which each of the primary photoreactions a, b and c occurs.

2) The extent to which each of the essential bio(macro)molecules is damaged.

3) The bioavailability of the given photoactive compound and its metabolites, not only in the organs exposed to light but also at cellular level.

For example, singlet molecular oxygen is a mutagenic species. However, although dyes are efficient singlet oxygen producers by energy transfer (c), mutagenesis is not an expected problem with the dye/visible light therapy. The fact is that the dyes commonly used in this therapy, accumulate in the lipid material of the membranes and remain outside the cellular nucleus.

Another example of bioavailability as determining factor is chlorpromazine whose photoreactivity to DNA is known from many in vitro studies: irreversible binding is one of the possibilities. However, as a result of poor bioavailability in the cellular nucleus, this photoreaction was not found in vivo. Contrary to this, photobinding to proteins and lipids of epidermal cells was observed.

These in vivo observations,[1] concerning the occurrence of photoreactivity with different biomacromolecules, correspond with the fact that photoallergy is a common side-effect of chlorpromazine but also that there are no clinical reports on (photo)carcinogenicity.

A comparable situation was found with furocoumarins:[1] photobinding to DNA got by far the most attention. However, it has been demonstrated that 8-methoxypsoralen, frequently used in PUVA therapy, photobinds in vivo not only to DNA but to a high extent also to proteins and lipid material of epidermal cells. This can be an important finding with regard to e.g. immunological effects from photopheresis.

2.3 Phototoxic versus Photoallergic Drugs

Toxic effects are subdivided in the discipline toxicology because of their different nature. Subdivisions are for example genotoxicity, hepatotoxicity, immunotoxicity and neurotoxicity. They give information about the kind of effects to be expected, the organs affected or about the biochemical end-points observed. Skin allergy caused by drugs or other chemicals is a part of immunotoxicity and belongs thus to a subdivision of toxicology.

Because of this organization of the well established discipline of toxicology, there is no reason to act differently with toxic effects resulting from the interaction of light and, for example, drugs. Thus, phototoxicology covers the whole field in which light plays a distinct role in the occurrence of the toxic effect; it includes all subdivisions.

The consequence of this is that each drug with light-induced side-effects is called phototoxic. Besides, it fits into the logic of toxicology and its subdivisions that photoallergic drugs are considered as a subdivision of phototoxic drugs.

From the above, it can also be concluded that reporting that a drug is toxic or phototoxic is a broad statement which is not very informative. More insight is obtained when a report more specifically deals with the kind of adverse photobiological effects connected with the use of a given drug or with the biochemical end-points observed. This more specific knowledge permits that a general characterization, namely that a compound is phototoxic, can be replaced by more precise ones such as photogenotoxic, photoimmunotoxic (e.g. photoallergic) or photohepatotoxic.[1]

2.4 Systemic Phototoxicity of Drugs

It is important to emphasize that there is no reason to assume that phototoxic effects of drugs remain restricted to the organs exposed to light: the skin and eyes. One should also take into account that in many cases biological effects do not concern the sunlight exposed area of the body at all, but exclusively internal organs. The latter statement is supported by natural photobiological processes in man. The conversion of 7-dehydrocholesterol into previtamin D3 and that of bilirubin into more water-soluble products, upon exposure of the body to UVB and to visible light respectively, are two examples of systemic photobiological effects of endogenous compounds. Vitamin D3 is essential for proper bone calcification and the photoconversion of bilirubin e.g. in the visible light therapy of neonatal jaundice, results in a decrease of brain-damaging effects. In both cases, photobiological effects do not concern the skin at all.

The last fifteen years, we published several reports which show that the interaction of light with a drug in the body can cause systemic biological effects.[1] This implies that the possibility of systemic effects should be considered while screening drugs for phototoxicity.

2.5 Structure-Phototoxicity-Relationship

Standard methods for the measurement of phototoxic effects, such as skin erythema, oedema, cancer, photoallergy, ocular opacity or those concerning the not light-exposed organs are useful. They permit to estimate the risks of the use of a drug in combination with exposure to light. This estimation can become even more meaningful if data are added concerning the damage to bio(macro)molecules and cells, for example in connection with photocarcinogenicity: damage to and changes in DNA, expression of proto-oncogenes and mutations.

However, knowledge about the nature of the phototoxic effects and the extent to which they can occur are not sufficient to drug research. This situation does not change if cell biological and biochemical data are added.

In drug research one needs to know why a given lead compound is phototoxic.

To answer the latter question, the molecular processes which a lead compound and its structural analogues undergo in vitro and in vivo after absorption of UV or visible light, have to be investigated. Only in this way, knowledge of the part of the molecular structure which makes a given drug phototoxic, can be obtained.

Medicinal Photochemistry, the discipline that aims at such a so-called structure-phototoxicity-relationship, is of great interest to industries which develop new drugs:

It provides the opportunity to alter the molecular structure in such a way that an analogue is obtained which still has the desired properties of the parent drug but is not phototoxic anymore. In the review on Medicinal Photochemistry,[1] examples are given which show that expertise in this field of research not only is profitable to the development of non-phototoxic drugs, but that it can also provide highly specific phototherapeutics.

To avoid the loss of a huge amount of money, such a so-called structure-phototoxicity-relationship must be obtained in an early stage and not for example during Phase III-IV studies. Severe phototoxic side-effects have been recently observed during such studies; for example, of new non-steroidal antiinflammatory drugs (propionic acid derivatives) and antibiotics (fluoroquinolones).

2.6 Research Approach for a Structure-Phototoxicity-Relationship

In view of a Structure-Phototoxicity-Relationship, research must include the drug itself (and/or its metabolites) as well as a number of its analogues.

Emphasis must be put on what occurs on molecular level with each compound. This can be learned from the molecular structure of (un)stable photoproducts and (short-lived) reactive intermediates derived from the compound under investigation.

Research should not be limited to the primary photoreactions but also concern the structure of photoproducts and reactive intermediates such as radicals or unstable ions formed in the presence of bio(macro)molecules. In particular, attention should be paid to the molecular structure of the photoadducts to bio(macro)molecules. The latter should include the remaining fragment of the drug and, with regard to the bio(macro)molecule, the part to which it is irreversibly bound. Besides, the possibility should be investigated whether the drug and its analogues are photosensitizers, which can cause phototoxicity by transfer of energy, e.g. to oxygen resulting in singlet oxygen.

The research aim as defined above can be achieved most efficiently by combining data from *in vitro* and *in vivo* investigations. The *in vitro* research may consist of the following steps:

A) Photoreactivity of the drug alone in e.g. phosphate buffered saline, in the presence of oxygen, with essential biomacromolecules such as proteins (enzymes) or DNA and with unsaturated fatty acids, as constituents of cellular membranes.

B) Photoreactivity and (photo)toxicity in microbiological or cellular test systems. If with step A, some evidence is found that the drug may have adverse photobiological effects, research can be continued with this kind of *in vitro* test systems. They give the opportunity to investigate the influence of cellular bioavailability and metabolism of the drug on the formation of its photoproducts, reactive intermediates and photoadducts to bio(macro)molecules. The use of different kind of cells permits to investigate the influence of the presence of special receptors.

The photoreactivity data of a compound can be linked to particular (photo)toxic effects; some examples:

-   Numerous microbiological systems are available for genotoxicity and mutagenicity testing, which are also applied in combination with light.

-   Red blood cells are used to assess membrane photodamage.

-   Lymphocytes are used to screen compounds with regard to possible immunophototoxicity.

-   Human skin cells (fibroblasts, keratinocytes and melanocytes) are used for *in vitro* phototoxicity testing such as cell viability, membrane damage and growth inhibition. Early markers for skin cell damage such as cytokines (IL-1a, IL-6, IL-8), prostaglandins (PGE2) and heat-shock proteins are also measured.

Dependent on the results from the *in vitro* testing under step A, the *in vitro* testing under step B can be more or less profoundly undertaken.

Only compounds coming out of the tests A and B as being potentially phototoxic, are further investigated. This may involve research of the photoreactivity data (as mentioned above in the introduction to this section) in ex vivo models such as porcine skin or eyes.

The *in vitro* and *ex vivo* investigations can be followed by *in vivo* experiments with animals and eventually with human volunteers. With this *in vivo* research, the following factors should be taken into account:

- Metabolism of the drug. Metabolites can differ in photoreactivity as compared to the parent compound.

- (Intracellular) distribution. The amount of the drug that reaches the site of exposure (e.g. skin or eyes), or specific sites within the cell (e.g. DNA or membranes of cell bodies).

- Absorption of radiation by skin components which determines the amount of light available to the drug.

- Repair processes by which the organism can respond to photochemically induced damage.

In this *in vivo* research, particular attention should be paid to the molecular structure of the photometabolites in blood, urine and bile. Besides, to the molecular structure of photoconjugates; with which is meant the remaining fragment of the drug and, with regard to the bio(macro)molecule, the part to which it is irreversibly bound. The study of irreversible binding to bio(macro)molecule should not be limited to the organs exposed to light but also concern the inner organs, because of possible systemic effects.

In the review on Medicinal Photochemistry,[1] an example is discussed in more detail, which illustrates the foregoing: how a structure-phototoxicity-relationship is obtained by a combination of *in vitro* and *in vivo* research.

The example concerns a class of phototoxic compounds, the imino-N-oxides, which find extensive application amongst others as drugs. By investigating not only the parent drug but also structure analogues, it was possible to show that the N-O group in the molecule is the cause of the phototoxic effects. Interesting result of this research was not only that an analogue lacking the N-O group was not phototoxic anymore, but also that this compound had the same desired properties as the parent drug from which it was derived.

3 KINETIC ASPECTS OF MEDICINAL PHOTOCHEMISTRY

3.1 Quantum Yield and Rate of Photochemical Reaction

The central parameter in organic photochemistry and photophysics is the quantum yield ϕ.

After the absorption of light, the excited molecule can undergo a number of processes[1] such as radiationless deactivation to the ground state, fluorescence, intersystem crossing to the triplet, phosphorescence, energy transfer and of course also photochemical reaction. Each of these processes has its own quantum yield, which expresses the chance that it occurs. The formula of the quantum yield is as follows (e.g. of a photochemical reaction):

$$\phi = \frac{\text{The number of moles of the compound which reacted}}{\text{The number of einsteins absorbed by the compound}} \qquad (1)$$

(One mole contains Avogadro's number, is 6.023×10^{23}, molecules. Similarly, one Einstein is equal to 6.023×10^{23} photons of a given wavelength)

The sum of all quantum yields is 1 and, for each of the processes mentioned, ϕ can have a value between 0 and 1. Sometimes, it seems that a photochemical reaction has a quantum yield higher than 1. However, this does not result from the photochemical reaction as such but from secondary thermochemical reactions. (For example, during the photochemical decomposition of the excited molecule, radicals are formed which thereafter are involved in a chain reaction.)

Just like the quantum yield of each of the physical processes mentioned, e.g. phosphorescence and energy transfer, ϕ of a photochemical reaction is characteristic for a given compound and independent of the wavelength absorbed.[1]

In a sample of volume V containing a photoreactive compound of which the concentration is c at a given moment, the total number of moles $N_T = c \times V$. The number of moles (N_R) that has reacted after a short period of time at a given wavelength (λ) is $n_\lambda \times \phi$; in which n_λ is the number of einsteins absorbed. The photochemical reaction rate (R) is the number of moles (N_R) which has reacted divided by the total number of molecules in the sample at the beginning of that short period of time:

$$R = \frac{N_R}{N_T} = \frac{n_\lambda \times \phi}{c \times V} \qquad (2)$$

As a consequence of the Beer-Lambert law, the following formula is obtained for the photochemical reaction rate (R) at a given moment and wavelength,

$$R = \frac{N_R}{N_T} = \frac{I_\lambda (1 - 10^{-\varepsilon_\lambda bc}) \phi}{c \times V} \qquad (3)$$

in which I_λ is the intensity of the incident light, ε_λ the molar absorptivity at wavelength λ, b the optical path length and c the molar concentration of the absorbing reactant and V the volume of the sample.

3.2 Factors Influencing the Rate of a Photochemical Reaction

Photochemical kinetics depends on a variety of factors which makes it difficult to compare results from different studies. The discussion of these factors below results in proposals which can facilitate comparison of data.

3.2.1 Concentration of the Drug. Photochemical kinetics is more complicated at higher concentrations of a drug for two reasons.

- Photodegradation usually proceeds through the formation of reactive intermediates. These reactive intermediates, e.g. free radicals, can induce chain reactions and thus increase the rate of decomposition; the higher the concentration of the drug, the higher the increase of its decomposition rate. At concentrations of $<10^{-5}$ M, the influence of chain reactions on the photodecomposition rate is small. To facilitate comparison of data, it is proposed to investigate the photoreaction at such a low starting concentration, e.g. 10^{-6} M.

- Most of the light, penetrating a volume containing a drug, is absorbed close to the surface. The higher the concentration of the drug, the more this is the case:

As can be seen in formula (3), 99.9% of the incident light is absorbed with $\varepsilon bc = 3$. If the concentration of the compound is 10^{-2} M and it has a moderate molar absorptivity of 3×10^3 l.mol.cm^{-1}, the absorption is almost complete at b = 0.1 cm. This means that the part of the drug more inside the volume is *protected from light* and does not participate in the photoreaction. If the concentration of the drug is 10^{-4} M instead of 10^{-2} M, the absorption is almost complete at b = 10 cm only.

Besides, the determination of the rate of the photoreaction is based on a change of concentration in the total volume and not on that found at the outside where the photoreaction occurs mainly. As a result of this, the photodecomposition rate seems to be smaller at high concentration than at low; with the above example, R is 100 times smaller at 10^{-2} M than at 10^{-4} M.

That part of the drug more inside the volume is protected from light and does not participate in the photoreaction is caused by the so-called inner filter effect. However, not only the drug itself but also light absorbing photoproducts or other compounds present in the sample can contribute to this inner filter effect.

The occurrence of the inner filter effect leads to much confusion about the photostability of a given drug and makes comparison between different drugs almost impossible. However, the influence of the inner filter effect can be neglected if the maximal absorption (εbc) by the drug and its products is lower than 0.01 in the spectral region of the light source.[2] The condition, absorption smaller than 0.01, is usually reached at concentration of $<10^{-6}$ M (sample diameter of 1 cm and molar absorptivity of $<10^4$ l.mol.cm^{-1}).

From the foregoing, it can be concluded that comparison of data will be facilitated if studies would be performed at concentrations of $<10^{-6}$ M. Besides, that there are no other absorbing substances present in the sample than the drug itself and its photoproducts.

3.2.2 Volume of the Sample. As can be seen in formula (3), the rate of a photoreaction depends on the volume of the sample. In view of standardization, it is proposed to irradiate in quartz tubes of 1 cm diameter which are filled with the same volume of sample. To ensure uniform irradiation, tubes can be placed in a merry-go-round system.

3.2.3 Quantum Yield of the Photochemical Reaction. The quantum yield ϕ of a photochemical reaction is characteristic for a given compound and independent of the wavelength absorbed.[1] However, the chemical and physical properties of the medium (including solutes) have an influence on the quantum yield, not only on that of the photochemical reaction but also on those of the physical deactivation processes of the excited drug molecule. The quantum yield of the physical and chemical processes for a given drug are interdependent and as soon as one of them changes this may have consequences for the others.

This means that ϕ of the photochemical reaction of a drug depends on the pharmaceutical formulation chosen or on the model used for the investigation of phototoxic effects e.g. the presence of lipophilic membranes or DNA. Also small molecules such as oxygen being present in low concentration can interfere with a photoreaction of the molecule in its triplet excited state and can change ϕ; in such a case oxygen acts as a quencher. Besides, ϕ of the photochemical reaction can depend on e.g. pH, kind of ions present and ionic strength.

To facilitate comparison of data the composition of the medium have to be well-defined. It is proposed that a solution of the drug is made in phosphate buffered saline (pH 7.4). If for solubilization, a small amount of organic solvent is needed this should be limited to 2%. This will be feasible if, as proposed above, the concentration of the drug is $<10^{-6}$ M.

The organic solvent should not interfere with the photochemical reaction (for this reason, DMSO and CH_3CN are often used in photochemistry).

In view of the research on phototoxicity, one can imagine that besides this solution in phosphate buffered saline, some other standard solutions are needed; e.g. one which additionally contains a defined amount of DNA or albumin. And if the drug is photostable but is suspected to be a photosensitizer, a defined amount of a singlet oxygen acceptor could be added to the standard solution in phosphate buffered saline (pH 7.4).

3.2.4 Intensity and Spectral Distribution of Light. If the photoreactivity of a drug is investigated at low concentration, e.g. 10^{-6} M and the dimensions of the sample are standardized, e.g. quartz tubes of 1 cm diameter which are filled with the same volume, a simplified formula (4) can be obtained which permits to study the photochemical reaction as a function of time and besides with first order kinetics:[2]

$$- \frac{dc}{dt} = 2.3\, I_\lambda \varepsilon_\lambda \phi c \qquad (4)$$

The quantum yield ϕ is independent of the wavelength and can be determined (for example at the absorption maximum of the drug for greater precision) using standard methods.

The quantum yield is an important parameter; yet for communication about photostability of drugs, it has limited value. In fact the actual photoreactivity of a given drug also depends on the overlap between its absorption spectrum and the emission spectrum of the light source used. The reaction rate constant k_λ ($= 2.3\ I_\lambda \epsilon_\lambda \phi$) depends on the wavelength and integration has to take place over the relevant wavelength range determined by the absorption spectrum of the drug (ϵ_λ) and the emission spectrum (I_λ) of the light source. The overlap integral $\int (I_\lambda \epsilon_\lambda)\ d\lambda$ is a constant for a given compound, provided that the light source is defined. If the latter is the case, a simple calculation with modern computer facilities provides the reaction rate constant for a given drug regarding the entire relevant wavelength region.

The problem of the uniformity of the light source in drug research can be solved by taking over the solution that has been found and accepted for a comparable problem. This problem concerned the photostability of paints plastics, textile etc.; products which are used outdoors and exposed to sunlight all the year round. To obtain a uniform test method which would make it possible to compare the photostability of commercial products e.g. different paints, the CIE (TCO2-2 Commission Internationale d'Eclairage) agreed in 1970 upon one spectral distribution and total intensity of sunlight for photostability testing.[3] Solar simulators which approximately meet the requirements of spectral distribution and total light intensity, have become available. Using such a well-defined solar simulator makes comparison of the photostability of drugs possible.

4 CONCLUSIONS

Summarizing the proposals made in the previous sections with regard to quantitative data concerning the photoreactivity of a drug:

a) Concentration of the drug of 10^{-6} M. Bimolecular reactions between the drug and its instable photoproducts (e.g. chain reactions) are avoided. Besides, the inner filter effect caused by the drug itself and by its photoproducts can be neglected.

b) As the rate of a photoreaction depends on the volume of the sample, it is proposed to irradiate in quartz tubes of 1 cm diameter which are filled with the same volume of sample. To ensure uniform irradiation, tubes can be placed in a merry-go-round system.

c) As the chemical and physical properties of the medium (including solutes) have an influence on the quantum yield of the photochemical reaction, it is proposed that a solution of the drug is made which is well-defined. For example, in phosphate buffered saline (pH 7.4). (If for solubilization, a small amount of organic solvent, e.g. DMSO or CH_3CN, is needed this should be limited to 2%. This will be feasible if, as proposed above, the concentration of the drug is $<10^{-6}$ M.)

In view of the research on phototoxicity, one can imagine that besides this solution in phosphate buffered saline, one or two other standard solutions are needed; e.g. one which additionally contains a defined amount of DNA or albumin. Might the drug show to be photostable, its photoreactivity can be studied as well in the above mentioned solution in phosphate buffered saline to which a given singlet oxygen acceptor is added in a defined concentration.

d) Using a well-defined solar simulator makes comparison of the photostability of drugs possible.[3]

The proposals are far from meeting the conditions found in most pharmaceutical formulations. However, both the complexity of the photochemical reactivity depending on so many factors and the large variety in pharmaceutical formulations and their composition, makes meeting the conditions an unreachable ideal.

Without denying the importance of the problem of the photostability of the drug in its formulation, my personal point of view is that more concern is required with regard to photoinstability of drugs after their administration to man because of the risk of phototoxicity. As soon as possible problems with phototoxicity of a particular drug are controlled, a solution for the photostability of its formulation can be found as well.

The proposals can be useful to get a first idea whether a drug may have a greater risk to cause phototoxicity as compared to other compounds. The reason for this is that the conditions mentioned in the proposals are not incompatible with those found in vivo; such as the low concentration, the use of phosphate buffered saline (pH 7.4) and a solar simulator.

However, the quantitative data obtained are only useful for comparison and as a first indication that photostability may be a problem. In this respect, it is advised that these data are considered as a part of the in vitro research aiming at a Structure-Phototoxicity-Relationship (as described under 2.6).

Only a Structure-Phototoxicity-Relationship, as provided by Medicinal Photochemistry can prevent the development of phototoxic drugs.

References

1. G. M. J. Beijersbergen van Henegouwen, *Adv. Drug Res.*, 1997, **29**, 79.
2. N. J. de Mol, G. M. J. Beijersbergen van Henegouwen and K.W. Gerritsma, *Photochem Photobiol.*, 1979, **29**, 7.
3.CIE Division 2, Report No. 20 (1970), Recommendations for the integrated irradiance and spectral distribution of simulated solar radiation for testing purposes; CIE Division 3, Report No. 16 (1970), Daylight; CIE Division 1, Report No. 15.2 (1986, 2nd Ed.), Colorimetry.

Photoreactivity of Selected Antimalarial Compounds in Solution and in the Solid State

H. Hjorth Tønnesen, S.Kristensen and K.Nord
Institute of Pharmacy
University of Oslo
P. O. Box 1068, Blindern, 0316 Oslo, Norway

1 INTRODUCTION

The photochemical stability of pharmaceutical substances is a matter of concern. Light-induced decomposition of more than 100 medicinal compounds has been reported in the literature and the European Pharmacopoeia prescribes light protection for about 250 medical drugs and a number of adjuvants. Although photostability testing has not been required by the regulating authorities up to now, this is about to change with the introduction of the ICH guidelines on photostability testing of drugs. The most obvious result of drug photodecomposition is a loss of potency of the product. In the final consequence this can result in a drug product which is therapeutically inactive. Although this is rarely the case, even less severe degradation can cause formation of (photo)toxic degradation products. The drug substance can also cause light-induced side-effects after administration to the patient by interaction with endogenous substances. This can be the case both for systemically and topically administered compounds. It is for instance, well established that parts of the sunlight spectrum (UV-A, visible light) penetrate the skin to a depth sufficient to reach the blood circulating in surface capillaries and thereby the dissolved drug molecules. Excitation of the drug molecule by absorption of irradiation in vivo may generate reactive oxygen species (ROS), free radicals or toxic photoproducts which can be biologically harmful.

It is essential to obtain information about the photoreactivity of a drug molecule as early as possible in the formulation process. Knowledge about the photochemical and photophysical properties of the compound is essential for handling, packaging and labelling of the drug substance and drug product but also in order to understand implications for lack of efficacy or possible photosensitizing activity in vivo. Several in vitro methods for phototoxicity studies are described but in many cases in vivo methods will also be required. A complete assay for photoreactivity screening is time and money consuming and requires a broad spectrum of techniques. A selection of the drugs to undergo a full screening can be made on certain criteria.[1] Large variations are found among molecules that can act as photosensitizers in biological systems, and photoreactivity is difficult to predict. A photostable drug may act as an efficient in vivo photosensitizer by a series of photophysical events while a photolabile compound may cause phototoxic reactions by formation of reactive intermediates or toxic photoproducts. It is important to keep in mind that the photoreactivity of a pure compound can change when the sample is introduced into a biological system. The most frequently used antimalarial compounds are structural analogues, i.e. they belong to the group of quinolines. They fulfil almost every criteria needed to subject a drug to a complete photoassay; they are sensitive to irradiation, contain photoreactive functionalities, accumulate in tissues exposed to light, have an absorption spectrum that overlaps with the transmission spectra of the actual tissues, are administered at high accumulative dosages, and are used in climatic zones where the solar irradiance is high. On the basis of these criteria the photoreactivity of selected antimalarial compounds has been investigated according to a previously described in vitro assay. [1] The compounds discussed belong to the groups of 4-aminoquinolines (chloroquine), 8-aminoquinolines (primaquine) and 4-quinolines (mefloquine) respectively.

1.1 In vitro Screening of Compounds in Solution

1.1.1 Reaction Medium. It is difficult to predict the photoreactivity of a drug molecule from structural considerations only.[2,3] Experiments have to be performed to elucidate the photochemical reaction-pattern of the substance, and the experimental design has to be planned thoroughly. In vivo conditions will always be very complex compared to what can be achieved in vitro. A certain in vitro - in vivo correlation may however be obtained by carefully designing the in vitro photoassay. It is important to mimic physiological conditions in the experimental set-up by use of aqueous solutions at physiological pH and ionic strength. Micelles, liposomes and cyclodextrines can be added to the medium to mimic the effects on membranes. Most photosensitized reactions in biological systems require the involvement of molecular oxygen. It is therefore natural to use media in equilibrium with air in photoreactivity studies. An elevated oxygen concentration or anaerobic conditions can, however, provide valuable information in mechanistic studies of photostability. An increase in oxygen concentration can lead to a decrease or an increase in degradation rate. If the drug molecule initiates the formation of singlet oxygen followed by a self-sensitized type II reaction an increase in degradation rate may be observed at higher oxygen levels. A decrease in degradation rate can be the result when quenching of the reactive triplet state of the drug molecule by oxygen is the process in favour. A change in the formation of degradation products can occur when the oxygen level is suppressed. Chloroquine is demonstrated to dechlorinate during anaerobic photolysis,[4] while dechlorination cannot be observed during aerobic photolysis.[5] There are also some exceptions in which oxygen-independent processes have an important place in the overall picture of photosensitization.[6,7] Photoreactivity studies carried out under elevated oxygen levels and under anaerobic conditions should therefore be included in the in vitro assay.

Many drug molecules, like the antimalarial compounds, are weak acids or weak bases. The pH of the reaction medium may in that case strongly influence the results obtained. The degree of protonation will affect the spectral characteristics of the drug in the medium as well as the photochemical reaction pattern. The pH of the medium can therefore have a considerable effect on the photochemical stability of a compound. It is obvious that valuable information can be lost if the experiments are carried out only in one medium, e.g. organic solvent that can not differentiate between protonated and deprotonated forms of the molecules. In mechanistic studies of drugs that are acids or bases experiments should therefore be carried out at a series of pH values covering all the protonation forms of the molecule.

1.1.2 Formation of Free Radicals and Singlet Oxygen. The photochemical decomposition of a drug substance will often involve radical processes and/or formation of singlet oxygen. The excited drug molecule can exchange electrons with another drug molecule, with oxygen or with other compounds present in the medium. Free radicals formed can subsequently react with new drug molecules, leading to degradation by several pathways. Excipients or impurities present in the drug product may also initiate radical chain reactions by absorption of radiation. The absorbed energy can further be transferred from the excited drug molecule to ground state oxygen leading to the formation of singlet oxygen which subsequently can participate in various reactions. Formation of radicals or singlet oxygen in vivo may cause damage to proteins, lipids, DNA and RNA, and/or cell membranes. Determination of the detailed photoreactivity characterisitcs of a drug molecule requires knowledge of the sensitizing and quenching properties of the molecule, e.g. quenching efficiency of the drug excited state by a substrate, the ability of the drug to transfer an electron to oxygen and the rate of reaction of the drug radical with ground state oxygen. A number of techniques have been developed to enable the detection of free radical intermediates, including electron paramagnetic resonance spectroscopy (EPR) and pulse radiolysis. An infrared luminescence technique is frequently used for determination of singlet oxygen.[8] These methods require specialized equipment. A more simple approach for studying radical formation is by addition of various free radical scavengers to the medium during irradiation. The rate of disappearance of the drug and the apparence of particular products is compared with that occurring without the scavenger. Polymerization reactions (e.g. polymerization of acrylamide) can be used to detect formation of free radicals in anaerobic media.[9] The photooxidation potential can be studied in terms of oxygen uptake measurements in the presence of oxidable substrates like histidine and 2,5-dimethyl furan which are substrates for singlet oxygen, and L-tryptophan which is a substrate for superoxide. The reactions should be confirmed by addition of suitable singlet oxygen quenchers and superoxide dismutase. The (lack of) specificity of the various scavengers and substrates should be taken into account, and a combination of scavengers or substrates should be used to obtain adequate information. It is important to remember that the relative reactivity of both the radicals and the scavenger will determine the outcome of the reaction. If the radical intermediates are extremely reactive, they may react with the solvent before any other reaction can occur and no change will be observed.

1.1.3 Characterization of Excited States. The lifetime of the excited singlet state of a drug molecule is generally of the order of nanoseconds. This time is normally too short for the excited molecule to react chemically even with neighbouring molecules, although photoionization can occur. Excited triplet states of drug molecules can have long lifetimes (up to several seconds). Long-lived intermediates with lifetimes around one second may diffuse between organelles or to neighbouring cells prior to a reaction with oxygen or endogenous compounds.[10] Excited drug triplets formed in vivo can therefore often reach several molecular targets prior to deexcitation, implicating a high probability for phototoxic reactions. The production of singlet oxygen has been reported to occur by energy transfer both from the singlet and triplet state of a sensitizer. The singlet-triplet interaction is, however, of very low probability and formation of singlet oxygen by energy tranfer from the triplet state of the sensitizer is highly preferred. Transient species like singlet and triplet excited states can be observed by use of flash photolysis.[9] Valuable information concerning the formation of singlet and triplet states can, however, also be obtained from fluorescence and phosphorescence measurements. The lifetime of a molecule in the excited singlet or triplet states is related directly to the fluorescence and phosphorescence lifetimes respectively. Information about the possible photoreactivity of the drug substance and its degradation products or metabolites can therefore be obtained from luminescence quantum yield measurements and life-time studies. Additionally, the singlet and triplet energies of the compound are of considerable interest in mechanistic studies especially with respect to energy transfer. Phosphorescence is usually too weak to be observed in solution at room temperature. The drug molecule should therefore be held in a transparent (glassy) matrix at low temperature during the experiment. Unless fluorescence is measured under the same conditions the quantum yields cannot be compared.

1.1.4 Photosensitized Damage of Endogenous Substances. A phototoxic drug can damage the body tissues by formation of radicals or by energy transfer to endogenous compounds. Oxygen is present in all body tissues, and the body therefore provides excellent conditions for formation of reactive oxygen species. Cell constituents as proteins, unsaturated fatty acids, cholesterol and DNA are likely targets for damage caused by ROS. The reactions are often leading to photomodification of the cell membranes. The tissues are normally equipped with antioxidants but accumulation of a phototoxic compound can overload the capacity of the protecting agents in a specific tissue.[11, 12] Photohaemolysis of red blood cells provides a simple technique to monitor membrane photomodification. Erythrocytes have no intracellular organelles, so the oberved effects are due to effects on the membrane.[13] Some of the antimalarial drugs are previously reported to accumulate in melanin-rich tissues (eye, skin, hair).[14] Accumulation of drugs in light exposed areas is important for their ability to act as photosensitizers. The turnover of melanin in the body is very low, except for epidermal melanin. Compounds with high melanin affinity can be retained in melanin-containing tissues for years.[15] Studies of in vitro interactions between the drug molecules and melanin is therefore of great importance in the evaluation of drug phototoxicity. The particle size and zeta potential of the melanin granules must be taken into account as the chemical composition of melanin isolated from different tissues will vary. Several of the aminoquinoline antimalarials are also known to induce ocular adverse effects. Both retina, lens and cornea are likely targets for photosensitized damage in the eye. Photosensitized reactions in the lens can lead to a modification of the amino acids in the cytosol proteins and/or a covalent binding of the sensitizer to the proteins. This leads to a change in the physical properties of the lens proteins which results in aggregation and opacification of the lens (lens cataracts).[16] Isolated proteins from calf lens can be used as an in vitro biological test system for estimation of ocular phototoxicity.

1.1.5 Pharmocokinetic Parameters and Dosage Regime. Most of the adverse effects associated with the use of antimalarials are related to the eye and skin. Drugs administered systemically like the antimalarial compounds must be distributed to the actual body tissues e.g. the eye, skin, or the outermost capillaries of the skin, to act as photosensitizers. Systemical distribution of exogenous substances to the eye is however limited. Although the retina is richly supplied with blood vessels, the blood-retinal barrier is normally very tight and therefore restricts the movements of substances from the capillaries. The lens has no blood supply and drugs cannot be distributed directly from the systemic circulation to this tissue. On certain occasions the permeability of the blood-retinal barrier can be altered. Drugs will then to a larger extent enter the retinal pigmented epithelium (RPE) and retina. Drugs accumulated in the RPE can further pass through the vitreous cavity and reach the lens. Phototoxicity is generally dose dependent, i.e. dependent on the concentration of the drug sensitizer at the site of action. The pharmacokinetic properties of the compound are therefore essential factors which have to be considered when in vitro results are used to predict in vivo photoreactivity. Mefloquine is used for the prophylaxis of malaria in areas of the world where there is a high risk of chloroquine-resistant falciparium malaria. Mefloquine is usually administered as a single dose or as two divided doses and is rarely used over a long period of time in order to prevent mefloquine resistance. This is in contrast to chloroquine and sometimes also primaquine, which can be used for periods of months or even years. The dosage regime for mefloquine is, however, about to change as mefloquine now also is used in long-term prophylaxis for up to one year.[17] As mentioned above most of the adverse effects reported after use of anti-malarial drugs seem to be dose-related. High accumulative dosages and long-term administration may be important factors for drug phototoxicity. The dosage regime should therefore also be taken into account when the phototoxic potential of the various drugs are to be compared.

1.2 In vitro Screening of Compounds in the Solid State

Pharmaceuticals intended for use in the tropics, like the antimalarial compounds, are required to maintain their stability under the most severe storage conditions. The bulk substance and drug product will often be distributed through local traders whom in the majority of cases are ignorant of the ideal storage requirements. The compounds may therefore be exposed to considerable amounts of humidity, heat and sunlight before use. The most frequently used dosage form for antimalarial drugs is tablets although chloroquine is also formulated as injection solutions and syrups. If the patient is seriously ill quinine should be given by long term intravenous infusion (8-12 hours).

Photodegradation of a compound in the solid state is quite different from degradation of the substance in solution. In the solid state the reactions take place only in a thin surface-layer of the compound. The particle size and shape and the structure of the crystal lattice (i.e. crystal modification) will strongly influence the photostability of the bulk material. The most obvious results of the irradiation of a solid sample is a change in appearance (i.e. change in colour) and photodegradation of the compound. Change in appearance does not necessarily relate to degradation of the compound and vice versa. This means that discolouration of e.g. a tablet in many cases not will affect the efficacy of the product but on the other hand there might be a loss in potency although the tablet looks unchanged. A discolouration of compounds stored under conditions as discussed above is often observed. It is then of great importance to evaluate the correlation between change in appearance and degradation of the parent compound. In many cases visual inspection will be the only "quality control" possible after the product has reached the local market.

2   IN VITRO PHOTOREACTIVITY SCREENING OF CHLOROQUINE, PRIMAQUINE AND MEFLOQUINE

The compounds discussed here represent the three main groups of quinolines used in malaria therapy, i.e. the group of 4-aminoquinolines consisting of chloroquine, hydrochloroquine and amodiaquine, the 8-aminoquinolines consisting of primaquine and the 4-quinolines consisting of quinine and mefloquine. Some of the antimalarial drugs e.g. chloroquine and hydroxychloroquine, have also proved to be useful in the treatment of other, noninfectious disorders.[18] All the compounds give adverse effects which are possibly phototoxic reactions; they are known to be photolabile in solution and a discolouration is often observed in the solid state. The photoreactivity screening in the present study was performed according to the methods described above.

2.1 Chloroquine Diphosphate

2.1.1 Chloroquine in Solution. Chloroquine (CQ) can exist as a single- and double-charged cation and in a neutral form. The pK_a values for CQ are reported to be 8.4 and 10.4 for the heterocyclic nitrogen atom and aliphatic nitrogen atom respectively. The pK_a value for the amino group bonded to the C_{10} atom on the ring has not yet been determined. At physiological pH chloroquine will exist both as the dication (91%) and monocation (9%) (Figure 1). The dication of CQ is shown to exist as the imine tautomer while the monocation exists as a mixture of the amine (8.1%) and the imine (0.9%) at physiological pH (19). The compound is demonstrated to decompose on exposure to irradiation both in aqueous solution and in organic solvent, and several degradation products are isolated and identified.[5, 20]

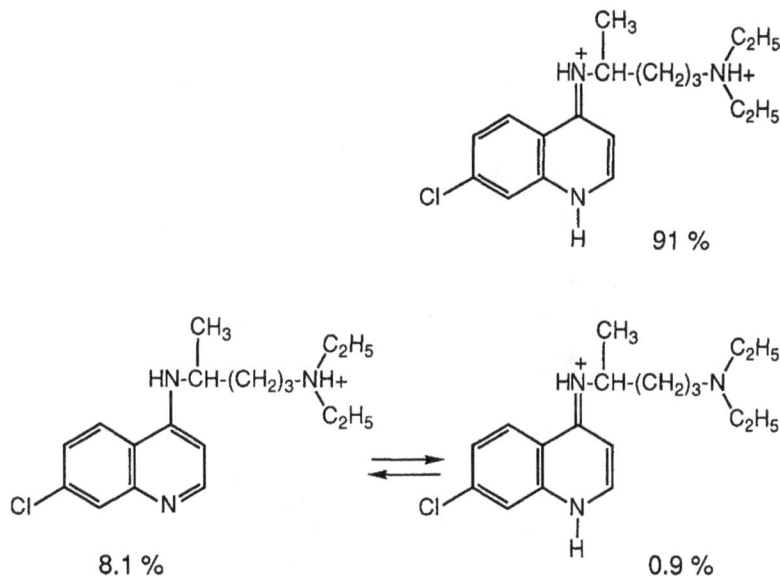

Figure 1 *Protonation of chloroquine at physiological pH.*

The photolysis quantum yield increases with an increase in pH indicating that the drug is most stabile as a dication.[20] The degradation process is also strongly dependent on the oxygen level of the sample. There is an increase in degradation rate with a decrease in oxygen concentration leading to the conclusion that the excited triplet of CQ is likely to play an important role in the degradation process. This is further emphasized by the observation that cobaltous ion, a triplet state quencher, has an inhibitory effect on the degradation of CQ. The degradation process seems to be initiated by alpha-C-H cleavage in the side chain.[20] The monocation of CQ is demonstrated to be a source of superoxide and hydroxyl radicals, but is a weak source of singlet oxygen and an inefficient quencher of this species.[20, 21] The monocation is strongly fluorescent with a fluorescence quantum yield of 0.14, while the dication is strongly phosphorescent with a triplet lifetime of approximately 1 second.[22] Chloroquine is a weak inducer of photohaemolysis of red blood cells (RBC), it induces photopolymerization of calf lens proteins and it binds strongly to melanin.[23-25]

2.1.2 Phototoxic Potential of Chloroquine. Based on the above results CQ should have the potential to induce phototoxic reactions in vivo. Chloroquine has a large distribution volume (200 l/kg) and a long elimination half-life (25-60 days). This indicates that the compound is widely distributed to- and accumulated in various tissues. Chloroquine medication can continue for years, both with respect to malaria prophylaxis and in the treatment of other diseases. The accumulative dose can be high, i.e. more than 200 grams. The absorption spectrum of CQ has a cut-off at 365nm, i.e. chloroquine will absorb radiation that penetrates skin, cornea and lens, but not retina. CQ is known to be retained in the eye for a long time.[26] The retinopathy observed after medication with this drug is, however, not likely to be caused by a reaction sensitized by chloroquine due to the fact that CQ does not absorb the wavelengths reaching this tissue. CQ is deposited in the cornea and has also been detected in the tear film, thus corneal adverse effects and lens cataracts resulting from CQ therapy[27] can be ascribed to photosensitized reactions. The drug accumulates in the epidermis of the skin, and the observed change in pigmentation[14] is also a possible phototoxic reaction. The monocationic form of CQ can penetrate lipid bilayers, e.g. of erythrocytes.[28] The RBC membrane is demonstrated to haemolyse after intravascular peroxidation reactions, thus haematologic photosensitization induced by CQ is a possibility.[29]

2.1.3 Chloroquine in the Solid State. The commercially available quality of chloroquine diphosphate is a hydrate (modification I) which can recrystallize to an anhydrous form on heating.[30, 31] The ratio of CQ/water in modification I is calculated to 3:1. The bulk substance (modification I) can take up water in humid air at room temperature forming a new hydrate (modification II) with a CQ/water ration of 2:1. Compression and grinding of modifiaction I will lead to the formation of two new forms, modification III and IV respectively. These four crystal modifications of CQ show different sensitivity to irradiation. The formation of modification II from the bulk substance seems to be catalyzed by irradiation. No significant changes seem to occur in the thermogram of modification II during exposure while the thermograms of modifications III and IV show great changes under the same conditions.[32] Formation of the degradation products 4-aminoquinoline and desethylchloroquine from the different modifications further emphasize that there is a difference in photostability between these forms in the order modification II > I > III > IV, the first being the most stable. This seems to correspond with a change in colour of the substance. After 50 hours exposure at 80W/m^2 in a sun simulating unit the colour of the powders changed from white to yellow described by an increase in Hunter b*-coordinate (measures yellowness) and a simultaneous decrease in the L*-coordinate (measures whiteness). Modifications I and II show less change in the Hunter b* and L* coordinates than modification III and IV.[32]

2.2 Primaquine Diphosphate

2.2.1 Primaquine in Solution. The pK$_a$ values for primaquine (PQ) are reported to be 3.2 and 10.4 for the heterocyclic nitrogen atom and aliphatic nitrogen atom, respectively.

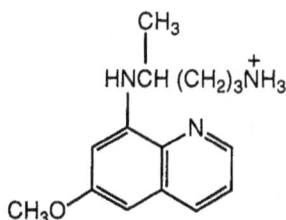

Figure 2 *Protonation of primaquine at physiological pH.*

PQ exists as a monocation at physiological pH (Figure 2). Primaquine is photoreactive as a monocation and forms several degradation products in aqueous solution at physiological pH.[33, 34] The influence of pH on the degradation process is not investigated at present. The photolysis is, however, strongly dependent on the oxygen content of the medium but in the opposite way compared to chloroquine, e.g. the degradation rate is increased by an increase in oxygen concentration.[33] The degradation process is postulated to be initiated by electron transfer between excited PQ and molecular oxygen, leading to the formation of superoxide and the cation radical of PQ.[33] In addition to superoxide, primaquine is a source of hydroxyl radicals but the formation of singlet oxygen is not detected.[21] On the other hand, primaquine is a very efficient quencher of singlet oxygen with a quenching rate constant of 2.6×10^8 M^{-1} s^{-1}.[21] PQ has weak fluorescence with a quantum yield of 0.00005 and a short-lived triplet state (lifetime approximately 4.8 μsec). It is a slightly more efficient inducer of phothaemolysis than chloroquine but the two compounds have about the same capacity to induce lens polymerization.[23, 24] Primaquine also binds to melanin.[25]

2.2.2 Phototoxic Potential of Primaquine. Like chloroquine, primaquine should also be regarded as a potentially phototoxic compound. PQ has, however, a low distribution volume (3-4 l/kg) and a short elimination half-life (7 hours) compared to chloroquine. It is also usually administered for a shorter period of time and at a lower accumulative dose. The absorption cut-off is 430nm which means that PQ absorbs visible light, i.e. light that reaches the retina, but the pharmacokinetic parameters indicate that primaquine is probably not distributed to the eye. Ocular phototoxicity is therefore quite unlikely. PQ will mainly be located in the blood where it is extensively bound to plasma proteins. Primaquine also penetrates cell membranes and concentrates in the erythrocytes.[35] Blood cells are therefore the targets which are most likely to be damaged by photochemical reactions induced by PQ, effectuated by light penetration through the outermost capillaries. This is consistent with the haematologic side effects observed after medication of this drug.

2.2.3 Primaquine in the Solid State. Commercially available primaquine exists only as one crystal modification. Elevated humidity and temperature, irradiation, high pressure or grinding do not lead to the formation of other crystal forms. Exposure of the samples in a sun-simulating unit does, however, lead to a change in colour measured by a change in the Hunter a*, b* - and L* - coordinates. In some cases a bleaching was observed while in some cases the samples became more yellow as a result of exposure. A change in colour was apparently not related to degradation of the drug substance as no decomposition was observed in any of the samples (to be published).

2.3 Mefloquine Hydrochloride

2.3.1 Mefloquine in Solution. Mefloquine consists of a mixture of the monocation (94%) and the neutral form (6%) at physiological pH (Figure 3). The pK$_a$ value for the quinoline N in MQ is in the range 4-5.5 while the corresponding value for the amine N is about 8.6.[36] Mefloquine in solution is sensitive to irradiation and several photodecomposition products are isolated and identified.[37] The photolysis shows the same dependency on pH and oxygen level of the medium as chloroquine, i.e. an increase in degradation rate by an increase in pH and by a decrease in oxygen concentration is observed (to be published). At physiological pH mefloquine is demonstrated to be an efficient source of singlet oxygen with a quantum yield of 0.38.[21] The substance is further known to induce the formation of superoxide. The ability to act as a source of reactive oxygen species is dependent on the pH of the medium, i. e. the state of protonation of MQ. This is also the case for the fluorescence quantum yield of this compound.[36] The phosphorescence lifetime of the monocation is quite short, about 3.5 milliseconds.[36] Mefloquine is a potent inducer of (photo)toxic reactions in the applied test systems. Incubation of RBC with mefloquine did result in a 100% dark haemolysis.[23] MQ is further a clearly more powerful inducer of lens polymerization than both chloroquine and primaquine.[24] The compound binds to melanin in vitro.[25]

2.3.2 Phototoxic potential of mefloquine. From the in vitro results mefloquine is a highly phototoxic drug compared to CQ and PQ. The compound has a fairly large distribution volume (13-29 l/kg) and a long elimination half-life (19.5 days). Since the substance is widely distributed to the tissues and slowly eliminated from the body it should also have the possibility of reaching the eye and accumulate in the retina. MQ does, however, not absorb irradiation above 330nm and is therefore not likely to act as a sensitizer in this part of the eye. Photosensitized reactions in the lens and cornea can occur assuming that the drug is distributed to this part of the eye. Mefloquine can accumulate in the melanocytes of the skin, and skin rashes reported to occur after medication with this compound can be due to photosensitized reactions. Unwanted effects are also possible in the blood as MQ is distributed to the erythrocytes and clearly affects the membrane. The phototoxic potential of mefloquine compared to CQ and PQ will be dependent on the applied dosage regime, e.g. single dose or one year prophylaxis, as discussed above.

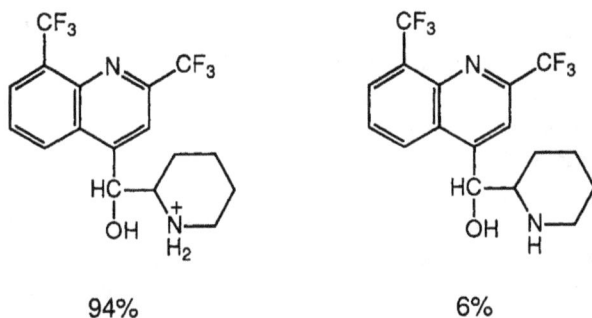

94%   6%

Figure 3 *Protonation of mefloquine at physiological pH.*

2.3.3 Mefloquine in the solid state. At least 8 different crystal modificaitons of MQ are described in the litterature.[38, 39] In the present work mefloquine bulk substance from two different suppliers has been investigated. The two qualities of MQ represent two crystal modifications. The two modifications respond differently to irradiation. One of the modifications decomposes without any decolouration while the other modification becomes yellow but is not decomposed as a result of exposure in a sun simulating unit. The crystal modifications and/or photostability are changed when the bulk substances are incorporated in a tablet formulation.[40]

3 CONCLUSION

The antimalarial compounds show different photoreactivity both in solution and in the solid state although they all belong to the group of quinoline antimalarials. The reactivity in solution is strongly dependent on ionization form of the substance and of the oxygen concentration of the medium. A discolouration of the bulk substance is not necessarily related to drug degradation. The formulation process can change the crystal modification and photostability of the drug substance. Both chloroquine, primaquine and mefloquine are sensitive to irradiation in solution and in the solid state and they have the potential to induce phototoxic reactions in vivo.

References

1. H. H.Tønnesen, S. Kristensen and K. Nord in 'Photostability of Drugs and Drug

Formulations', H. H.Tønnesen, Ed., Taylor and Francis, London, 1996, Chapter 12, p.2 67

2. J. V. Greenhill in ref. 1, Chapter 5, p. 83

3. T. Oppenländer in ref. 1, Chapter 11, p. 217

4. D. E. Moore and V. J. Hemmens, *Photochem. Photobiol.*, 1982, **36**, 71

5. K. Nord, J. Karlsen and H. H.Tønnesen, *Int. J. Pharm.*, 1991, **72**, 11

6. I. E. Kochevar and A. A. Lamola, *Photochem. Photobiol.*, 1979, **29**, 791

7. G. Rodighiero and F. Dall'Acqua, *Photochem. Photobiol.*, 1976, **24**, 647

8. J. E. Roberts in ref. 1, Chapter 10, p. 189

9. D. E. Moore in ref. 1, Chapter 2, p. 9

10. H. Bayley, F. Gasparro and R. Edelson, *Trends Pharmacol. Sci.*, 1987, **8**, 138

11. J. Dillon, *J. Photochem. Photobiol.: B: Biol.*, 1991, **10**, 23

12. J. P. Kehrer and C. V. Smith in 'Natural Antioxidants in Human Health and Disease', B. Frei, Ed., Academic Press, USA, 1994, p. 53

13. D. P. Valenzeno in 'Photobiological Techniques' D. P. Valenzeno, R. H. Pottier, P. Mathis and R. H. Douglas, NATO ASI Series. Series A: Life Sciences, Plenum Press, New York, 1991, Vol.216, p.99

14. L. Tannenbaum and D. L. Tuffanelli, *Arch. Dermatol.*, 1980, **116**, 587

15. N. G. Lindquist, *Uppsala J. Med. Sci.*, 1986, **91**, 283

16. J. E. Roberts in 'Photosensitisation', G. Moreno, Ed., NATO ASI Series, Springer-Verlag, Berlin-Heidelberg, 1988, Vol. H 15, p.325

17. 'British National Formulary', Brit. Med. Ass., The Pharmaceutical Press, London, 1997, **33**, 281

18. D. J. Wallace, *Lupus*, 1996, **5**, S59

19. L. S. Rosenberg and S. G. Schulman, *J. Pharm. Sci.*, 1978, **67**, 1770

20. K. Nord, A-L. Orsteen, J. Karlsen and H. H. Tønnesen, *Pharmazie*, 1997, **52**, 8

21. A. G. Motton, L. J. Martinez, N. Holt, R. H. Sik, K. Reszka, C. F. Chignell, H. H. Tønnesen and J. E. Roberts, *Photochem. Photobiol.*, 1997, **65**, 27S

22. K. Nord, J. Karlsen, H. H. Tønnesen, *Photochem. Photobiol.*, 1994, **60**, 427

23. S. Kristensen, J. Karlsen and H. H.Tønnesen, *Pharm. Sci. Comm.*, 1994, **4**, 183

24. S. Kristensen, R-H. Wang, H. H. Tønnesen, J. Dillon and J. E. Roberts, *Photochem. Photobiol.*, 1995, **61**, 124

25. S. Kristensen, A-L. Orsteen, S. A. Sande and H. H. Tønnesen, *J. Photochem. Photobiol., B: Biol.*, 1994, **26**, 87

26. N. G. Lindquist and S. Ullberg, *Acta Pharm. Toxicol.*, 1972, **31**, 1

27. F. T. Fraunfelder and S. M. Meyer, 'Drug-Induced Ocular Side Effects and Drug Interactions', Lea and Febiger, USA, 1989, p.58

28. C. A. Homewood, D. C. Warhurst, W. Peters and V. C. Baggaley, *Nature*, 1972, **235**, 50

29. D. Chiu, F. Kuypers, B. Lubin, *Semin. Hematol.*, 1989, **26**, 257

30. A-K. B. Bjåen, K. Nord, S. Furuseth, T. Ågren, H. H. Tønnesen and J. Karlsen, *Int. J. Pharm.*, 1993, **92**, 183

31. S. Furuseth, J. Karlsen, A. Mostad, C. Rømming, R. Salmén and H. H. Tønnesen, *Acta*

Chem. Scand., 1990, **44**, 741

32. K. Nord, H. Andersen and H. H. Tønnesen, *Drug Stability*, in press.

33. S. Kristensen, K. Nord, A-L. Orsteen and H. H. Tønnesen, *Pharmazie*, in press.

34. S. Kristensen, A-L. Grislingaas, J. V. Greenhill, T. Skjetne, J. Karlsen and H. H. Tønnesen, *Int.J.Pharm.*, 1993, **100**, 15

35. E. Kennedy and H. Frisher, *J. Lab. Clin. Med.*, 1990, **116**, 871

36. H. H. Tønnesen and D. E. Moore, *Int. J. Pharm.*, 1991, **70**, 95

37. H. H. Tønnesen and A-L. Grislingaas, *Int. J. Pharm.*, 1990, **60**, 157

38. S. Kitamura, L-C. Chang and J. K. Guillory, *Int. J. Pharm.*, 1994, **101**, 127

39. A. Kiss, J. Repasi, Z. Salamon, C. Novak, G. Pogol and K. Tomor, *J. Pharm. Biomed. Anal.*, 1994, **12**, 889

40. H. H. Tønnesen, G. Skrede and B. K. Martinsen, *Drug Stability*, in press.

Photochemistry of Diuretic Drugs in Solution

Douglas E. Moore
Department of Pharmacy
The University of Sydney
Sydney 2006, Australia

1 INTRODUCTION

Drugs that facilitate diuresis are in widespread use in the community for the treatment of oedema and hypertension, and for other conditions for which the increase in urinary flow can relieve symptoms. There are several mechanisms by which this action is achieved, so the drugs are generally grouped in terms of their particular mechanism of action. There are some similarities of chemical structure of the drugs within such a grouping, for example, the thiazides, which are inhibitors of sodium and chloride ion reabsorption. Of interest here are those diuretic drugs which have an absorption spectrum in the sunlight region above 280 nm, a necessary requirement for the possibility of participation in photochemical reactions in the normal environment. Several of these have been reported to give rise to adverse photosensitivity responses *in vivo*, and many have been the subject of photodegradation and other photochemical studies *in vitro* which will be described below. A selection of the more important photoactive diuretic agents is shown in Figure 1 with their longest wavelength absorption maxima. Acetazolamide has been included since its absorption when in alkaline solution does extend into the UVB region, even though no photochemical studies have been reported in the literature for this drug.

The most frequently used diuretic drugs are hydrochlorothiazide and frusemide (furosemide) and consequently these have been the most studied in all aspects of stability. Hydrochlorothiazide can be used alone but is mostly co-formulated with a potassium-sparing agent such as amiloride or triamterene to yield a better balanced diuretic effect. Recent developments of new diuretics have tended to concentrate on the thiazide structure. The first of the thiazides was chlorothiazide with a double bond at the 3,4 position. Saturation to produce the 3,4-dihydro-derivative (i.e., hydrochlorothiazide) gives a compound with 10 times more activity than the parent. These compounds are weakly acidic with the H atom on N-2 being the most acidic because of the electron withdrawing effects of the neighbouring sulphone group. An electron withdrawing group is necessary at C-6 for the diuretic action. Hence the chlorine atom is a feature of the structure of many diuretics, but is supplanted by the trifluoromethyl group in many of the newer compounds. The sulphonamide group at C-7 is also necessary for effective diuretic action.[1]

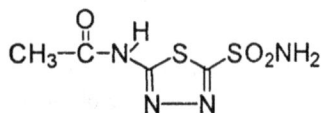

Acetazolamide
291 nm (alkaline solution)

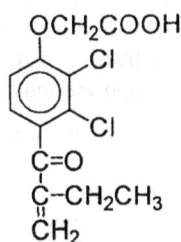

Ethacrynic acid
270 nm

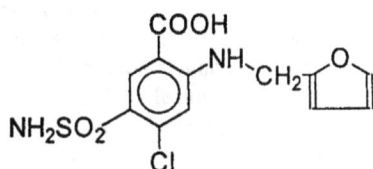

Frusemide
330 nm

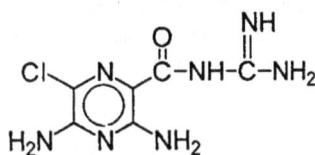

Amiloride
360 nm

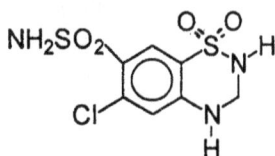

Hydrochlorothiazide
318 nm

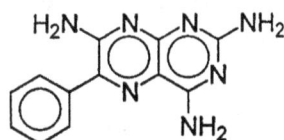

Triamterene
366 nm

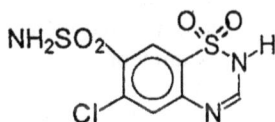

Chlorothiazide
315 nm

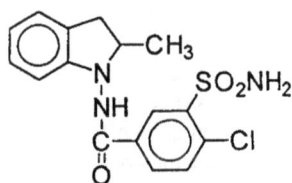

Indapamide
288 nm

Figure 1   *Chemical structures of photoactive diuretic drugs.*

The chlorine and sulphonamide substituents are also components of the frusemide structure, but the presence of the free carboxyl group makes it a stronger acid than the thiazides. Frusemide is a high-ceiling or loop diuretic with a saluretic effect 8 to 10 times stronger than the hydrothiazides. Indapamide has some similarities to the thiazides and frusemide with its chlorine and sulphonamide substituents.

The main potassium-sparing diuretics, amiloride and triamterene, are both basic substances which interfere with the process of cation exchange in the distal tubule, blocking the reabsorption of sodium and the secretion of potassium ions. These two compounds have the strongest UV absorption at the longest wavelength of all the diuretics. Nonetheless, the major photochemical activity of the diuretics has been shown by frusemide and hydrochlorothiazide.

A survey of the photochemical and photobiological studies relevant to the understanding of the photostability of the major diuretics has been made. In examining the literature, those studies are considered relevant in which the irradiation conditions have been chosen to simulate sunlight, as this is considered to be the most relevant condition that would be experienced in practice. Also important in comparing photochemical studies is the fact that measured rate constants and half-lives are related to the particular irradiation conditions used. The wavelength range absorbed by the sample will depend on the type of photon source and filter used. The use of sunlight as the irradiating source will also show significant variations according to the latitude, time of day and season. The quantum yield should be used as an absolute standard for reporting the efficiency of a photochemical reaction, thereby taking account of variations in irradiating source and experimental positioning of samples.[2] However, for photostability testing of formulations in an industrial context, a standardized sunlight-simulating source consisting of a xenon arc lamp with special filters is recommended.[3]

2 PHOTOCHEMISTRY OF FRUSEMIDE

There have been a number of studies of the photodegradation of frusemide with some apparent variation in the results reported.[4-9] The differences in outcomes from the various studies are due to differences in irradiation conditions, such as the wavelength range, the concentration of frusemide and the solvent used, as well as the presence or absence of oxygen in the system. In our laboratory the approach has been to irradiate solutions which have an absorbance of about 1 at the λ_{max} of the absorbing compound. Such a concentration usually permits direct spectrophotometric and chromatographic analysis of the irradiated solution. The presence of oxygen frequently leads to a more complex array of photoproducts due to the fact that oxygen is an efficient radical scavenger. Thus initial studies performed under a nitrogen atmosphere may provide a less complex product mixture which is easier to analyse.

The chlorine substituent is a major factor in the photochemistry of the diuretics, as dehalogenation occurs quite efficiently upon irradiation of the drugs in solution. The concomitant spectral changes are very slight and are not very useful for following the progress of the photoreaction, as shown for frusemide in Figure 2. The release of the chloride ion and the concomitant increase in acidity can be detected by potentiometric titration as

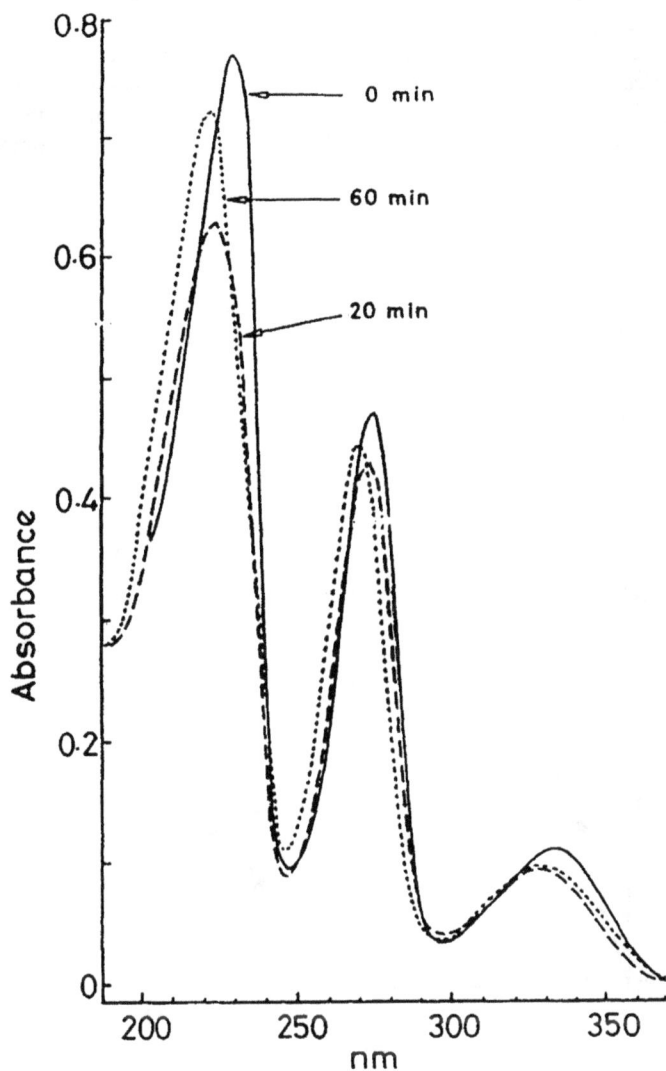

Figure 2 *UV absorption spectra of frusemide solution in oxygen-free methanol after various times of irradiation.*

shown in Figure 3. The solvent clearly affects the nature of the reaction when aerated solutions are irradiated.[5]

Oxygen-free solutions of frusemide in methanol produced hydrogen and chloride ions with a quantum yield of 0.40 ± 0.08. In aqueous solution (containing 25% methanol because of the low solubility of frusemide in water) complete dechlorination was also observed with the same quantum yield. In oxygenated solutions, chloride production was inhibited more than H^+ in methanol. Accompanying studies on oxidation photosensitized by frusemide show that it is an effective generator of singlet oxygen in methanol solutions but not in aqueous buffer at pH 7. Free radicals production is observed in deaerated solutions by the addition of a polymerizable monomer such as acrylamide, and is associated with the dechlorination process.

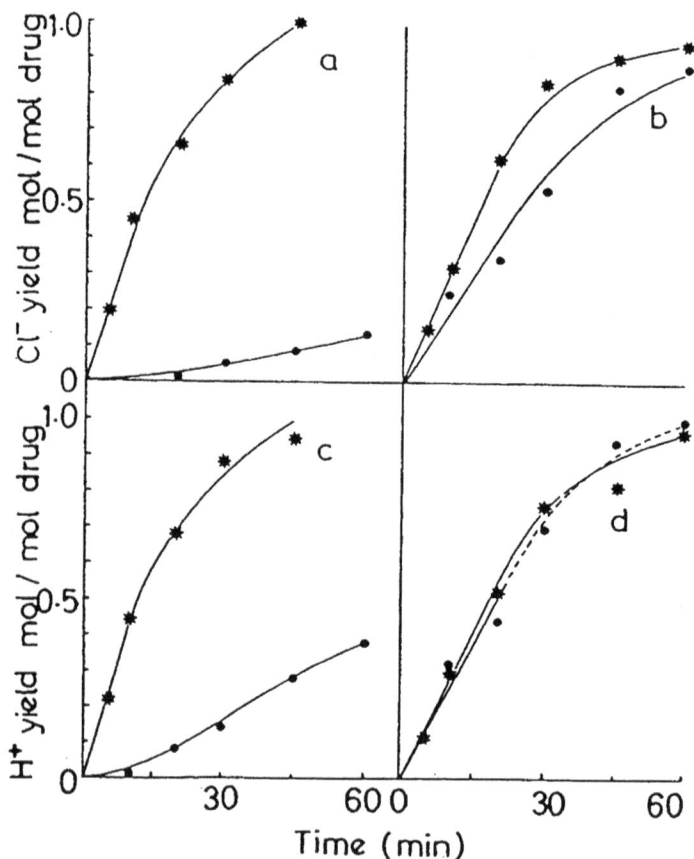

Figure 3 *Yield of chloride and hydrogen ions in the photolysis of frusemide (5×10^{-4} M) in solution saturated with nitrogen (✳) or oxygen (●). Solvents: for (a) and (c) methanol; for (b) and (d) water-methanol 3:1.*

Further investigation of the photosensitizing behaviour of frusemide in aqueous solution revealed that it is the neutral molecule which is effective in generating singlet oxygen and free radicals. With a pK_a of 3.9 at 20°C, the drug is completely in the anionic form at pH 7.0 and both its photosensitizing ability and photodegradation are greatly diminished. The variation of photooxidation rate with pH followed a sigmoidal curve with an inflection corresponding to the apparent pK_a of the photosensitizer in the system. This study was performed in surfactant solutions to overcome the poor solubility of frusemide, but it also demonstrated that the apparent pK_a of the drug can be shifted according to the charge carried by the surfactant.

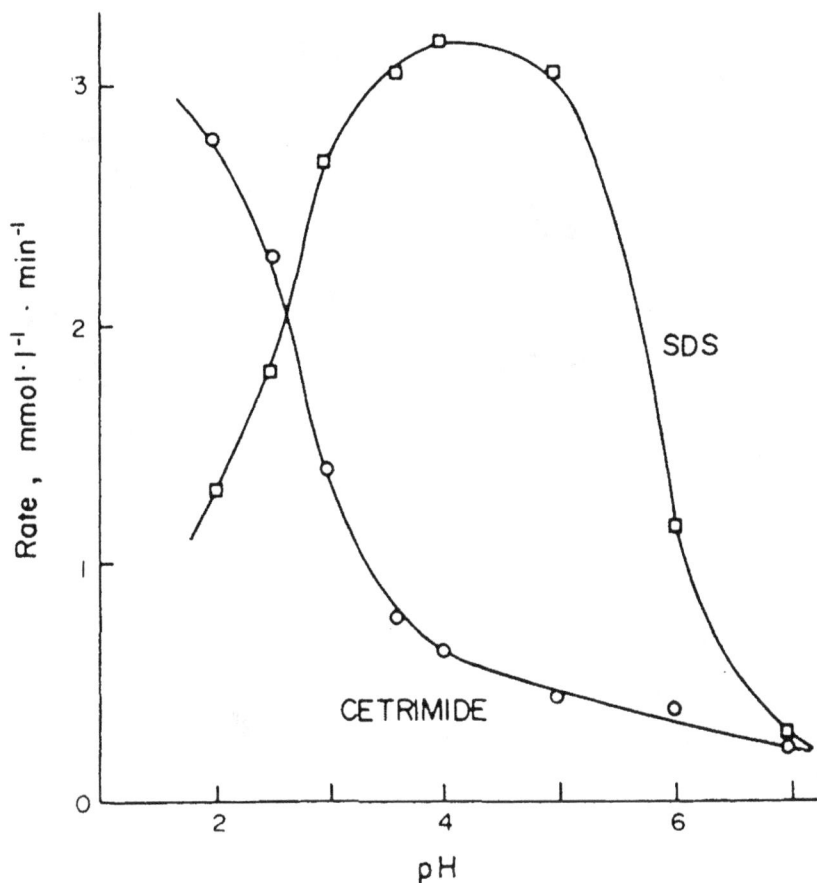

Figure 4 *Rates of photopolymerization of acrylamide (0.125 M) as a function of pH for frusemide (0.12 mM) at 30°C: (O-O) in cetrimide (0.01 M); (□-□) in sodium dodecyl sulphate (0.05 M).*

Using sodium dodecyl sulphate as surfactant suppressed the ionization of frusemide such that the apparent pK_a became close to 6 as shown for the photosensitized polymerization of acrylamide in Figure 4.[6]

Flash photolysis studies on frusemide (unpublished observations) revealed a strong triplet state formation from the neutral molecule in aqueous solution as well as methanol, connecting with the efficient generation of singlet oxygen in aerated solutions. Degradation would thus appear to occur from the triplet state of frusemide.

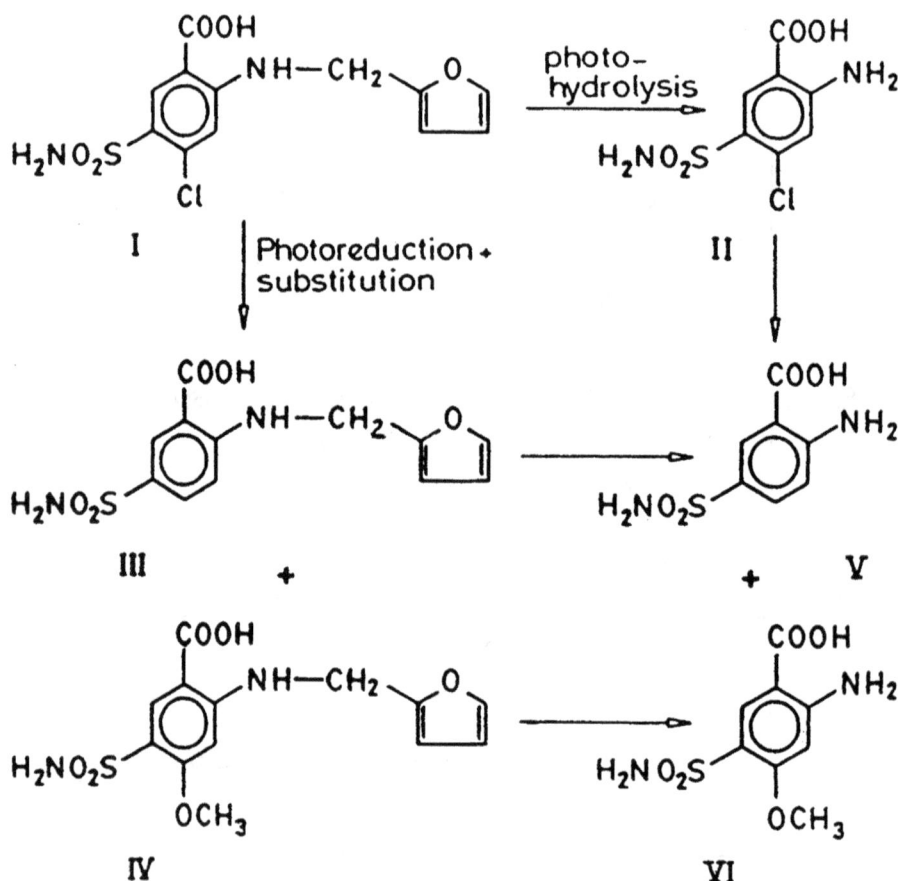

Figure 5   *Pathways of photodegradation of frusemide in oxygen-free methanol.*

Analysis of the product mix after irradiation of frusemide in deaerated solution shows that photodegradation leads to a combination of dehalogenation and hydrolysis reactions, as shown in Figure 5.[7] Dehalogenation leads to a mixture of substitution and reduction products. Photodechlorination has been postulated to occur through the formation of a pair of radical ions {Aryl-Cl[+]} {Aryl-Cl[-]} following electron transfer from the excited state molecule to a ground state molecule.[10] The radical cation is suggested to be the precursor of the substitution product {Aryl-OR} while the radical anion leads to the reduction product {Aryl-H} where OR and H are abstracted from the solvent ROH. The only difference is in terms of the substitution of -OH or -OR for -Cl depending on solvent.

The occurrence of a proportion of photohydrolysis in the photodegradation is similar to the situation observed following irradiation of some drugs susceptible to solvolysis, such as pentobarbitone[11] and others in the diuretic group mentioned below. Frusemide has a secondary amine functional grouping and is therefore susceptible to acid-catalysed hydrolysis and slow uncatalysed hydrolysis in neutral and alkaline solutions to form saluamine as major product.[12] These dark reactions are considerably slower than the photoinitiated ones.

In an earlier study of the photodegradation of frusemide in alkaline solution[4] the irradiation was continued for 48 h and the only product isolated and identified was 4-amino-5-carboxy-2-chlorobenzenesulphonic acid, representing the photooxidation of the sulphonamide group as well as hydrolysis of the furylmethyl group. This result serves to illustrate the need to follow the course of the photodegradation by chromatographic analysis from the beginning of the irradiation, as it is not known whether the product identified is the only product, and whether it is formed in a primary or secondary process. Additionally, from the perspective of the stability of the drug in a pharmaceutical formulation, the irradiation conditions used in the research and development laboratory are usually far more intense than is likely to be experienced in storage and use of the formulation. Thus the early stages of photodegradation must be considered as the more important in this regard, also serving to provide information to assist in elucidating the pathway(s) of the degradation.

As an example of the practical consideration of formulation stability, it was shown that frusemide infusion solutions of pH 8-10, stored in burette administration sets, are stable over a 48 h period when exposed to diffuse daylight/fluorescent strip room lighting.[8] However, the solutions decomposed rather quickly with a half-life of 4 h when exposed to direct sunlight. While validating an HPLC assay, it was found that frusemide alkaline solutions show the presence of small amounts of photoproducts after being exposed to sunlight for extended periods.[13]

The stability of several frusemide esters was examined in aqueous solutions of varying pH under different lighting conditions as a means of determining the importance of the carboxyl group of frusemide in the photodegradation.[9] These esters were very susceptible to photodegradation in aqueous solutions exposed to artificial light or diffuse sunlight. Frusemide itself is readily degraded in acidic solution but is relatively stable in alkaline solutions, with the pH profile of the degradation reflecting the change in ionization at its pK_a of 3.9. As shown in Figure 6, the relative rate constant k is a function of a_H the activity of hydrogen ion in solution (eq. 1).

$$k = k' a_H / (a_H + K_a) \qquad (1)$$

where k' is the first order rate constant for the photodegradation of the neutral form of frusemide under the same conditions of irradiation. This agrees with the stronger photosensitizing capacity of the neutral molecule compared with the frusemide anion (*vide supra*), and indicates that the degradation occurs through the long-lived triplet state species.

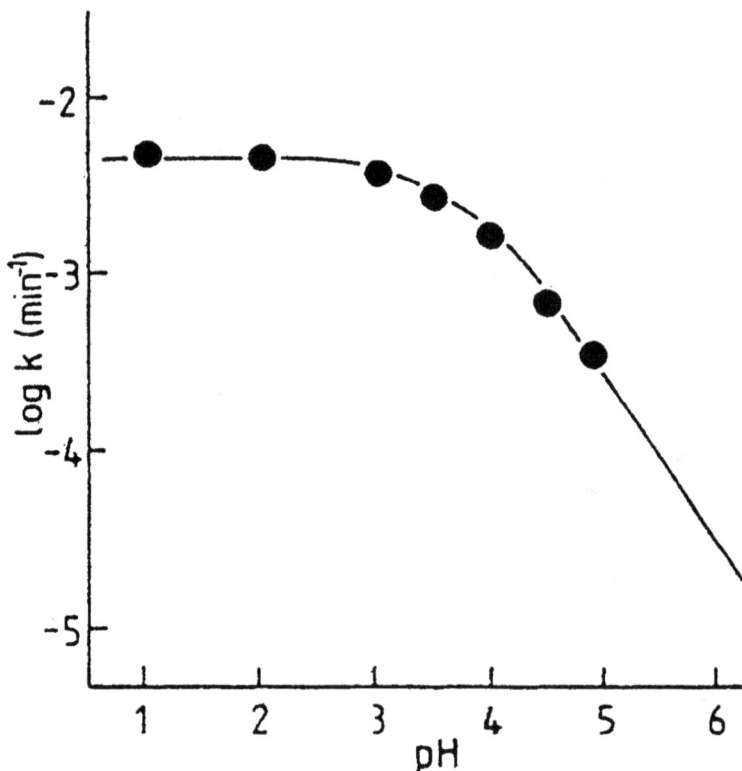

Figure 6 *The pH-rate profile for the photodegradation of frusemide in aqueous solutions at 20°C. The points are experimental and the solid curve calculated from Equation 1.*

The photolability of the frusemide esters has implications to the analysis of frusemide in urine samples, which may be acidic. A major pathway for the biotransformation of frusemide is conjugation with glucuronic acid forming esters, so adequate precautions should be taken to protect such samples from exposure to UV light.

3 PHOTOCHEMISTRY OF HYDROCHLOROTHIAZIDE

Hydrochlorothiazide (λ_{max} 318 nm, ε_{max} 3200 l mol^{-1} cm^{-1}) is a weak absorber compared with frusemide in the region above 300 nm. Nevertheless, it is completely dechlorinated upon irradiation with a medium pressure Hg arc lamp through a glass filter for 5 hours, in deaerated aqueous or alcohol solutions. The quantum yield for dechlorination is 0.18 ± 0.05 under deaerated conditions.[5] The thiadiazine ring is liable to hydrolyse in aqueous solution, and the pH-rate profile for the dark hydrolysis has a complex bell shape with inflections caused by the two ionizations occurring with pK values of 8.6 and 9.9.[14]

The photodegradation mechanism follows the same pattern as for frusemide with dechlorination and hydrolysis both occurring, as shown in Figure 7.[15] The dechlorination and hydrolysis occur almost equally so that the product mixture contains all the products from the beginning of the irradiation. The pH dependence of the photodegradation of hydrochlorothiazide has not been reported.

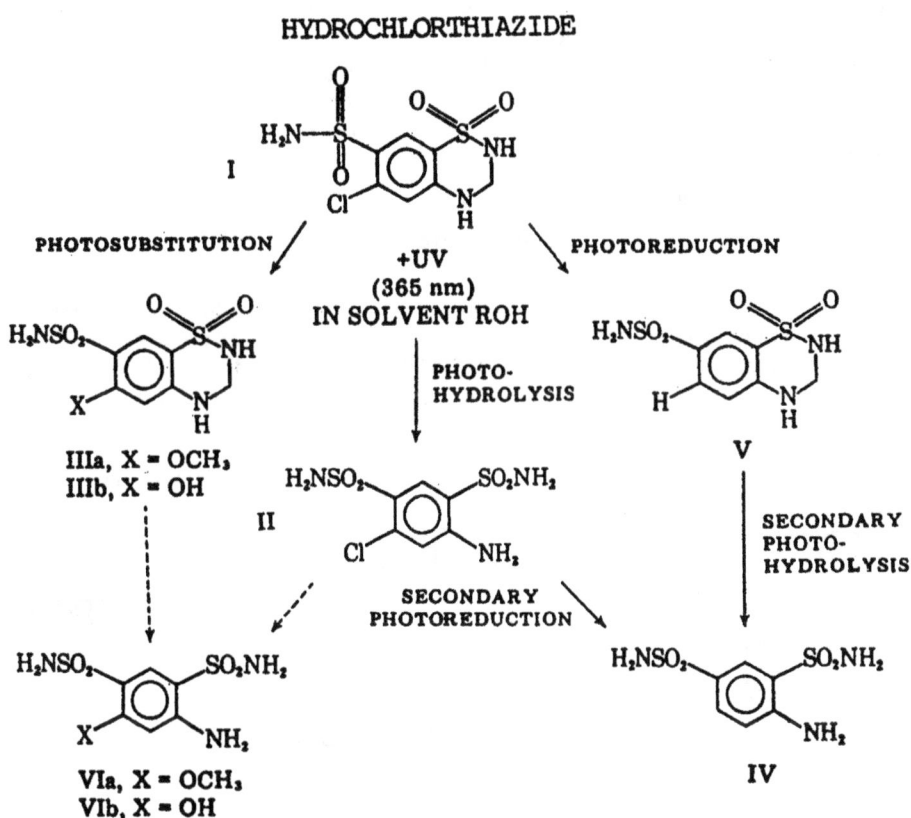

Figure 7 *Pathways of photodegradation of hydrochlorothiazide in oxygen-free methanol.*

Flash photolysis of hydrochlorothiazide has not provided a clear-cut identification of the primary photochemical species. The flash photolysis sources currently in use are Nd-YAG lasers which give the capability of irradiating at 355 nm or 266 nm. This is a problem for studying the effect of sunlight on drugs as the important UVB region is more energetic than 355 nm, but use of 266 nm may lead to results corresponding to excitation processes, e.g., photoionization, which are normally not achieved with sunlight. Because hydrochlorothiazide has only a very weak residual absorbance at 355 nm, it did not produce a response to flash photolysis at that wavelength. When 266 nm excitation was used for hydrochlorothiazide, a small broad absorption was observed between 350 and 450 nm, suggestive of a triplet state rather than a cation radical. There was no evidence of solvated electron production.[16]

The photosensitizing ability of hydrochlorothiazide and chlorothiazide were compared with frusemide in photooxidation reactions with 2,5-dimethylfuran, a singlet oxygen substrate, and free radical photopolymerization of acrylamide.[5,17] In methanol solutions, the thiazides were equally poor as sensitizers of photooxidation, whereas frusemide was more than 10-fold more effective. On the other hand, in the photopolymerization experiments, the thiazides demonstrated a modest level of free radicals production, about a third that of frusemide. This suggests that the most likely mechanism by which hydrochlorothiazide would participate in generating a photobiological response is through free radicals formed in the dechlorination.

4 PHOTOCHEMISTRY OF INDAPAMIDE

Indapamide was irradiated in methanol solution under nitrogen with a medium pressure mercury lamp through a copper sulphate filter solution which cut off wavelengths below 300 nm. The products were separated by preparative TLC and identified using MS and NMR as resulting from cleavage of the N-NH-CO-C linkage (Figure 8).

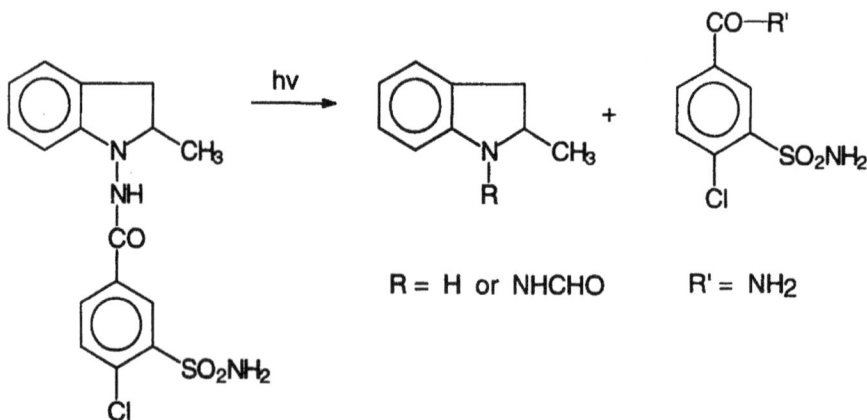

Figure 8 *Pathways of photodegradation of indapamide in oxygen-free methanol.*

After irradiation under oxygen, the same products were augmented by others formed by esterification and oxidative cleavage. Since the major cleavage reactions were not inhibited by oxygen (a triplet statequencher) the photolytic degradation was presumed to occur from the excited singlet state.[18] No photosensitization experiments have been reported for indapamide.

5 PHOTOCHEMISTRY OF AMILORIDE

The potassium-sparing diuretic, amiloride, is a basic compound with pK_a of 8.7 whereas the major diuretics are weakly acidic compounds. Although it is used almost exclusively in combination with hydrochlorothiazide, there are some reported instances of photosensitivity responses from use of the drug alone.[19,20] Amiloride has a strong absorption at 360 nm in the UVA giving it the potential for photochemical action. Additionally, it is susceptible to hydrolytic degradation at elevated temperatures with a complex pattern of products depending on the pH of the medium.[21] In alkaline solution the guanidine group is hydrolysed while one of the NH_2 substituents on the pyrazine ring is converted to OH in acidic solution. At intermediate pH, the guanidine side chain cyclizes forming a pteridine ring system. Although regarded as stable in the solid state, solid amiloride hydrochloride was reported to darken upon exposure to strong UV light, although no degradation products could be detected.[21]

The potential for amiloride to participate in photochemical processes was revealed by a laser flash photolysis and pulse radiolysis study.[22] The primary process occurring upon photoexcitation of amiloride in aqueous solution is photoionization yielding a solvated electron and the amiloride cation radical. On the other hand, a semireduced radical was observed in isopropanol solution. The yield of solvated electron was approximately 40% higher for the neutral form of amiloride (pH 10) than for the cation (pH 6). This effect is expected as the ejection of an electron from the cation is clearly energetically less favourable than from the neutral form. A relatively small yield of amiloride triplet state was also detected.

As expected from the flash photolysis results, the photochemistry of amiloride shows a distinctly pH-dependent behaviour, with the neutral molecule existing at high pH being 3-4 fold more reactive than the protonated species. The photodegradation of amiloride has been studied in aqueous solution at 30°C over the pH range 4.5 to 11 (Li, Moore and Tattam, manuscript in preparation). HPLC analysis and actinometry were used to determine the quantum yields of 0.23 ± 0.02 and 0.09 ± 0.01 for the photodegradation of neutral amiloride and its conjugate acid, respectively. Isolation of the photolysis reaction mixture revealed the presence of only one photoproduct which was identified as arising from the replacement of the chlorine substituent by OH. The stable form of the product appeared to be the tautomer shown in Figure 9. This very simple product pattern differs markedly from that of frusemide and hydrochlorothiazide where the reduction product was dominant after dechlorination, and hydrolytic products were also found. As pointed out above for frusemide, the precursor of the substitution product is postulated to be a cation radical in an ion-pair exciplex formed from the excited state and ground state molecules. A cation radical has been clearly identified in flash photolysis of amiloride. It is postulated therefore that the degradation occurs directly from the cation radical of amiloride, i.e., an exciplex ion-pair is not involved as has been proposed for frusemide and hydrochlorothiazide.

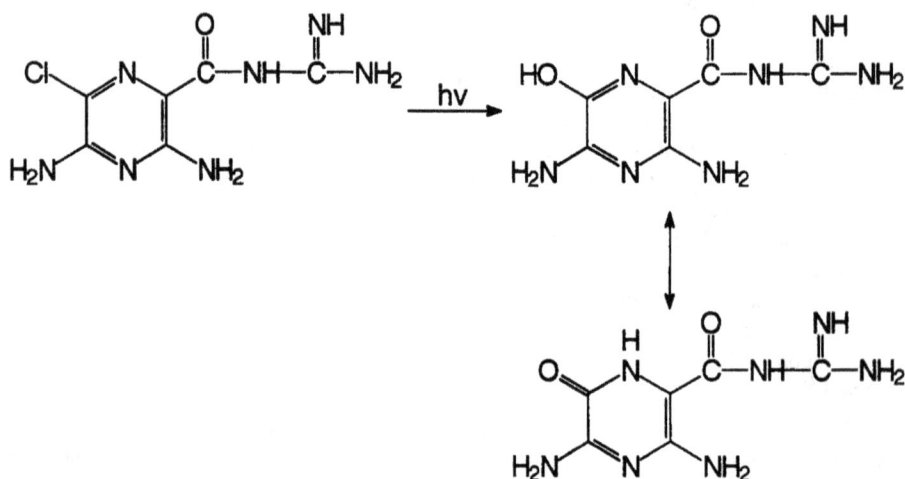

Figure 9 *Pathways of photodegradation of amiloride in oxygen-free aqueous solution.*

6 PHOTOCHEMISTRY OF TRIAMTERENE

Photosensitivity responses have been reported for triamterene[23] but even though it absorbs strongly in the UVA region at 360 nm, this compound has been found to degrade relatively slowly under UV irradiation. This may be due to the strong fluorescence providing an efficient means of dissipating the absorbed energy. The photodegradation of triamterene led to small amounts of several products which have not been identified at this time.

Since triamterene is primarily used in a combination formulation with hydrochlorothiazide, a photochemical study was performed to determine whether there might be any type of interaction between the components leading to degradation or other reactions detrimental to the formulation or the patient.[16] HPLC analysis was used to determine the degradation profile of triamterene and hydrochlorothiazide when irradiated together under anaerobic conditions. The photodegradation of triamterene proceeded very slowly compared with that of hydrochlorothiazide, but neither appeared to be affected significantly by the presence of the other. In oxygenated solutions, both drugs acted individually as weak sensitizers of the photooxidation of histidine. When both sensitizers were present in the same solution, the overall rate of photooxidation was 50% greater than the sum of the individually sensitized processes, an indication of a degree of interaction between the two sensitizers.

Further evidence of an interaction was found in a laser flash photolysis study. Using 355 nm excitation, photolysis of triamterene in buffer solution at pH 7.0 revealed a strong solvated electron signal as well as a transient identified as a cation radical, indicating photoionization to be the principal photochemical process. As noted above, hydrochlorothiazide does not respond to excitation at 355 nm, so it was possible to add hydrochlorothiazide to the triamterene flash photolysis solution and observe the effect on the decay of the solvated electron signal. The significant effect of hydrochlorothiazide on triamterene was seen as a doubling of the rate of decay of the solvated electron by the addition of 12 μM hydrochlorothiazide. Thus ground state hydrochlorothiazide appears to scavenge the electron formed on photoionization of triamterene, even though this does not lead to an increase in the rate of degradation. Overall, the results of both the transient and steady state experiments show there is only a minor interaction or cross-sensitization between hydrochlorothiazide and triamterene. This is not sufficient to suggest that the combination formulation might be markedly less stable to the effects of UV light compared with the individual drugs.

7 PHOTOTOXICITY OF DIURETIC DRUGS IN CELL CULTURES

The diuretic drugs have been tested for their photoxic potency to a limited extent only. It is generally believed that photosensitizers may induce phototoxic damage at different subcellular locations, but that they have one preferred site of action.[24] Yeast and fungus contain only the nuclear target, and were not damaged by UV irradiation with chlorothiazide, hydrochlorothiazide or acetazolamide.[25] Photohaemolysis, a test of phototoxic membrane damage, gave positive responses with hydrochlorothiazide and several other thiazides, but not with frusemide.[25] Tissue cultures may be regarded as more appropriate test models because both membranes and the nucleus are present as potential targets. Only one significant study has been reported involving the diuretics and a tissue culture system consisting of NHIK 3025 cells derived from a human cervical carcinoma, as shown below (see Table 1).[26]

Curiously, frusemide showed no phototoxic action in this system, a result which may be ascribed to the drug being in its relatively unreactive anionic form under the test conditions.

The addition of ascorbic acid and α-tocopherol reduced the phototoxic cell death caused by hydrochlorthiazide, bendroflumethazide, bumetanide and trichlormethiazide, while no significant results were obtained on addition of β-carotene or ubiquinone. These positive effects of free radical scavenging antioxidants suggest that free radical mechanisms play the major role in the phototoxic action of the photoactive diuretic drugs. In agreement with that conclusion, it can be stated from the above survey of the photochemistry of the diuretic drugs that they demonstrate a preponderance of free radical activity in their photodegradation pathways.

Table 1 *Phototoxicity of diuretic drugs.*

Test Substance	Phototoxic Cell Death at Test Concentration		
	0.5 mM	0.25 mM	0.05mM
bemetizide	+	+	
bendroflumethazide	+	+	+
benzylhydrochlorthiazide	+	+	
bumetanide	+	+	
butizide	+		
hydrochlorthiazide	+		
hydroflumethazide	+	+	
piretanide	+		
polythiazide	+		
trichlormethiazide	+		
chlorthalidone	-		
furosemide	-		
indapamide	-		
xipamide	-		

References

1. W. O. Foye, T. L. Lemke and D. A. Williams. 'Principles of Medicinal Chemistry', 4th edition, Williams and Wilkins, Baltimore, 1993, Chapter 21, p.405.
2. D. E. Moore, in 'Photostability of Drugs and Drug Formulations', H. H. Tonnesen, Ed., Taylor and Francis, London, 1996, Chapter 4, p. 63.
3. J. Boxhammer, in ref. 2, Chapter 3, p. 39.
4. P. C. Rowbotham, J. B. Stanford and J. K. Sugden, *Pharm. Acta Helv.*, 1976, **51**, 304.
5. D. E. Moore and S. R. Tamat, *J. Pharm. Pharmacol.*, 1980, **32**, 172.
6. D. E. Moore and C. D. Burt, *Photochem. Photobiol.*, 1981, **34**, 431.
7. D. E. Moore and V. Sithipitaks, *J. Pharm. Pharmacol.*, 1983, **35**, 489.
8. A. M. Yahya, J. C. McElnay and P. F. D'Arcy, *Int. J. Pharm.*, 1986, **31**, 65.
9. H. Bundgard, T. Norgaard and N. M. Nielsen, *Int. J. Pharm.*, 1988, **42**, 217.
10. J. Grimshaw and A. P. de Silva, *Chem. Soc. Rev.*, 1981, **10**, 181.
11. H. Barton, J. Mokrosz, J. Bojarski and M. Klimezak, *Pharmazie*, 1980, **35**, 155.
12. J. E. Cruz, D. D. Maness and G. J. Yakatan, *Int. J. Pharm.*, 1979, **2**, 275.
13. J. M. Neil, A. F. Fell and G. Smith, *Int. J. Pharm.*, 1984, **22**, 105.
14. J. A. Mollica, C. R. Rehm, J. B. Smith and H. K. Gowan, *J. Pharm. Sci.*, 1971, **60**, 1380.
15. S. R. Tamat and D. E. Moore, *J. Pharm. Sci.*, 1983, **72**, 180.
16. D. E. Moore and J. L. Mallesch, *Int. J. Pharm.*, 1991, **76**, 187.
17. D. E. Moore, *J. Pharm. Sci.*, 1977, **66**, 1282.

18. R. Davis, C. H. J. Wells and A. R. Taylor, *J. Pharm. Sci.*, 1979, **68,** 1063.

19. Adverse Drug Reactions Advisory Committee, 'Adverse Drug Reactions Bulletin', Australian Department of Health, Canberra, February, 1987.

20. K. Thestrup-Pedersen, *Danish Med. Bull.*, 1987, **34,** (suppl. 1), 3.

21. D. J. Mazzo, *Analyt. Profiles Drug Substances*, 1986, **15,** 1.

22. H. I. Hamoudi, P. F. Heelis, R. A. Jones, S. Navaratnam, B. J. Parsons, G. O. Phillips, M. J. Vandenburg and W. J. Currie, *Photochem. Photobiol.*, 1984, **40,** 35.

23. J. E. F. Reynolds, Ed., 'Martindale, The Extra Pharmacopoeia', 31st edition, Pharmaceutical Press, London, 1996, p. 957.

24. I. E. Kochevar, *Photochem. Photobiol.*, 1987, **45,** 891.

25. B. E. Johnson, E. M. Walker and A. M. Hetherington, in 'Skin Models', R. Marks and G. Plewig, Eds., Springer-Verlag, Berlin, 1986, p. 264.

26. E. Selvaag, H. Anholt, J. Moan and P. Thune, *J. Photochem. Photobiol. B: Biology*, 1997, **38,** 88.

New Results in the Photoinstability of Antimycotics

K. Thoma and N. Kübler
Dept. of Pharmacy and Food Chemistry
Ludwig-Maximilian University
Sophienstr. 10, D-80333 Muenchen, Germany

1 INTRODUCTION

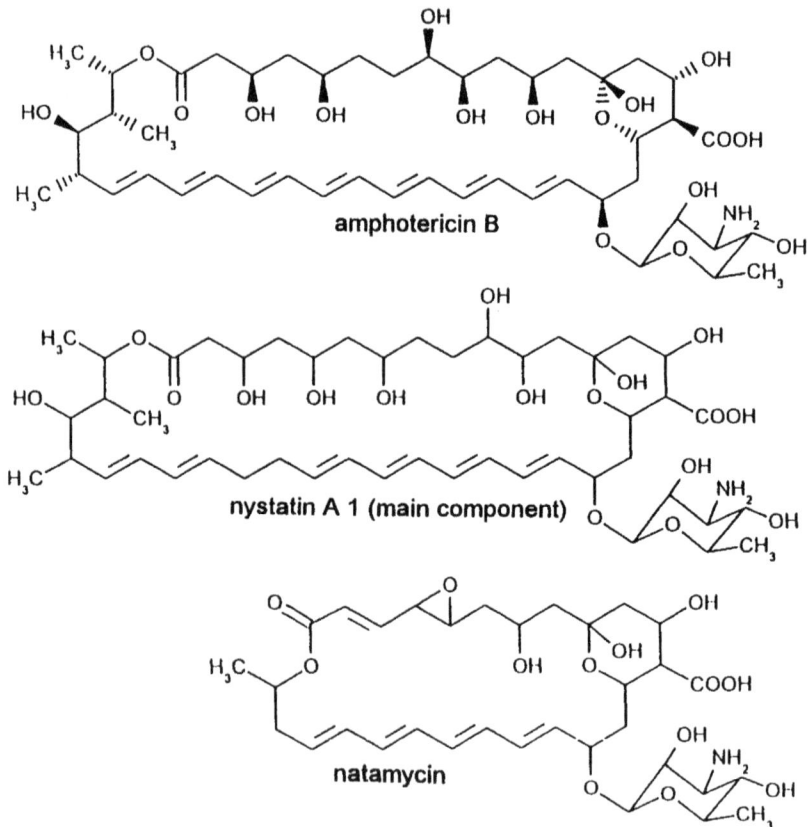

Figure 1 *Polyenantibiotics*

Antimycotics are a therapeutic group of drugs with a broad range of local and systemic applications. Photostability is known only for some drug substances as polyenantibiotics and griseofulvin. However, the knowledge is restricted to qualitative data.[1,2,3,4,5]

The reason for this restriction must be sought in their composition. Polyenantibiotics are not defined single substances but a mixture of various substances. That is why until now their potency is determined by microbiological methods. Investigations of photostability were therefore limited to qualitative results (Figure 1).

For the topical application various substances are used. One big substance group is the azolantimycotics. Clotrimazole and miconazole are well known representatives of this substance group (Figure 2).

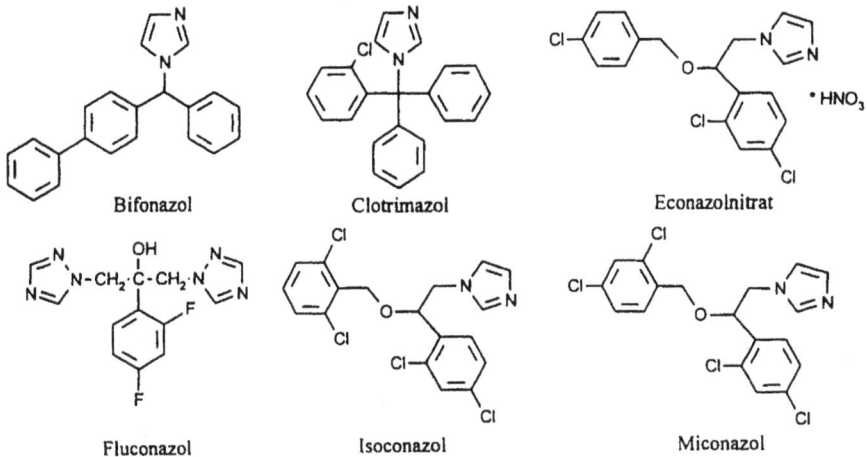

Figure 2 *Local azolantimycotics*

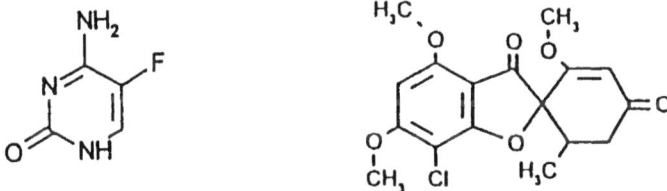

Figure 4 *Griseofulvin (left) and flucytosine*

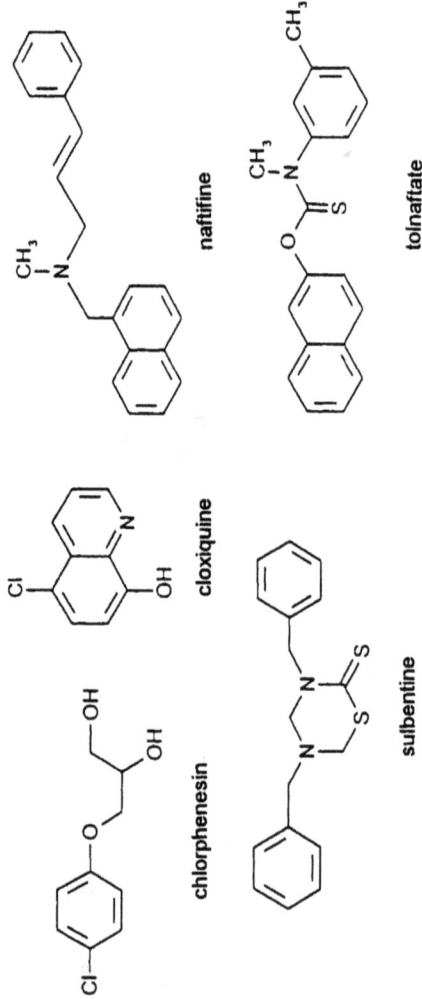

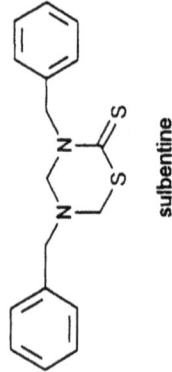

Figure 3   *Local antimycotics of other substance groups*

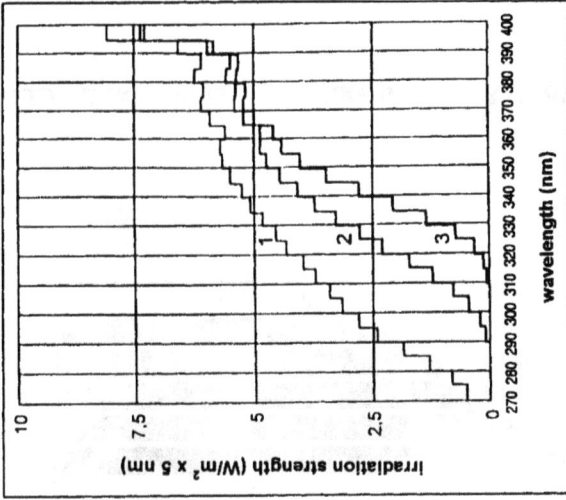

Spectral distribution of different light filters

1: without filter
2: UV special glass filter
3: window glass filter

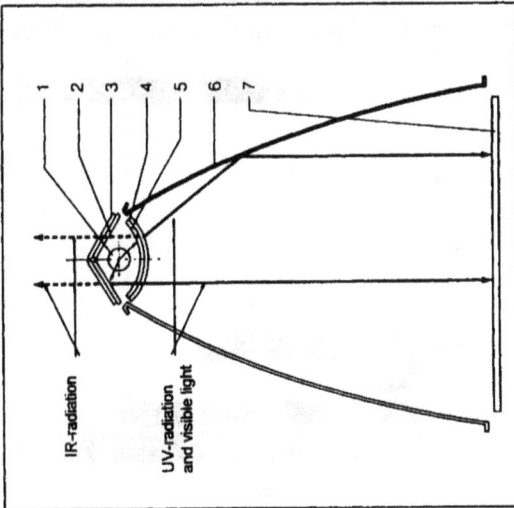

Suntest CPS: 1.Xenon lamp, 2.UV mirror, 3.Light mirror,
4. Quarz glass shell, 5. Extra UV special glass/window glass filter,
6. Parabolic reflector, 7. sample level

Figure 5 *Suntest CPS+*

120   *Drugs: Photochemistry and Photostability*

As examples for other substance groups we would like to mention tolnaftate and naftifine (Figure 3).

In the systemic therapy beside amphotericin B and certain azolantimycotics particularly griseofulvin and flucytosine are used (Figure 4).

In this paper we want to present new results in the photostability of different antifungal agents.

2 PHOTOINSTABILITY OF DIFFERENT SUBSTANCE GROUPS OF ANTIMYCOTICS

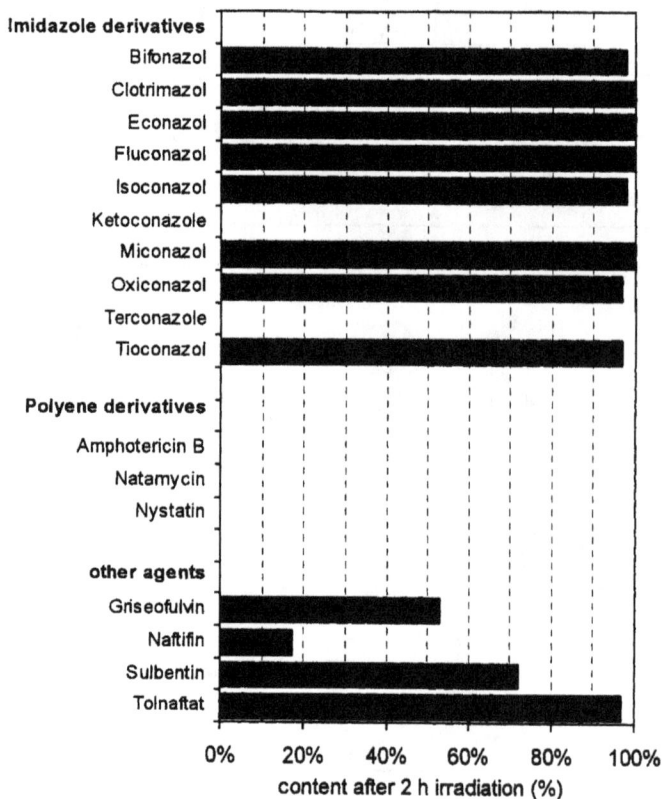

Figure 6 *Photoinstability of antimycotics in methanolic solution (2 mg/100 ml)with UV-filter*

The photostability tests have been carried out with the Suntest CPS.[6] This light testing cabinet is equipped with a xenon lamp as light source. To adopt its spectrum to that of natural sunlight a UV filter is set in. By these means the unnatural UV range below 290 nm can be eliminated (Figure 5).

In Figure 6 of the photoinstability testing results of the various antimycotics in this work are compared: The black bars symbolise the residual drug content after 2 h irradiation in a methanolic solution. The HPLC methods are described.[7, 8, 9, 10]

The azolantimycotics show no uniform behaviour to irradiation. 4 out of 10 investigated imidazole derivatives are photostable like clotrimazol, 4 of them are photolytically degraded only to 1-3% like tioconazole. However, ketoconazole and terconazole are completely destroyed after this light stressing,[7] (Figures 7 and 8).

The polyenantibiotics are a very light sensitive substance group. Amphotericin B, natamycin and nystatin were completely degraded after 2 h of irradiation.[8]

Other antifungal agents show strong differences of their photostability, varying from 5% photodecomposition of tolnaftate to 85% degradation of naftifine. Griseofulvin with a loss of about 47 % and sulbentine with about 30 % show significant photodegradation, too.[9, 10]

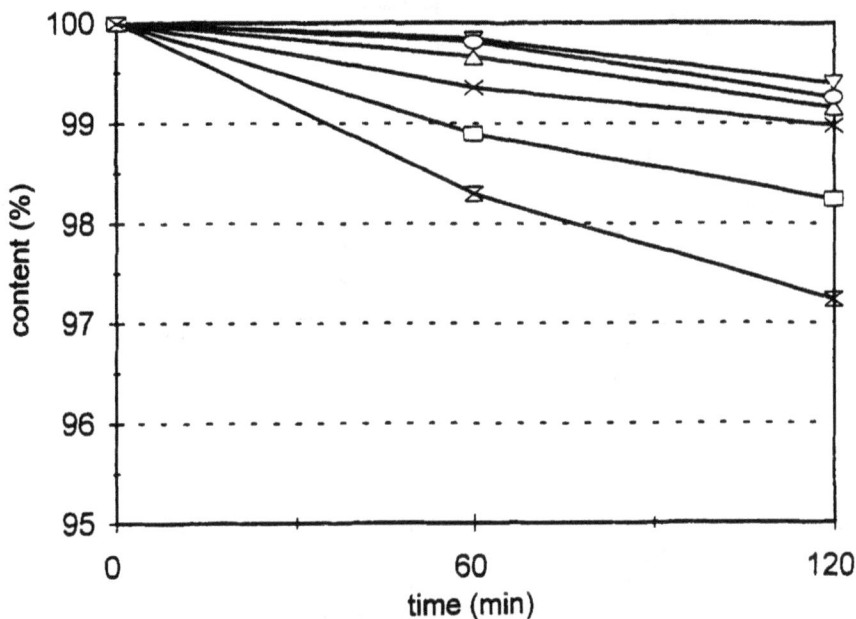

Figure 7 *Photodegradation of 6 azolantimycotics(2 mg/100 ml in methanolic solution, Suntest CPS, UV-filter). Bifonazol* ⚹ *, Fluconazol* ○ *, Clotrimazol* ▽ *, Isoconazol* □ *, Econazolnitrat* △ *, Miconazol* ✳ *.*

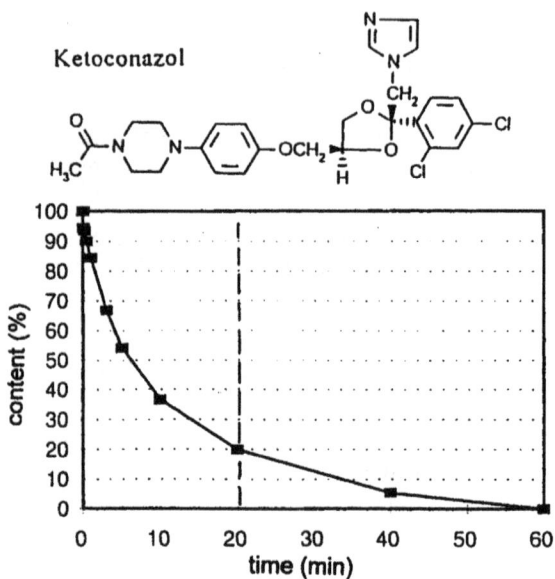

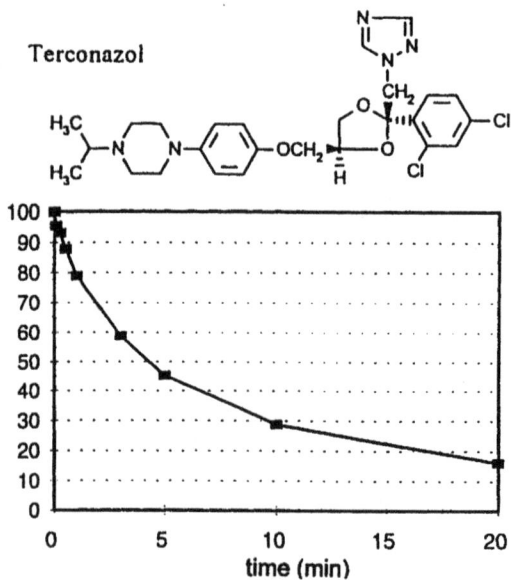

Figure 8 *Photodegradation of the azolantimycotics ketoconazol and terconazol (10 mg/100 ml in methanolic solution, Suntest CPS, UV-filter)*

3 COMPARATIVE INVESTIGATION OF THE PHOTOINSTABILITY OF LOCAL ANTIMYCOTICS

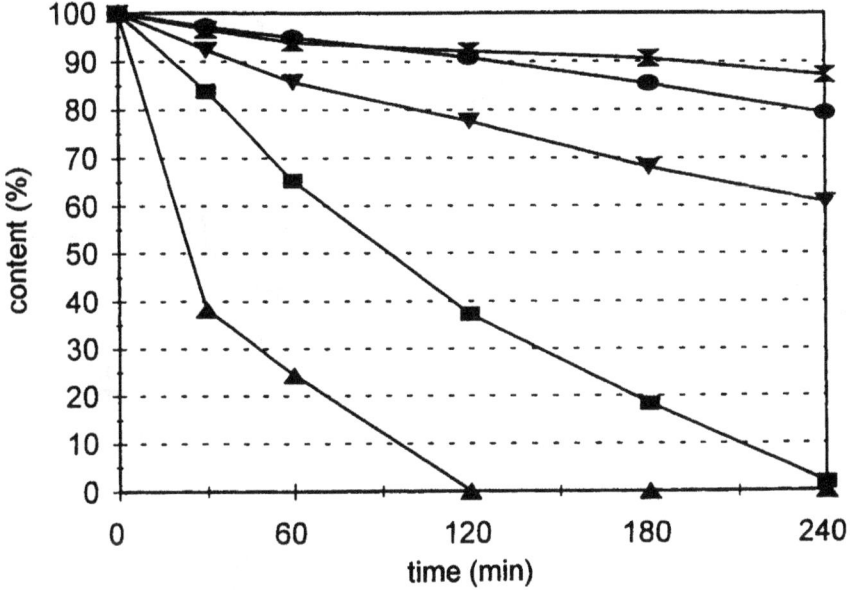

Figure 9 *Photodegradation of topical antimycotics (in 10mg/100ml methanolic solution, Suntest CPS, UV filter): chlorphenesin* ✖ *, sulbentine* ◼ *, cloxiquine* ▼ *, tolnaftate* ● *, naftifine-HCl* ▲

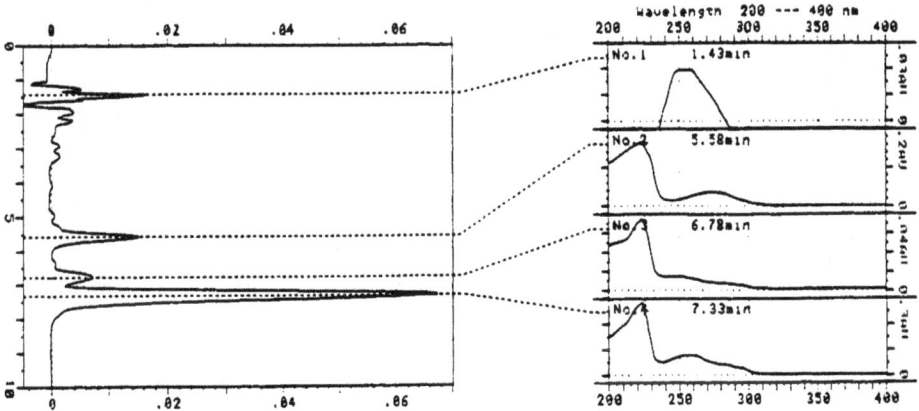

Figure 10 *Degradation products of naftifine HCl in methanolic solution after 30 min irradiation:1.43 min: injection peak; 6.78 min: photoproduct (2); 5.58 min: photoproduct (1); 7.33 min: naftifine.*

To compare the photoinstability of various local antimycotics these have been irradiated in methanolic solution (Figure 9).

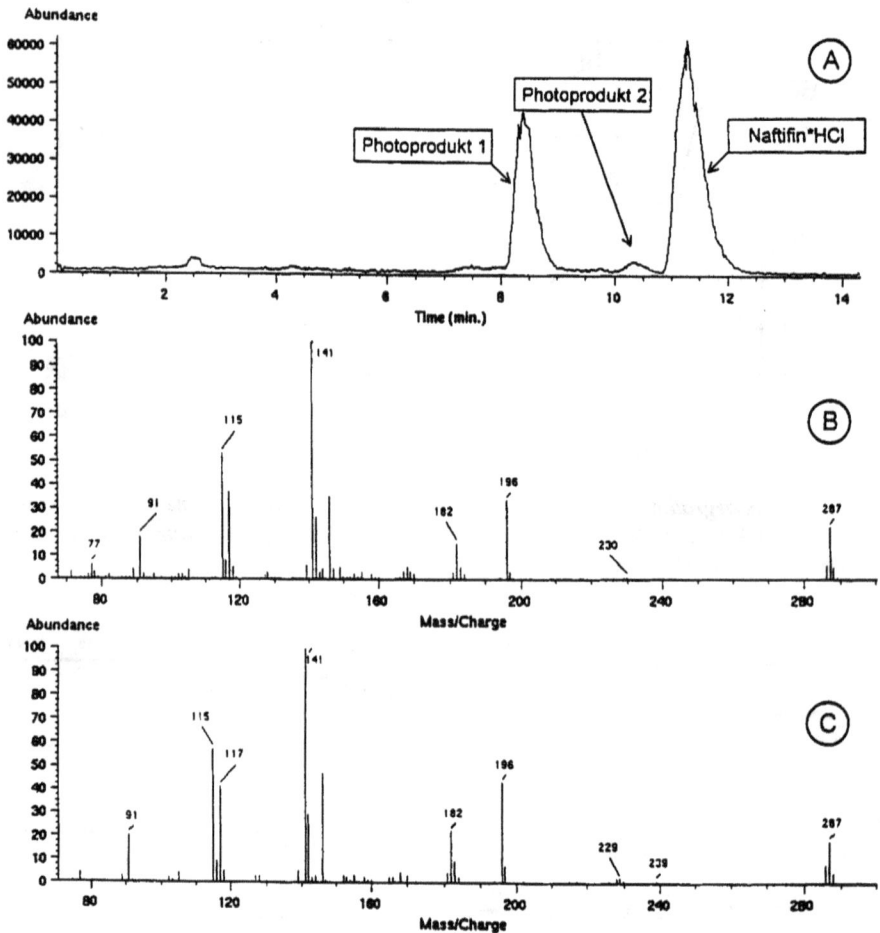

Figure 11 *Mass spectroscopic determination of the photodegradation products of naftifine (methanolic soluton, 10mg/100ml, Suntest CPS, UV filter, 30 min)*
A Total ion-chromatogramm after 30 min irradiation
B mass-spectrum of naftifine (RT 11.3 min)
C mass-spectrum of photoproduct (1) (RT 8.4 min)

Large differences have been observed between the photoinstability of the topical applied antimycotics. The fastest photodegradation has been observed for naftifine. After 2 hours it was completely decomposed.[9] Sulbentine showed high light sensitivity, too. After 4 hours of irradiation only 1.7% of the initial drug content remained intact. The other local antimycotics showed significant, but not that pronounced photodegradation.

Not only the degree of photodegradation has been of interest in these investigations, but also the identification of the degradation products. Figure 10 shows the HPLC chromatogram of an irradiated solution of naftifine. A photodiode array detector has been used for detection. As can be seen the two main photodegradation products (1) and (2) (retention times 5.58 min and 6.78 min, respectively) show spectral deviations to naftifine.

For further information about the nature of the degradation products the irradiated solution has been investigated by a combined HPLC-MS device. It consists of HPLC system connected by an interface to a mass spectrometer.[6, 11]

In the total ion chromatogram the two photodegradation products of naftifine (1) and (2) can be detected. The mass spectra of photodegradation product (1) (C) and (2) (not shown on this Figure) are identical with the mass spectrum of naftifine (B) (Figure 11).[9]

Photodegradation product (1) could be identified as the cis-isomer of naftifine by comparison with a standard solution (Figure 12). Isolation and nuclear magnetic resonance spectroscopic (NMR) investigations of the photodegradation product 2 proved its identity as the dimer of naftifine.

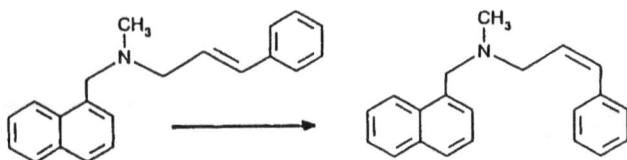

Figure 12 *Chemical structures of trans- naftifine and its photoproduct cis-naftifine*

4 DETECTION AND IDENTIFICATION OF PHOTODEGRADATION PRODUCTS OF FLUCYTOSINE AND GRISEOFULVIN

Degradation products of drug substances can be responsible for unwanted side effects. Therefore the detection and identification of the degradation products of flucytosine and griseofulvin are of interest (Figure 13). Both are some of the few drug substances used for the systemic therapy of mycotic infections.

Flucytosine is orally or parenterally applied for generalised mycoses. 5-Fluorouracil is known as a toxic degradation product of flucytosine. Therefore pharmacopoeias like the Ph. Eur. 97 and the USP 23 demand for flucytosine the testing of the absence of 5-fluorouracil. For griseofulvin, it has been known since 1960 that degradation products can cause toxic side effects.[12] Nevertheless, until now, the structures of these photolytic degradation products are not clear

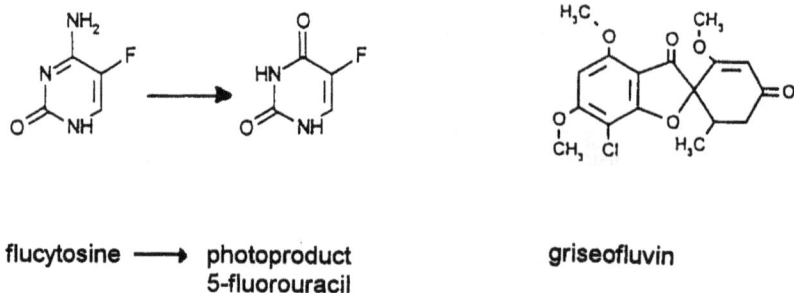

flucytosine ⟶ photoproduct
5-fluorouracil

griseofluvin

Figure 13 *Chemical structure of flucytosine, its degradation product 5-fluorouracil and of griseofulvin*

After 6 min of irradiation 80% of flucytosine was degraded in methanolic solution. After a longer irradiation of 16 h of an aqueous solution 5 degradation products were detected in the HPLC chromatogram. The chromatogram of a standard solution of 5-fluorouracil corresponds exactly with the peak of the photodegradation product W3 (retention time 4.18 min) (Figure 14).

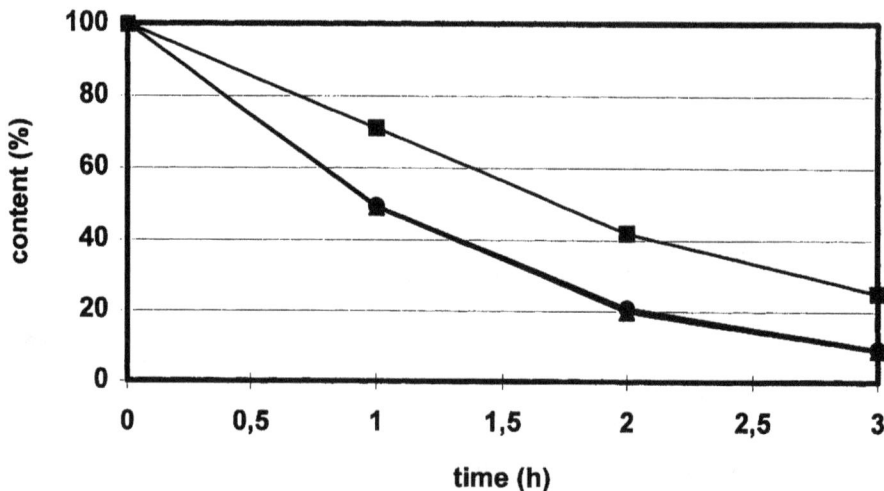

Figure 15 *Influence of the solvent on the photodegradation of griseofulvin (10mg/100ml, Suntest CPS, UV filter).*

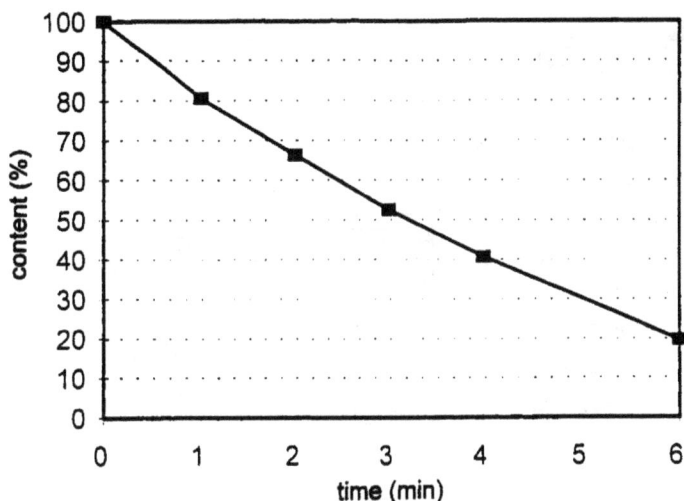

Photodegradation of flucytosine in methanolic solution
(10 mg/100 ml, Suntest CPS, UV-filter)

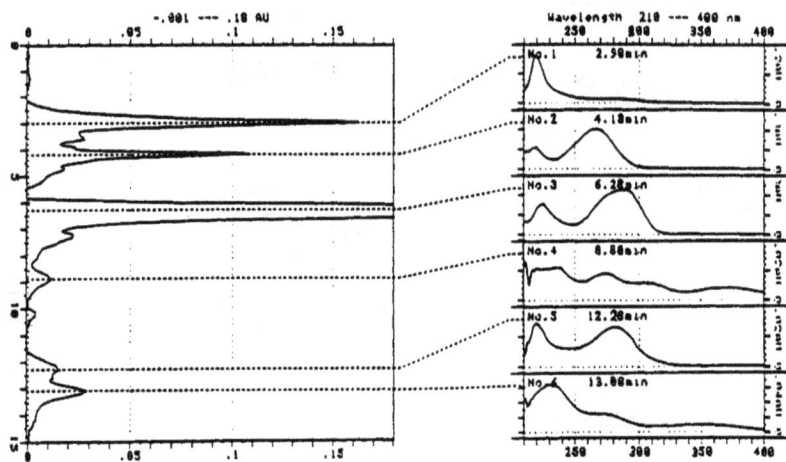

Degradation products of flucytosine in H_2O
(16 h, Suntest CPS, UV-filter)

2,98 min: W1	8,88 min: W6
4,18 min: W3	12,28 min: W8
6,28 min: flucytosine	13,08 min: W9

Figure 14 *Photodegradation of flucytosine in methanolic solution (10 mg/ 1oo ml, Suntest CPS, UV-filter) and detection of 5 photodegradation products of flucytosine after 16 h irradiation (aqueous solution 100 mg / ml, Suntest CPS, UV filter).*

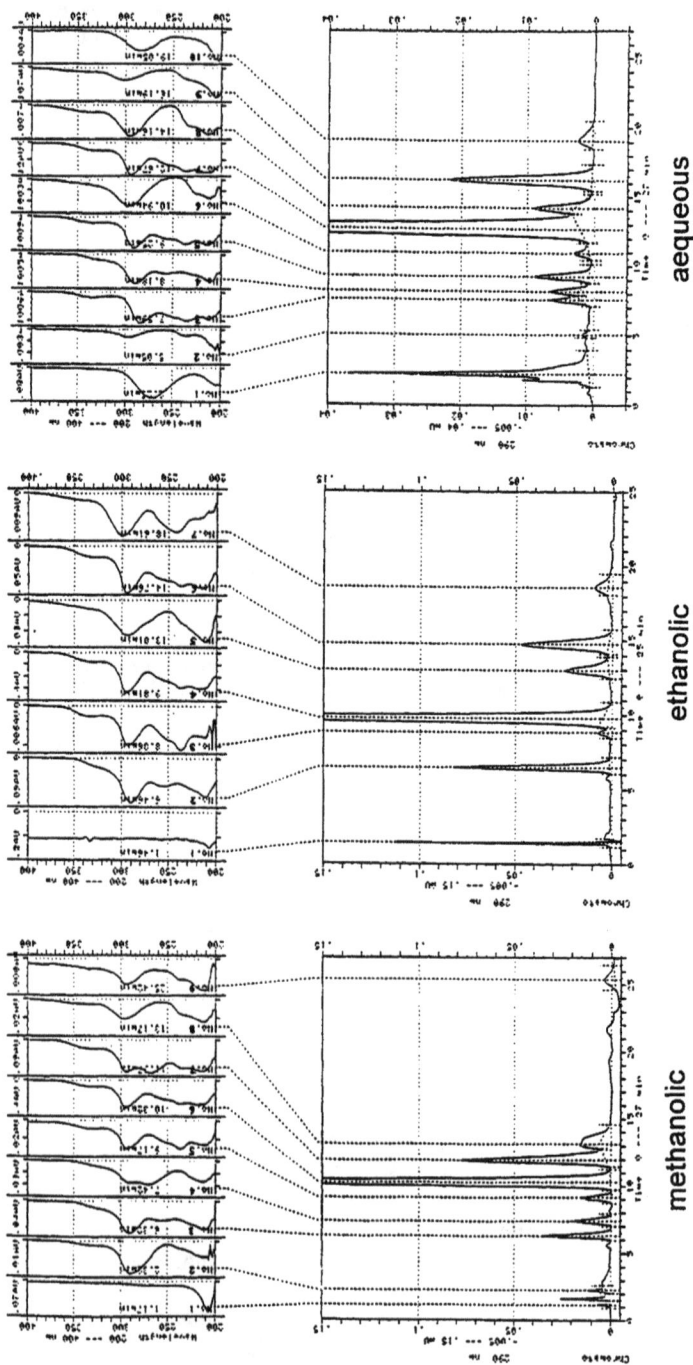

Figure 16   *HPLC-chromatograms of irradiated griseofulvin solutions: in methanolic (100mg/100ml, 8h), in ethanolic (100mg/100ml, 6h) and in aqueous (2mg/100ml, 2h). (Suntest CPS, UV-filter).*

For griseofulvin it was found that its photodegradation occurs faster in alcoholic solutions than in water. The photolysis of griseofulvin in methanolic solution resulted in a degree of about 50% of the initial drug content after 1 h irradiation. After 3 h the residual drug content was below 10%. The decomposition in an ethanol solution was similar to that of a methanolic solution. In aqueous solution the drugs substance was significantly more photostable. After 1 h only 30% were destroyed, after 3 h the residual drug content still amounted to 25% (Figure 15).

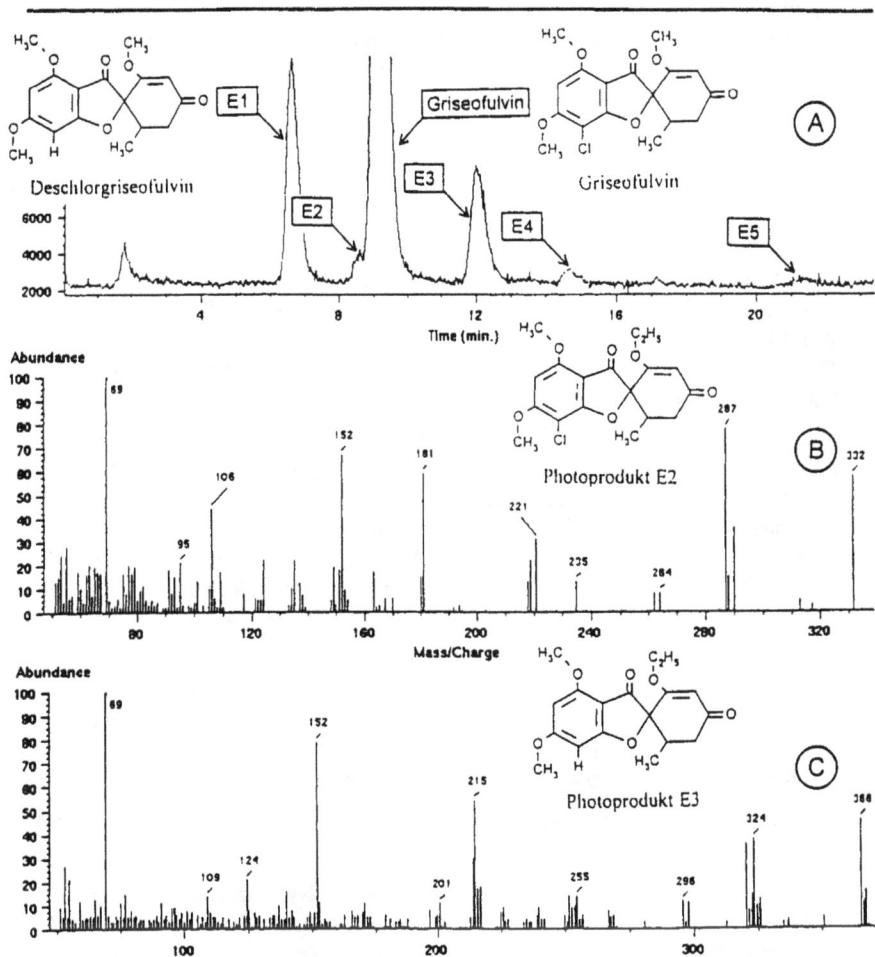

Figure 17 *Mass spectra of irradiated griseofulvin in ethanol solution: (100mg/100ml, 16h) (Suntest CPS, UV filter)*
A Total ion-chromatogram after 16 h irradiation
B mass-spectrum of photoproduct E2 (RT 8.5 min)
C mass-spectrum of photoproduct E3 (RT 12.0 min)

Because of this fact it was of interest as to whether the different degradation kinetics are caused by different degradation pathways.

It is conspicuous, that the number of degradation products in the methanolic solution is higher than in the other solutions (Figure 16).

By means of the HPLC-MS technique the photodegradation products after irradiation of an ethanol solution were partly identified. The MS spectrum of the degradation product E1 proves its identity as dechlorogriseofulvin. This spectrum is not shown here. Two of the photodegradation products are formed only by irradiation of griseofulvin in ethanol solution, because ethanol takes part in the photochemical reaction as a reaction partner. The MS spectra of these two degradation products are shown in Figure 17. E2 is the ethyl ester of griseofulvin, E3 the corresponding dechlorinated product.

5 PHOTOSTABILITY PROBLEMS OF TOPICAL DOSAGE FORMS OF ANTIMYCOTICS

The determination of the photoinstability of the drug substance is the first step of the photostability testing procedure of drugs. The next one is the investigation of the light sensitivity of the drug substance in its preparation. Antimycotics are often used externally. In many cases it is necessary to determine the photostability of such substances in their topical dosage form. As examples results with natamycin and nystatin will be presented.

In methanolic solution the photolysis of natamycin is very fast. Already after 10 sec more than 10% of the drug was destroyed, after 10 min no drug substance is detectable any more. Therefore, natamycin belongs to the drugs with extremely high light sensitivity. Its tetraene structure, which could be detected in 5 degradation products, is more and more transformed to a triene structure found in other photodegradation products. Natamycin is part of two topical dosage forms, shown in Figure 18. In both it is suspended. The diameter of the particles reached 30 μm in the cream, in the eye ointment the particle diameter was mainly below 10 μm.

Of course both preparations are protected from light in tubes during storage. However, sunlight exposure of the ointments after application results in a fast photodegradation of the drug. The preparations were applied in a layer thickness of 1 mm. After 30 min about 50% of the drug substance was decomposed in both preparations.

Photodecomposition of drugs after application on the skin has been demonstrated for other drug substances, too, e.g. for corticosteroid gels.[13]

The second drug which will be to presented is nystatin. Nystatin consists of a mixture of various polyenes. Its photolysis in solution is very fast. After 1 min irradiation more than 50% of the initial drug content was destroyed. Already after 30 sec 3 photodegradation products with tetraene structure are detectable. The fourth degradation product has got a different structure. Longer lasting irradiation resulted in the formation of further degradation products.

In the topical preparations, two creams and three paraffin/polyethylene-ointments, nystatin is suspended as thin crystal needles of about 10 μm.

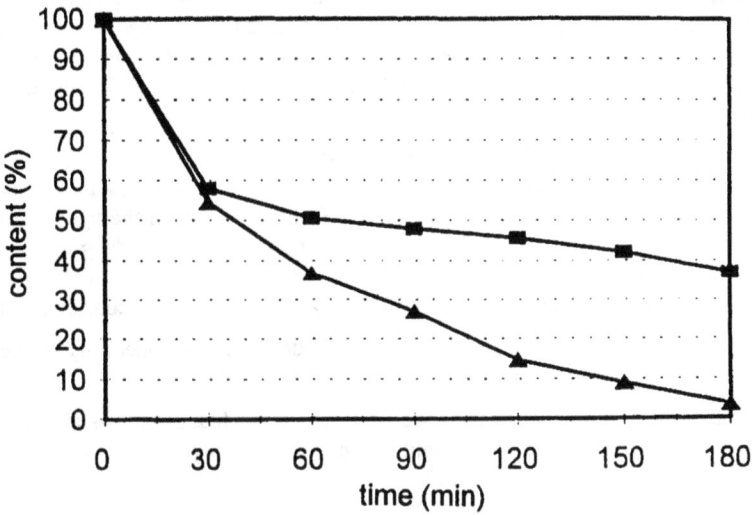

Figure 18 *Photodegradation of natamycin in topical preparations (Suntest CPS, UV filter):*
cream (20 mg/g) ■ *eye-ointment (10 mg/g)* ▲

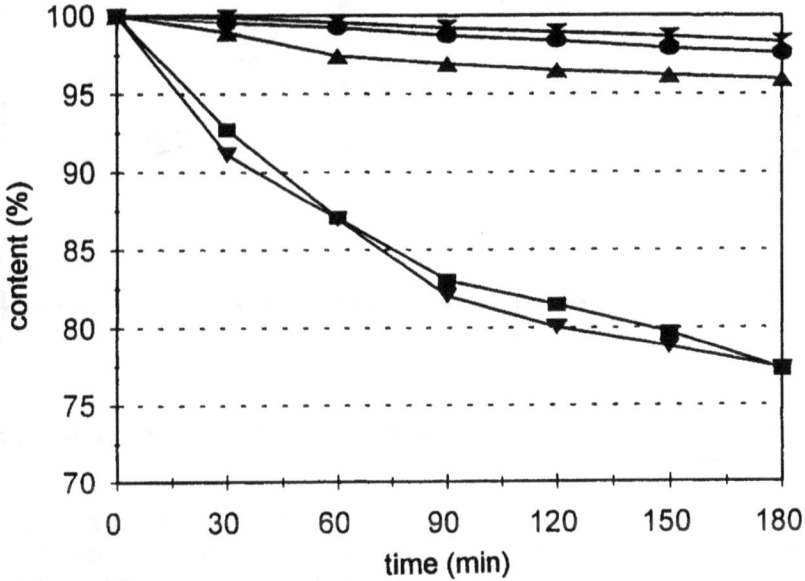

Figure 19 *Photodegradation of nystatin in topical preparations after irradiation (Suntest CPS, UV filter):* *creams* ● ✖ *ointments* ■ ▼ ▲

The topical preparations were applied as layers of 1 mm thickness and irradiated. After 60 min a degree of 13% of the initial drug content was found in two of three ointments. The third ointment showed only a loss of 3% nystatin because of the presence of zinc oxide. Zinc oxide is a pigment with good light reflective properties and therefore can be regarded as a light protective agent in this preparation. Both of the creams were significantly more stable than the ointments (Figure 19).

Because of these results changes for the monographs in the European Pharmacopœia should be taken into consideration for both drug substances, natamycin and nystatin. They have shown photodegradation kinetics similar to nifedipine, of which the light sensitivity is very well known. Therefore light protection during the analytical handling should be demanded in the monographs. For topical application on uncovered parts of the body, light protection by pigments or light absorbing excipients should be taken into consideration.

How is it possible to photostabilise light sensitive drugs? As a model drug the very light sensitive substance terconazole has been chosen.

A methanolic solution of terconazole was stabilised by adding excipients as antioxidative substances and substances which absorb light in a similar wavelength range like the drug itself (vanillin, *p*-aminobenzoic acid) (Figure 20).

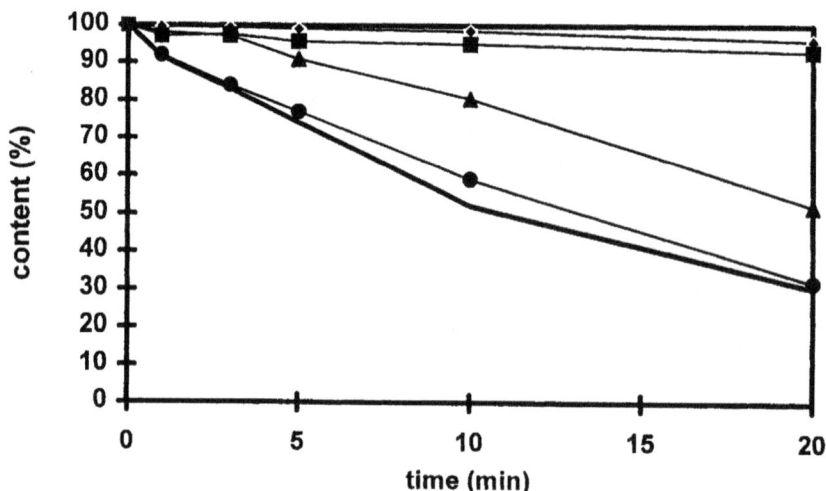

Figure 20 *Influence of excipients on the photostability of terconazole in methanolic solution (10mg/100ml, Suntest CPS, UV filter)*

Without the addition of a protective excipient 70% of terconazole were degraded within 20 min of light exposure. The redox stabilisers sodium bisulfite and ascorbic acid had only little effect on the photostability of terconazole solutions. Whereas excipients showing spectral overlay of the absorption spectrum of terconazole had good light protective properties: The addition of p-aminobenzoic acid and the addition of vanillin (100mg/100ml) to the terconazole solutions resulted in a good photostabilisation of the drug. Only 5% to 6% were degraded after 20 min of irradiation. The photodegradation kinetics were reduced by the factor of 20. The photostabilisation principle of spectral overlay is often successfully applied.[14, 15, 16, 17]

References

1. I. M. Asher and G. Schwartzman, *Anal. Prof. Drug Subst.*, 1977, **6**, 1.
2. H. Brik, *Anal. Prof. Drug Subst.*, 1981, **10**, 513.
3. G. W. Michel, *Anal. Prof. Drug Subst.*, 1977, **6**, 341.
4. K. Thoma, T. Strittmatter and D. Steinbach, *Acta Pharm. Technol.*, 1980, **26**, 269.
5. E. R. Townley and P. Roden, *J. Pharm. Sci.*, 1980, **69**, 523.
6. K. Thoma and N. Kübler, *Pharmazie*, 1996, **51**, 919.
7. K. Thoma and N. Kübler, *Pharmazie*, 1996, **51**, 885.
8. K. Thoma and N. Kübler, *Pharmazie*, 1997, **52**, 294.
9. K. Thoma and N. Kübler, *Pharmazie*, 1997, **52**, 362.
10. K. Thoma and N. Kübler, *Pharmazie*, 1997, **52**, 455.
11. K. Thoma and N. Kübler, *Pharmazie*, 1996, **51**, 940.
12. N. J. Goldberg and M. B. Sulzberger, *Arch. Dermatol.*, 1960, **81**, 859.
13. K. Thoma and R. Kerker, *Pharm. Ind.*, 1992, **54**, 551.
14. K. Thoma and R. Klimek, *Inter. J. Pharm.*, 1991, **67**, 169.
15. K. Thoma and R. Kerker, *Pharm. Ind.*, 1992, **54**, 359.
16. K. Thoma, in 'Photostability of Drugs and Drugs Formulation', H. H. Tønnesen (Ed.), Taylor and Francis, London, 1996, p. 111.
17. K. Thoma and R. Kerker, *Pharm. Ind.*, 1992, **54**, 630.

Photoreactivity *versus* Activity of a Selected Class of Phenothiazines: A Comparative Study

Beverley D. Glass, Michael E. Brown and Patricia M. Drummond
School of Pharmaceutical Sciences
Rhodes University
Grahamstown, 6140 South Africa

1 INTRODUCTION

The phenothiazines, a class of drugs widely used as neuroleptics in the therapy of schizophrenia, organic psychoses, the manic phase of manic depressive illness and other acute or chronic idiopathic psychotic illnesses[1] although divided into the aliphatic, piperidine and piperazine subclasses, based on similarities in their chemical structure exhibit different pharmacological activity and potency. The piperazine subclass (Figure 1, Table 1) including, prochlorperazine (1), perphenazine (2) trifluoperazine (3) and fluphenazine (4) possess potent antipsychotic activity with more pronounced extrapyramidal but fewer anticholinergic and sedative effects than the other groups.

The phenothiazines known to cause a variety of adverse effects in individuals[2] with high doses producing ocular opacity, tremors and deposition of melanin in the skin,[3,4] have also been associated with photosenstivity effects in individuals exposed to light. These photosensitising properties have been assessed using two physical models.[4,5] Both the phospholipid spherule and the lecithin monolayer model concur with the involvement of the chlorine containing derivatives in photosensitivity reactions and explain a mechanism involving the liberation of HCl with subsequent formation of the 2-hydroxy derivatives contributing to a decrease in pH which explains the sensitivity, inflammation, or dermatitis that usually precedes the deposition of melanin in the skin of patients treated with large doses of the drug. In contrast to the increased surface activity observed with the chloro derivatives the trifluoromethyl-containing derivatives, trifluoperazine and fluphenazine are essentially non-photosensitising.

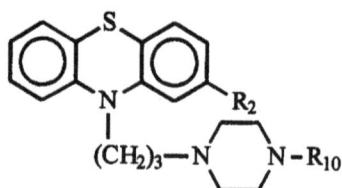

Figure 1. *General Structure.*

Table 1 *Propyl piperazine-substituted phenothiazines*

Phenothiazine	R_2	R_{10}
Prochlorperazine (1)	-Cl	-CH$_3$
Perphenazine (2)	-Cl	-CH$_2$CH$_2$OH
Trifluoperazine (3)	-CF$_3$	-CH$_3$
Fluphenazine (4)	-CF$_3$	-CH$_2$CH$_2$OH

The nature of the R_2 and R_{10} substituents affects the neuroleptic activity of the phenothiazine with the presence of the trifluoromethyl and hydroxyethyl groups contributing to enhanced activity as compared to the chloro and methyl groups respectively, while the influence of structure on stability has been noted by Pawelczyk et al.[6,7,8] where trifluoperazine is found to be more photostable than prochlorperazine. However, although the effect of the R_{10} substituent on activity has been reported this is not so in terms of its effect on photostability. Pilpel et al.[3] explain the adverse effects associated with the phenothiazines on exposure to light to be due to the formation of the hydroxy derivatives following dechlorination only by the action of UV irradiation, while the sulphoxides which are natural metabolites of the phenothiazines in the body and therefore in low doses are unlikely to be responsible for the adverse effects of these drugs associated with sunlight. The effect of structure on the development of adverse effects must therefore be taken into account in the comparative study on the activity versus the stability of these drug molecules.

2 STRUCTURE-ACTIVITY

2.1 Introduction

The mechanism of action of these phenothiazine drugs involves blockade of the dopamine receptors in the brain by virtue of their ability to mimic the *trans alpha* conformation of dopamine while the distances of the nitrogen atoms of the piperazine ring from the centres of the aromatic rings are more or less constant and correspond to those occurring at the dopamine receptor as illustrated in the preferred conformation of dopamine and the propyl piperazine derivative, prochlorperazine[9] (Figure 2).

In this preferred conformation, the side chain is tilted toward ring A allowing for favourable v.d. Waals interaction of the side chain with the R_2 substituent. A CF$_3$/piperazine/ hydroxyethyl group provides a greater number of favourable v.d. Waals contact with the side chain than does a Cl/alkylamino/methyl group.[10] This model allows several predictions to be made about the structure of the phenothiazines that might be expected to lead to enhanced interaction with the dopamine receptor.

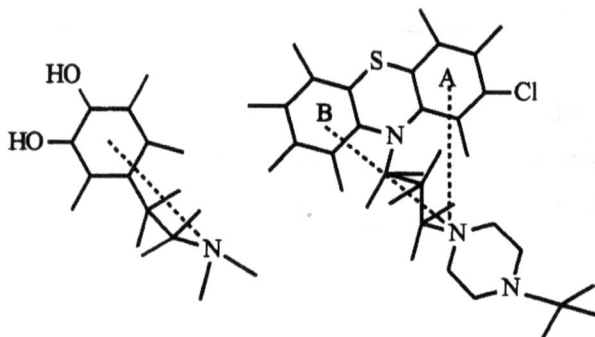

Figure 2. *Conformation of dopamine and prochlorperazine.*

2.2 Activity

Structure-activity describing the relationship between molecular structure or physical properties and activity proposes to explain the effect of structure on the interaction of the drug with the relevant receptor. A widely used theoretical method based on the work of Hansch considers relationships between biological activity and common physical properties, namely lipid solubility, degree of ionisation or molecular size. The octanol-water system allows for the measurement of the preference of drugs for the hydrophilic or lipophilic phase.[11] This preference is described by Hansch as the partition coefficient P and the parameter for the phase distribution relationship is standardised as log P or π. Thus some optimum value of log P for a particular drug in a given biological system would give rise to a maximum probability of reaching the receptor in a given time. It has been shown that such an optimum value for those drugs active in the CNS such as the phenothiazines is 2. The two benzene rings present in all phenothiazines confer sufficient lipid solubility for penetration of the brain; however, the implication of the R_2 substituent in lipophilicity is supported by Hansch where π is a measure of the contribution of the substituent to the lipophilicity of the molecule. π Values for the trifluoromethyl (1.07) and chloro (0.76) groups confirm the greater contribution of the former group to enhanced lipophilicity.[11]

This contribution of the trifluoromethyl group to the hydrophobicity of the molecule is further confirmed by the surface activity determinations of 0.58 for prochlorperazine and 0.12 for trifluoperazine, where values of less than 1 indicate greater surface activity than chlorpromazine. Nightingale et al.[12] have demonstrated a correlation of pharmacological effect with physicochemical properties of the phenothiazines, and the role of the absorptive process in modifying such a response in apparent water/dodecane partition coefficients is effective in predicting phenothiazine absorption efficiency, which suggests a relationship between phenothiazine absorption and hydrophobicity showing the greater contribution of the trifluoromethyl group. The idea of "dissecting" drug activity into physical contributions (hydrophobic, electronic, steric) is the central theme of the Hansch approach with the enhanced activity of trifluoperazine and fluphenazine as compared with the chloro derivatives being explained in terms of their hydrophobicity.

Thus activity of these propyl piperazine derivatives is determined by the strucural features contributing to lipophilicity and enhanced interaction with the receptor resulting in the reporting of a decreased order of activity being as follows: fluphenazine, trifluoperazine, perphenazine and prochlorperazine.[1]

3 STRUCTURE-PHOTOREACTIVITY

3.1 Introduction

As a result of reported phototoxicity of the phenothiazines, their photoreactivity has been extensively investigated.[6,7,8]

Pawelczyk et al.[6,7] report the rate and type of photodegradation of aqueous solutions of various perazine derivatives to be dependent upon the nature of the R_2 substituent. The methodology used involves irradiation by UV light of 254nm (low pressure TUV 30 Philips lamp) of an aqueous solution of the salts of perazine derivatives in a phosphate buffer at pH 3 in the presence of air and nitrogen. Results indicate the absence of sulphoxides in the presence of nitrogen and for prochlorperazine and trifluoperazine, the degradation proceeds due to reversible first-order photooxidation. The photochemical degradation of perazine and thioethylperazine is complex consisting of parallel reactions of photolysis and photooxidation, while substituents Cl and CF_3 prevent photolysis such that the rate of degradation is determined by the rate of the photooxidation process. Based on these findings, this group continued the investigation on some physicochemical parameters in order to explain the chemical reactivity of these compounds. Many observations can be explained in terms of the volume (Å) including the fact that the basic properties increase with an increase in the substituent volume. In the Hammett type plot showing the relationship between the basicity of the compound and the rate of degradation prochlorperazine (Å = 29) exhibits an increased degradation rate on irradiation as compared with trifluoperazine (Å = 88). The double-lined Hammett plot due to different R_2 substituents suggests two different mechanisms of reaction confirming degradation by photooxidation for those derivatives with Cl and CF_3 substituents and photooxidation and photolysis for the H and SC_2H_5 containing derivatives.

The fact that both Moore et al.[13] and Sharples[14] report on the photolabile nature of chlorine in chloro-aromatic compounds can be applied to prochlorperazine and perphenazine which would then be capable of undergoing both Type I (free radical) and Type II (singlet molecular oxygen) reactions. Prochlorperazine yielded chloride and hydrogen ions at half the rate observed for chlorpromazine and thus appears to be an effective photosensitiser of 2,5-dimethylfuran oxidation. In both studies, methanolic solutions of the phenothiazines were irradiated over a period of time under nitrogen or oxygen as desired using in the case of Moore and coworkers[13] a medium pressure mercury lamp (Hanovia, 125W), and in the other study an Allen type A 409 fixed wavelength (365nm) UV lamp.[14] Sharples reports that the 2-chlorophenothiazines give rise to a dechlorinated product, a dimer and the corresponding sulphoxide. The free radicals formed are believed to explain the high phototoxicity of the chloro-substituted phenothiazines. These results concur with those reported by Moore et al.,[13] who also found that the HCl yield is independent of the solvent used. When considering the mechanism whereby chloropromazine photoinitiates the polymerisation of acrylamide, it is possible that either the promazine radical arising from the direct homolysis of the triplet chlorpromazine or the chlorpromazine cation radical may be implicated. This cation radical on reaction with oxygen gives rise to the sulphoxide. Irradiation of 2-chlorophenothiazine in methanol gives rise to phenothiazine and 2-methoxy phenothiazine. Photodechlorination of the same compound occurs in acetonitrile-water implicating the *N*-alkyl substituent in the acceleration of chlorine removal by an intramolecular electron transfer mechanism.[15]

The work published by Underberg[16] and Abdel-Moety[17] et al. represents important findings in respect of the R_2 trifluoromethyl-substituted phenothiazines. Consideration of the thermal degradation of selected phenothiazines attempts to ascertain whether there is a relationship between oxidative degradation and the nature of the R_2 substituent. In the case of the trifluoromethyl derivatives, it was found that the degradation was pH dependent yielding the sulphoxide at pH 3 while at pH 6.3 an additional product, *N*-monomethyl-nortrifluopromazine was isolated. This latter product occurs in the degradation profile due to the cleavage of the side chain of these molecules only if the R_2 substituent is electron-withdrawing and if the dimethylamino group is unprotonated.[16] Results from the accelerated trifluoperazine photolysis carried out on the aqueous solution of the drug using a 60W UV (254nm) lamp indicate the development of reddish brown solution and a new photoproduct 3-trifluoromethyl(biphenylthiophen)sulphoxide characterised by mass spectrometry. In the UV spectrum of the irradiated aqueous solutions an increase in the light absorption in the visible region at about 523nm caused a red colouration. The observed red colour in the case of trifluoperazine can be attributed to the stable red radical. Because of the distribution of the π electrons in the trifluoperazine molecule, radical forms may develop at the S atom, the N_{10} and between the S and the N atom on the phenothiazine ring. Owing to the effect of the short-lasting UV irradiation, a radical at the S atom is expected. Decomposition of this sulphoxide dimer results to give the sulphoxide, which is confirmed by the presence of relevant absorption bands in the UV spectrum.

While the metabolism of these four piperazine-substituted derivatives is proposed to occur via *N*-oxidation, *N*-demethylation, sulphoxidation and hydroxylation, the degradation of the piperazine ring is independent of the presence of the methyl or β-hydroxyethyl substituent and the R_2 substituent with no further mention of the role of these substituents in their metabolism. However, there certainly is a relationship between phototoxicity and the R_2 substituent where the chloro derivatives are photosensitising with the trifluoromethyl derivatives showing few adverse effects. In the stability studies it is reported that the degradation of those compounds with Cl and CF_3, R_2 substituents occurs via photooxidation. It can thus be seen that these R_2 and R_{10} substituents which contribute to activity also play a role in the metabolism, development of adverse effects and stability of these compounds.

3.2 Photoreactivity - Kinetics

A study of instability problems in pharmaceutical products is important since there are at least six possible results of drug instability i.e. loss of the active drug, vehicle and content uniformity, reduction of bioavailability, impairment of pharmaceutical elegance and production of potentially toxic materials.[18] The loss of active drug and formation of degradants on irradiation can be used not only as a means of predicting stability and thus relating structure to photoreactivity but the nature of the degradants serves to evaluate their potential as toxic photoproducts based on previous stability and metabolic studies.

Drug stability is affected by both the intensity and the spectral character of radiation; thus, in order to evaluate the photostability of these related phenothiazines various light sources need to be utilised.[19] Photochemical and oxidative decomposition reactions are a stability liability with respect to certain liquid dosage forms, particularly small volume parenterals. Although both of these reactions are free radical mediated, the photochemical reaction can occur in the absence of oxygen, and oxidative reactions can occur in the absence of the catalytic effect of light. In ampoules the large surface area to formulation volume ratio allows maximum impingement on the relatively dilute drug solution and the short path length. In addition, the head space gas to formulation offers a presence of molecular oxygen. Although irradiation stress testing is widely used qualitatively the quantitative estimate of photolytic decomposition based on stress testing has proved to be difficult. From a practical point of view for the pharmaceutical scientist the first 10-20% degradation on irradiation of a commercial dosage is most significant. Although zero-order behaviour has been predicted in many cases for dilute solutions a first-order relationship is apparent.

Although the rate of photodegradation is quantified in forms of a rate constant, the difficulty associated with the comparisons of photochemical as opposed to thermal reactions is due to the dependence of photochemical reactions on the wavelength and intensity of the irradiating source and the shape and distance of the reaction vessel from the source. Thus, although the rate of photodegradation of a dilute solution of a drug may approximate to first-order kinetics, these mixtures may only be compared if exactly the same irradiation conditions are applied.

Exposure of the four propyl piperazine-substituted derivatives to three different irradiation sources (30W Philips UV lamp, sunlight and a 55 W fluorescent/diffuse light) allow the degradation rates of the different derivatives in the presence of each light source to be compared.

The pseudo first-order rate constants, K_{app}, will be complex constants containing contributions from factors other than the chemical reaction itself, i.e.:

$$K_{app} = f \text{(reaction, irradiation conditions, temperature)} \qquad (1)$$

The irradiation contribution may be further factored into the contributions from the intensity and the wavelength of the radiation as previously mentioned[19] while the reaction contribution will include the influence of the substituents.

Figure 3 represents the logarithm of the residual phenothiazine derivative (D) in solution (KH_2PO_4/NaOH, pH 6.4) irradiated in the presence the three light sources expressed as a percentage of the initial concentration (D_0).

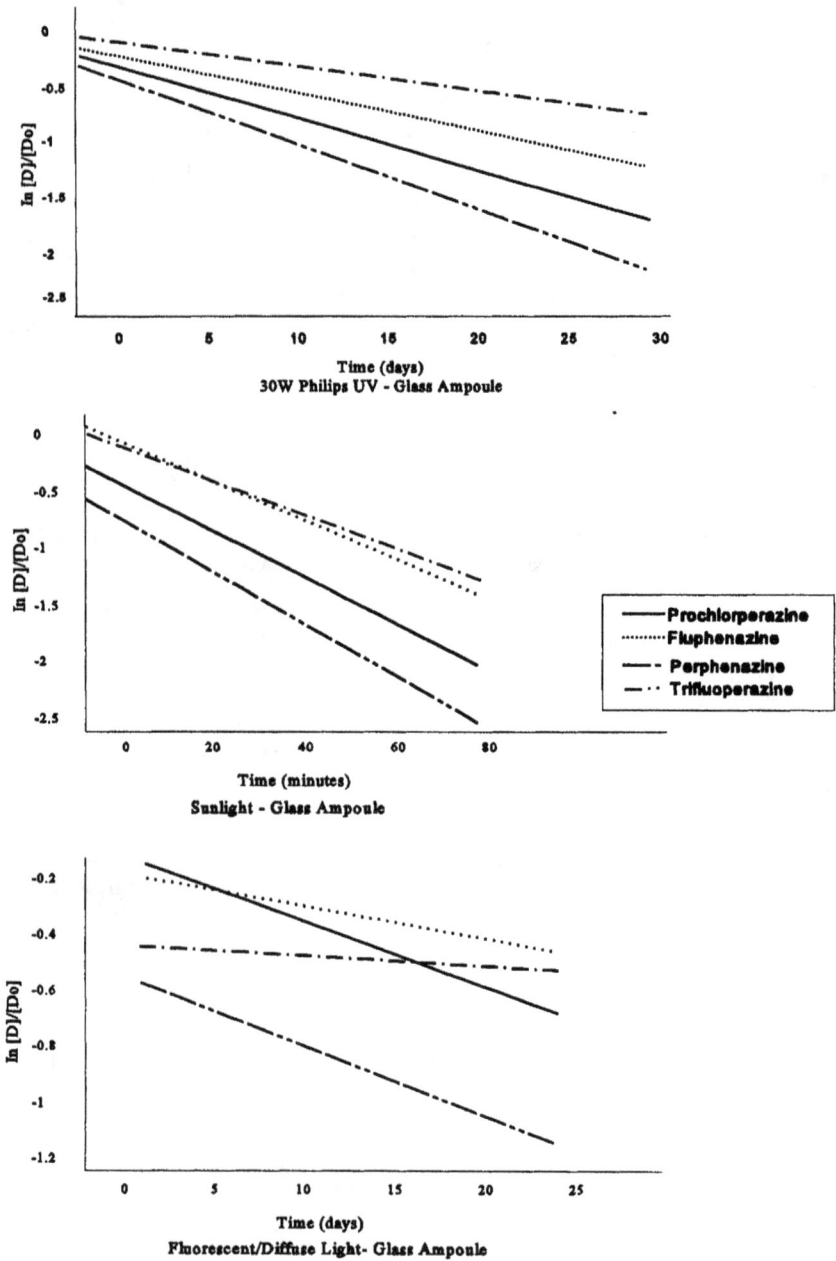

Figure 3. *Degradation profiles of prochloperazine, perphenazine, trifluoperazine and fluphenazine*

The order of degradation (phenothiazine % remaining) on exposure to the 30W UV source and sunlight at t ½ is from the most to the least stable: trifluoperazine (46.88/39.54), fluphenazine (40.55/41.77), prochlorperazine (30.04/26.23) and perphenazine (19.70/14.11). Irradiation under fluorescent/diffuse light further highlights the instability of perphenazine with 59% degradation occurring after 16 days while for the remaining three derivatives the degradation ranged between 40-50%

The degradation of these compounds displayed first-order kinetics with the pseudo first-order rate constants determined from the slopes of the linear plots (Figure 3) and correlation coefficients for these apparent first-order kinetic reactions are presented in Table 2.

Table 2 *First-order rate constants (correlation coefficients)*

Phenothiazine Derivative	30W UV (day^{-1})	Fluorescent/diffuse light (day^{-1})	Sunlight (min^{-1})
Prochlorperazine	0.0334 (0.9437)	0.0214 (0.8983)	0.0217 (0.9404)
Perphenazine	0.1314 (0.9647)	0.0732 (0.9203)	0.0618 (0.9955)
Trifluoperazine	-	0.0279 (0.9237)	0.0181 (0.9895)
Fluphenazine	0.0309 (0.9694)	0.0263 (0.9155)	0.0154 (0.9622)

Although Rhodes[18] comments on the limitations placed on the kinetic interpretation of data due to the lack of accuracy, precision, sensitivity and specificity of the assay, a high performance liquid chromatography (HPLC) method confirms precision with a % RSD at a concentration of 5×10^{-3} mg/mL of not greater than 1.86% and at 4×10^{-2} mg/mL of not greater than 0.91%.

In these stability studies, although the rate of degradation presented as a semi-logarithmic plot of the percentage remaining phenothiazine derivative expressed as a percentage of the original versus time produces a series of straight lines, as mentioned by Moore,[20] these rate constants can only be compared if these reaction mixtures of different phenothiazines are studied under the same irradiation conditions. The fact that Gu[21] reports that drugs with absorption of greater than 280nm have potential for decomposition in sunlight may account for the accelerated degradation of these phenothiazines in sunlight, with further observation by Bundgaard[22] that rate acceleration can be due to the higher temperatures of these solutions exposed to sunlight. Colouration of the solution hindering UV absorption of the reacting molecules and slowing down the rate of reaction may explain the increased degradation from zero time to the first sampling as compared to subsequent readings.

These stability findings concur with those of Pawelczyk et al.[6,7,8] in that trifluoperazine is more stable than prochlorperazine, confirming the contribution of the R_2 substituent to stability; however, since perphenazine is more susceptible to degradation than fluphenazine the R_2 substituent is a primary influence with perphenazine being less stable than prochlorperazine highlighting the contribution of the R_{10} substituent to stability.[6,7,8]

3.3 Photoreactivity - Photoproducts

Current regulations of drug purity control within the International Conference on Harmonisation (ICH) Forum require that impurities exceeding the 0.1% level are examined and preferably identified. Since these impurities may result from degradation, impure starting materials or the synthetic procedure, characterisation of photoproducts following solution and solid-state stability studies allowing the proposal of mechanisms and degradation profiles compared to existing photosensitivity information and metabolic pathways could relate structure to loss of stability and the development of adverse effects.[23]

Mass spectrometry (MS) provides information based on molecular mass and thus the combination of HPLC and MS (LC-MS) is a powerful and rapid method for the analysis of drugs and biological substances. These analyses combine optimised HPLC separation conditions on-line with an electrospray-MS interface to obtain molecular mass information from LC-MS chromatograms and structural information from LC-MS-MS spectra. This technique is extensively used in the pharmaceutical industry and related laboratories for drug identification as well as screening and quantification.[24]

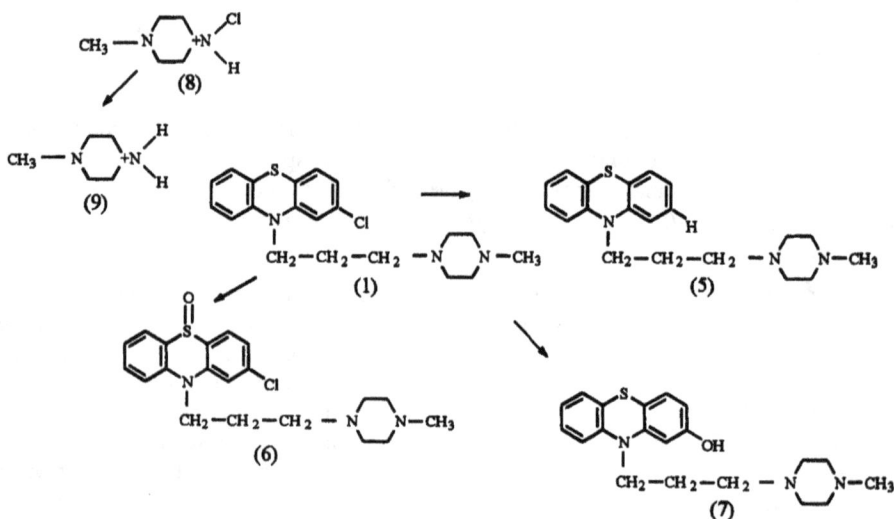

Figure 4. *Proposed degradation pathways for prochlorperazine.*

Figure 4 represents the common degradation pathway for prochlorperazine under all light conditions (30W UV {254nm}, sunlight and 55W fluorescent/diffuse light). In addition to photodechlorination and subsequent photoreduction and substitution, a common *in vitro* degradation product and *in vivo* metabolite which results due to oxidation is the 5-sulphoxide.[25]

Table 3 contains the relative intensities of degradants formed from prochlorperazine under the various light conditions.

Although the presence of the β-hydroxyethyl group as the structural difference between perphenazine and prochlorperazine affects the rate of photodegradation, this group remains intact during the degradation process with the identification of the sulphoxide, the dechloro sulphoxide and dechloro derivatives of perphenazine (Figure 5) and the relative intensities of the degradants from perphenazine expressed as a percentage in Table 4 showing photodechlorination and the resulting photoreduced product as the major degradant. Although similar pathways for the chloro derivatives involve the formation of the dechloro derivatives, perphenazine appears more susceptible to photodechlorination. This can be explained in terms of the involvement of the *N*-alkyl substituent in accelerating chlorine removal.[15]

Table 3 *Relative intensities as a % of the degradants from prochlorperazine*

Degradant	30W UV (254nm)	Sunlight	Fluorescent/diffuse Light
9	6.57	5.11	0.43
8	11.95	8.55	9.72
5	11.32	18.54	3.89
7	8.54	8.1	0.44
6	8.62	14.47	0.66

Figure 6 proposes a pathway for the formation of the degradants from trifluoperazine and fluphenazine. While similarities are evident in respect of the nature of the degradants formed in the irradiated solutions of trifluoperazine and fluphenazine, the relative intensities differ (Table 5), with the amount of the sulphone which has undergone cleavage of the trifluoro group from the 2-position between 16 and 55% for trifluoperazine, the major degradant and although present in the irradiated solutions of fluphenazine, the relative intensity is only between 2.39 - 8.27 %

Common to all irradiated samples is the occurrence of the corresponding sulphoxides; however, there is a considerable variance in the relative intensities due to structure and light sources (see Table 6).

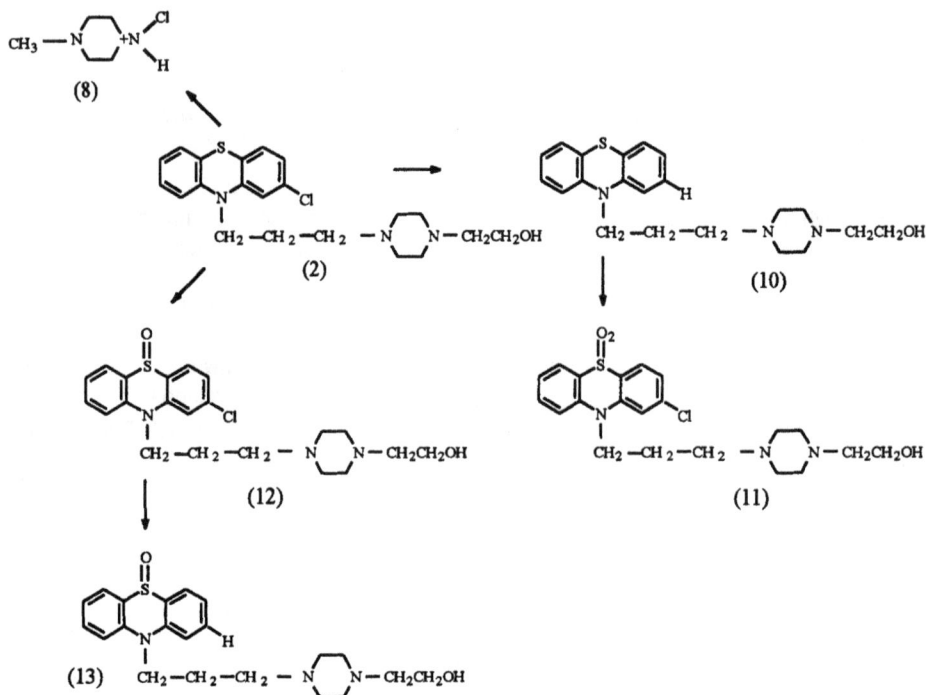

Figure 5. *Proposed degradation pathways for perphenazine.*

Table 4 *Relative intensities as a % of the degradants from perphenazine*

Degradant	30W UV (254nm)	Sunlight	Fluorescent/diffuse Light
8	10.26	14.23	10.64
10	23.71	33.95	18.9
13	4.42	9.02	3.41
11	6.37	7.56	5.31
12	3.69	-*	5.36

* less than 0.1%

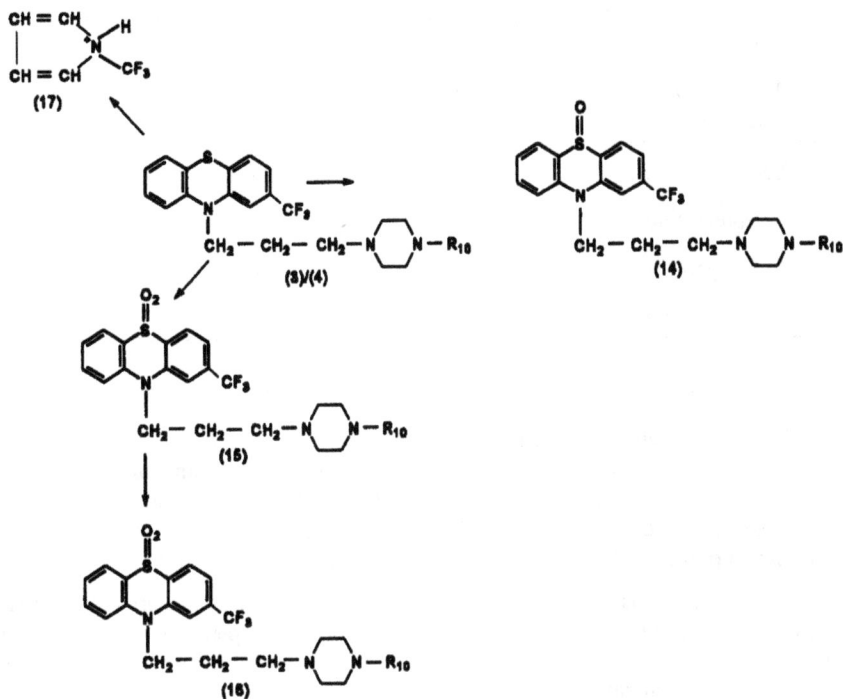

R_{10} = -CH$_3$ (trifluoperazine) / -CH$_2$CH$_2$OH (fluphenazine)

Figure 6. *Proposed degradation pathways for trifluoperazine and fluphenazine.*

Table 5 *Relative intensities as a % of the degradants from trifluoperazine and fluphenazine*

Phenothiazine Derivative	Degradant	30W UV (254nm)	Sunlight	Fluorescent/ diffuse light
Trifluoperazine/Fluphenazine	17	1.37/34.67	3.92/2.04	0.10/3.35
	16	33.04/34.67	41.33/44.45	16.9/2.39
	14	2.73/1.51	3.09/0.47	-*/2.19
	15	0.63/2.29	1.71/1.11	0.60/1.49

* less than 0.1%

Table 6 *Relative intensities as a % of sulphoxides*

Phenothiazine Derivative	30W UV (254nm)	Sunlight	Fluorescent/ diffuse light
Prochlorperazine	8.62	14.47	0.66
Perphenazine	3.69	-*	
Trifluoperazine	2.73	3.09	-*
Fluphenazine	1.51	0.47	2.19

This LC-MS method applied to solid-state photostability studies was undertaken to establish not only the nature of the degradants but also any commonality between major metabolic pathways of *N*-dealkylation, *N*-oxidation, sulphoxidation and hydroxylation. The rate of these *in vitro* transformations is dependent on the concentration of the substrates, and the formation of the sulphoxides has been proved to be a slow reaction *in vitro*.[25]

In spite of previous reports,[26] sulphoxidation is identified as the principal degradation pathway for irradiated samples of prochlorperazine and trifluoperazine giving rise to the 5-sulphoxide in the presence of each respective parent compound, while the relative intensities indicate the prochlorperazine sulphoxide to be 1.93% and trifluoperazine 0.1%, confirming the instability of the chlorine containing derivatives. For fluphenazine *N*-dealkylation, sulphoxidation and dehalogenation are the proposed degradation pathways in Figure 7.

Solid-state perphenazine gave rise to three degradants including the dechloro derivative which upon subsequent fragmentation showed identical product ions to those observed for the *N*-dealkylated derivative of fluphenazine, suggesting a pattern of dehalogenation and *N*-dealkylation for the β-hydroxyethyl derivatives. Similar degradation pathways exist for the chloro derivatives with dehalogenation and subsequent photoreduction and photosubstitution most prominent with sulphoxidation of the parent compounds of secondary importance. Perphenazine is more susceptible to dehalogenation with levels of between 18.90 and 26.85% for the dechloro derivative while prochlorperazine results in significantly lower levels of between 3.89 and 18.54 %.

Sulphoxidation is evident throughout the solution studies at levels greater than 0.1% with the exception of trifluoperazine / perphenazine under fluorescent-diffuse light/sunlight. However, in the case of perphenazine the sulphoxide is formed after dechlorination, a phenomenon not observed for prochlorperazine. While solid-state irradiation of trifluoperazine and prochlorperazine yields only the sulphoxides, fluphenazine, in addition to the sulphoxide (20.23%), loses the trifluoro group from the sulphone and undergoes *N*-dealkylation as additional pathways.

It can thus be concluded that sulphoxidation appears to be a common route of degradation for these four derivatives on exposure to light while evidence of dehalogenation, sulphonation and dealkylation are amongst the other degradation pathways observed in the photostudies. These results concur with the routes of degradation which have been well documented in previously reported *in vitro* studies.[26] In order of decreasing stability the following is proposed: trifluoperazine, fluphenazine, prochlorperazine and perphenazine.

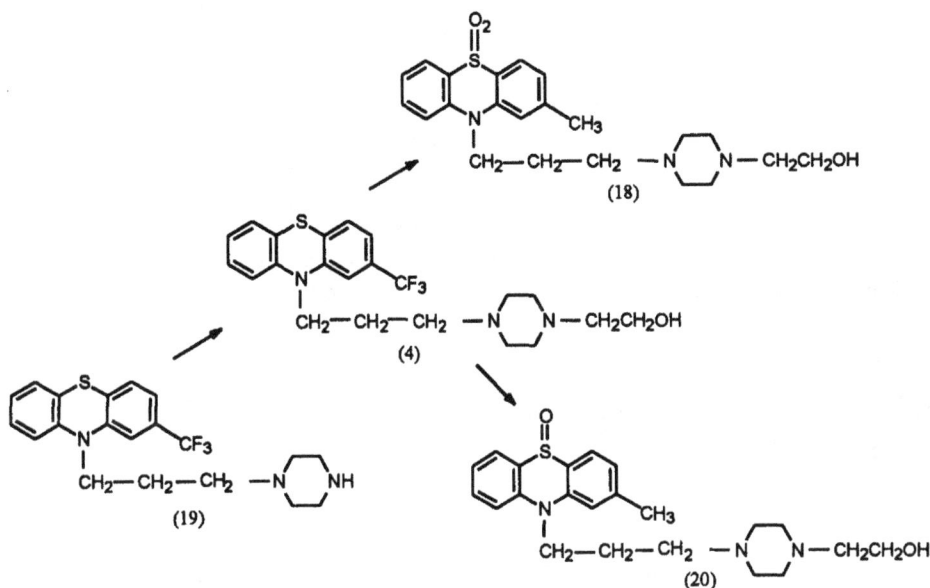

Figure 7. *Proposed degradation for fluphenazine (solid state).*

4 CONCLUSION

These propyl piperazine substituted phenothiazines are an important subclass of the phenothiazines due to their increased potency over the prototype chlorpromazine, with fluphenazine displaying the greatest neuroleptic potency due to increased affinity of the β-hydroxyethyl side chain for the trifluoromethyl group enhancing interaction with the dopamine receptor.[9,10] This is supported by the fact that the 2-substituent affects the lipophilicity of the molecule with the trifluoromethyl group making a greater contribution than the chloro group thus facilitating penetration of the CNS.[11]

Consideration of these structural differences and their effect on activity is important in terms of stability and the development of adverse effects. Kinetics, although useful in predicting the relative stability of the derivatives and thus decreased activity are insufficient due to the need to relate the nature of the degradants to the occurrence of these adverse effects. Although the sulphoxides lack antipsychotic activity, the 2-hydroxy derivatives are implicated in photosensitivity effects.

Stability studies in solution showed perphenazine to be the least stable of all the derivatives. Although dechlorination and sulphoxidation occurred under all light conditions studied, dechlorination with subsequent photoreduction and substitution proved to be a major degradation pathway. For the trifluoromethyl derivatives creation of a radical centre at position 2 did not occur and the sulphoxide and sulphone were identified. In the solid-state studies, sulphoxidation appeared to be the major degradation pathway with the dechloro derivative isolated from perphenazine.

Thus the design of these derivatives to include the trifluoromethyl group to improve activity over the chloro derivatives has merit, especially taking into account the instability of chloroaromatic compounds[14] and the involvement of the photosubstituted product in photosensitivity reactions, while other pathways e.g. *N*-demethylation, suphoxidation and aromatic hydroxylation are in common with those reported in *in vivo* metabolic pathways.[18] However, the inclusion of the β-hydroxyethyl group, as opposed to the methyl group, in the side chain to improve activity is questionable in terms of its effect on stability.

Acknowledgements. The authors wish to acknowledge the Foundation for Research and Development (R.S.A.), Pharmacare-Lennon and Rhodes University for financial support, and Professor Marciniec for the early work in this area and the privilege of meeting her at 'Photostability' 1997.

References

1.   W. O. Foye, T. L. Lemke and D. A. Williams. ' Principles of Medicinal Chemistry', Williams & Wilkins, Baltimore, 1995, 4th Edition, Chapter 12, p. 205.
2.   G. Brookes, *Pharm. J.,* 1990, **7**, 19.
3.   N. Pilpel, M. Nejmeh and W. H. K. Sanniez, *Int. J. Pharm.,* 1980, **5**, 273.
4.   M. Nejmeh and N. Pilpel, *J. Pharm. Pharmacol.,* 1978, **30**, 748.
5.   A. D. Bangham, M. M. Standish and J. C. Watkins, *J. Mol. Biol.,* 1965, **13**, 238.
6.   E. Pawelczyk and B. Marciniec, *Pol. J. Pharmacol. Pharm.,* 1977, **29**, 137.
7.   E. Pawelczyk and B. Marciniec, *Pol. J. Pharmacol. Pharm.,* 1977, **29**, 143.
8.   E. Pawelczyk, B. Marciniec and B. Matlak, *Pol. J. Pharmacol. Pharm.,* 1975, **27**, 317.
9.   A. S. Horn, M. L. Post and O. Kennard, *J. Pharm. Pharmacol.,* 1975, **27**, 553.
10. R. J. Miller, A. J. Horn and L. L. Iverson, *Mol. Pharmacol,* 1974, **10**, 759.
11. C. Hansch, P. G. Sammes and J. B. Taylor, 'Comprehensive Medicinal Chemistry', Pergamon Press, Oxford, 1990, Vol 4, p. 6/371.
12. C. H. Nightingale, M. Tse and E. I. Stupak, *J. Pharm Sci.,* 1972. **61**, 1498.
13. D. E. Moore and S. R. Tamat, *J. Pharm. Pharmacol.,* 1980, **32**, 172.
14. D. Sharples, *J. Pharm. Pharmacol.,* 1981, **33**, 242.
15. N. J. Bunce and B. Marciniec, *J. Med. Chem.,* 1979, **22**, 202.
16. W. J. M. Underberg, L. G. S. Ten Veen and W. P. L. E. Gottgens, *Pharm. Acta Helv.,* 1979, **54**, 239.

17. E. M. Abdel-Moety, K. A. Al-Rashood and N. A. Khattab, *J. Pharm. Biomed. Anal.*, 1996, **14**, 1969.

18. C. T. Rhodes, *Drug Dev. Ind. Pharm.*, 1984, **10**, 1163.

19. D. C. Monkhouse, *Drug Dev. Ind. Pharm.*, 1984, **10**, 1373.

20. D. Moore, *Int. J. Pharm.*, 1990, **63**, R5.

21. L. Gu, H. Chiang and D. Johnson, *Int. J. Pharm.*, 1988, **41**, 105.

22. H. Bundgaard, T. Norgaard and M. Nielsen, *Int. J. Pharm.*, 1988, **42**, 217.

23. M. Erickson, K. Karlsson, B. Lamm, S. Larsson, L. A. Svenson and J. Vessman, *J. Pharm. Biomed. Anal.*, 1995, **13**, 567.

24. X. Qin, D. P. Ip, K. H. Chang, P. M. Dradransky, M. A. Brooks and T. Sakuma, *J. Pharm. Biomed. Anal.*, 1994, **12**, 221.

25. U. Breyer, *Biochem. Pharmacol.*, 1971, **20**, 3341.

26. H. J. Gaertner, U. Breyer and G. Liomin, *Biochem. Pharmacol.*, 1974, **23**, 303.

Photoprocesses in Photosensitising Drugs Containing a Benzophenone-like Chromophore

Sandra Monti,[a] Salvatore Sortino,[a,b] Susana Encinas,[a,c] Giancarlo Marconi,[a] Guido De Guidi,[b] Miguel A. Miranda[c]

[a]Istituto di Fotochimica e Radiazioni d'Alta Energia, CNR, Area della Ricerca, Via P. Gobetti 101, I-40129 Bologna, Italy

[b]Dipartimento di Scienze Chimiche, Università di Catania, Viale A. Doria 8, I-95125 Catania, Italy

[c]Departamento de Quimica/Instituto de TecnologiaQuimica UPV-CSIC, Camino de Vera, s/n, Apartado 22012, E-46071 Valencia, Spain

1 INTRODUCTION

Non steroidal antiinflammatory drugs may cause phototoxic effects both *in vitro* and *in vivo*. The sensitisation reactions are initiated by photochemical processes leading to intermediates and stable photoproducts which attack the biological substrates. A key parameter in these processes is the environment. Location of the drugs either in polar, hydrophilic or in apolar, hydrophobic sites may determine completely different excited state deactivation pathways and intermediates, so that control of the photoreactivity and preparation of protective strategies can only be achieved if the structural/environmental factors determining the photobehaviour of the drugs are clarified.

Ketoprofen (KPF), suprofen (SPF) and tiaprofenic acid (TPA) belong to the aryl-propionic acid family and all contain a benzophenone-like aromatic moiety (Chart 1). They were reported to cause photohemolysis,[1,2] lipid peroxidation,[3] as well as damage to proteins and nucleic acid chains.[4]

In aqueous medium at neutral pH, where these compounds are present almost exclusively as the carboxylate anions, the main photoreaction is decarboxylation leading to the corresponding ethyl derivatives and, in the presence of oxygen, additionally to a distribution of (decarboxylated) oxidation products.[1,3,5] The damage to the biological substrates is caused by the formation of radicals and singlet oxygen and the accumulation of the final products, still containing the photoactive ketone functional group.[5,6]

Despite the structural analogy and the basic similarity of the photoproducts, important peculiarities were found in the photodegradation mechanisms. This paper illustrates the relationship between the excited state character and the photoreactivity pathways. The role of the other key parameter, i.e. the environment, is also discussed in the light of the results obtained in drug-β-cyclodextrin (β-CD) inclusion complexes. The apolar cavity of cyclodextrins was shown to be capable of modifying the photoreactivity of a large variety of molecules[7] and, interestingly, β-CD is known to attenuate the toxic effects of KPF and SPF *in vitro.*[8,9] Moreover, the association to CD via non covalent binding can represent a useful model to mimic the interaction of the drug with hydrophobic biological sites. The photodegradation of KPF, SPF and TPA (carboxylate forms), was investigated in aqueous medium[10,11] and in the β-cyclodextrin cavity[12,13] and the photochemical intermediates were characterised by time resolved emission and absorption studies in the picosecond, nanosecond and microsecond time domains. Quantum mechanical calculations contributed to assign the electronic nature of the lowest energy electronic states.[10,11] The main results of these studies are presented and the structural/environmental factors determining the features of the photoreactivity of these drugs are highlighted.

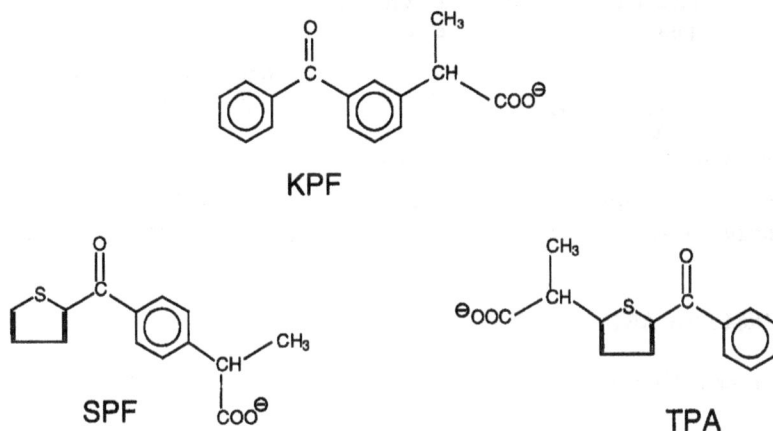

Chart 1

2 METHODS

2.1 Theoretical

The geometries of the isolated molecules were optimised using either the MM3-92 Allinger[14] or MM+ molecular mechanics routines. The excited states were investigated by means of the Zerner Intermediate Neglect of Differential Overlap/Spectroscopic (ZINDO/S) method (Hyperchem software package). Electronic population analysis was performed by the Complete Neglect of Differential Overlap/Spectroscopic (CNDO/S) method with singly excited Configuration Interaction.

2.2 Experimental

Irradiations were performed with light at $\lambda \geq 310$ nm. The incident light intensity was measured by using the ferrioxalate actinometer. Quantitative analysis of the degradation and decarboxylation yields was performed by High Performance Liquid Chromatography. All the details on the steady state experiments and on the analytical procedures can be found in refs.10 and 11.

Induced Circular Dichroism (i.c.d.) of racemic mixtures KPF, SPF and TPA in the presence of β-cyclodextrin was performed by means of a Jasco J715 spectropolarimeter. The inclusion geometry of KPF was investigated by theoretically interpreting the i.c.d. signal. Conformational calculations based on a Dynamic Monte Carlo procedure including the solvent were used to find low energy associated structures.[15-17] The rotational strength was calculated by using the Tinoco-Kirkwood approach.[18] The most likely inclusion geometry was taken as that best reproducing the sign and the intensity of the i.c.d. bands. This procedure, detailed in ref. 13, proved to be very effective in interpreting the i.c.d. spectra of several chromophores included in cyclodextrins.[19]

Transient absorption spectra in the subnanosecond time domain were obtained by a pump-probe technique already described.[20] Pump pulses of 35 ps (FWHM) at 266 nm were obtained from an actively-passively mode-locked Nd-YAG laser source (Continuum PY62-10). Time "zero" was assumed to be in the middle of the excitation pulse.

The setup for the nanosecond absorption measurements has been described previously.[21] The excitation pulse, 20 ns FWHM, at 266 or 355 nm was from a Nd-YAG laser (JK-Lasers).

3 RESULTS AND DISCUSSION

3.1 The Lowest Excited States

The absorption spectra of the three drugs in neutral phosphate buffer are reported in Figure 1. The spectrum of KPF is similar to that of benzophenone. It is characterised by a band with $\lambda_{max} = 260$ nm attributed to the $S_0 \rightarrow S_2$ transition of π,π^* nature, and by a lower energy shoulder in the 300-350 nm region corresponding to the $S_0 \rightarrow S_1$ forbidden transition of n,π^* parentage. The lowest singlet (calculated at 360 nm, f = 0.0108) participates of the π,π^* nature. Two very close triplets are located at 71.0 and 72.6 kcal/mol *in vacuo*. The solvent effect, investigated through the method of the virtual charges,[22] shows that they are almost unaffected in energy by solvation, their electronic distribution being very close to that of S_0. The lowest one, if compared to the lowest n,π^* triplet state in benzophenone, participates more of the π,π^* character.[10] The absorption spectrum of SPF is characterised by a broad band with maximum at 300 nm, corresponding to the $S_0 \rightarrow S_2$ (π,π^*) transition, a shoulder at 270 nm, attributed to $S_0 \rightarrow S_3$, the latter also of π,π^* nature, and a weak tail extending beyond 360 nm, identified as the transition to S_1 (n,π^*) state. This latter was calculated at 74.4 kcal/mol. As far as the triplet states are concerned, T_1 is of π,π^* nature, while T_2 (n,π*) is higher in energy by ca. 11 kcal/mol. T_3 (π,π*) lies close to S_1 (n,π*) (~2.5 kcal/mol below).

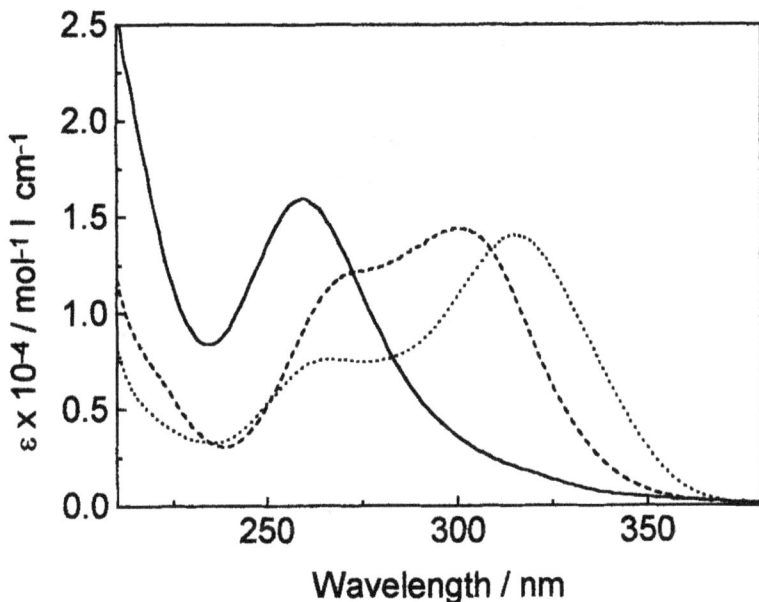

Figure 1 *Absorption spectra of KPF (—), SPF (----) and TPA (····) in phosphate buffer 10^{-2} mol l^{-1}, pH 7.4.*

The absorption spectrum of TPA is characterised by intense bands at 314 and 266 nm and a weak shoulder extending over 350 nm. The assignment of these bands parallels that given for the corresponding ones in SPF. S_1 was calculated at 71.3 kcal/mol. The lowest state of the triplet manifold, T_1 was calculated at 49 kcal/mol and is of π,π* character, while T_2 (n,π*) is located ca. 15 kcal/mol higher than T_1. T_3 (π,π*) lies close to S_1 (less than 1 kcal/mol higher). The existence of a large energy gap between T_1 and T_2 is confirmed to be a characteristic feature of the benzoylthiophene moiety. The actual value of the energy separation in TPA is likely lower than that calculated, being the energy of T_1 (π,π*) rather underestimated with respect to the experimental value (ca. 58-60 kcal/mol), and is thought to be similar to that in SPF.[11]

3.2 The Photoreactivity

3.2.1 Aqueous Environment. In neutral aqueous medium KPF, SPF and TPA are present as the carboxylate anions (pK_a <5). The photodegradation of the drugs in these conditions is almost exclusively due to decarboxylation. For KPF in deoxygenated phosphate buffer at pH

7.4 the quantum yield of the photoprocess (Φ_r) is very high, 0.75,[1] and does not depend on the concentration,[#] the presence of oxygen and the temperature (in the range 0-50 °C). For SPF Φ_r is considerably lower, 0.07, for a solution 5 x 10^{-5} mol l^{-1} at room temperature.[4] In this case the quantum yield increases by decreasing the concentration, and is lower in aerobic conditions. Moreover, it is remarkably temperature dependent between 10 and 50 °C. An activation energy $\Delta E^{\neq}$ ~ ca. 9 kcal/mol is derived from an Arrhenius-like plot.[12] Similar features were observed in the photodegradation of TPA: the reaction is affected by oxygen[3] and the quantum yield, ca. 0.25 at 25 °C in anaerobic conditions, depends on the temperature with $\Delta E^{\neq}$ ~ 8-10 kcal/mol.[11] The photodegradation quantum yields of TPA and SPF at different temperatures are shown in Figure 2.

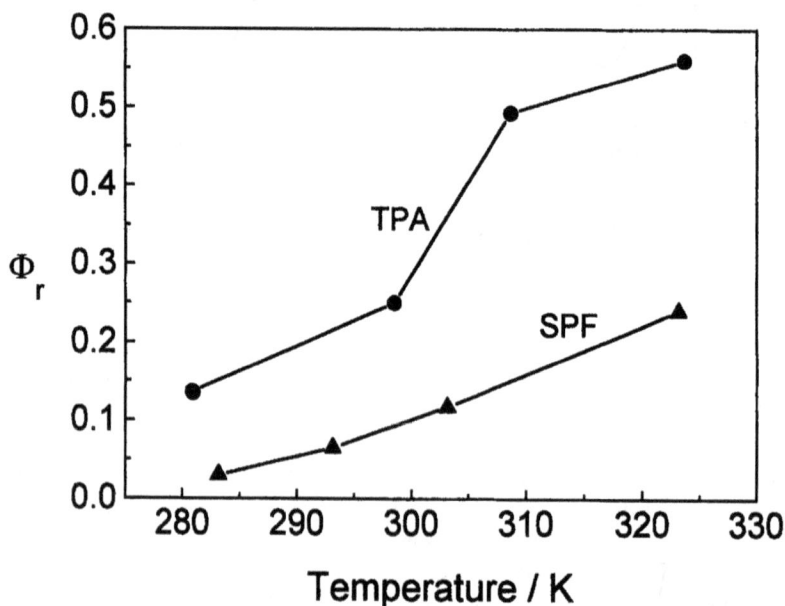

Figure 2 *Quantum yield of photodegradation (Φ_r) in deoxygenated phosphate buffer 10^{-2} mol l^{-1} pH 7.4: [SPF] = 5.0 x 10^{-5} mol l^{-1}; [TPA] = 6.5 x 10^{-4} mol l^{-1}.*

3.2.2 *β-Cyclodextrin Cavity*. The three drugs form inclusion complexes with β-cyclodextrin characterised by 1:1 stoichiometry: the association constants were determined by studying the induced circular dichroism (i.c.d.) of the complexes and found to be 2700,[8]

[#] At high concentration (10^{-2} mol l^{-1}) additional photoproducts are obtained in which the benzophenone chromophore has participated. However, this photodegradation pathway is minor.

1900^{12} and 3000 l mol^{-1} for KPF, SPF and TPA, respectively. The KPF molecule deeply includes in the host cavity with the carbonyl and the aromatic part, while the propionic group remains outside the torus.[13] In the case of SPF, ^{1}H NMR spectra are in agreement with an inclusion geometry with both the thienyl and the phenyl rings located inside the cavity.

The insertion of the drugs in the apolar cavity of the cyclodextrin has marked consequences on their photoreactivity. In the KPF-β-CD inclusion complex, in the absence of oxygen, the quantum efficiency of decarboxylation is 0.42, remarkably lower than in aqueous solution; correspondingly, the photodegradation quantum yield is 0.85, i.e. somewhat higher than in water. A fraction of ca. 50% of the photoreacted compound leads to a product with an absorption λ_{max} = 205 nm, characteristic of the reduced carbonyl group, and with NMR properties typical of an adduct containing the macrocycle as covalently linked moiety.[13] The β-CD-SPF inclusion complex exhibits similar behaviour: the photodegradation and the photodecarboxylation quantum yields are different, 0.23 and 0.11 respectively in the absence of oxygen at 20°C, and negligibly concentration dependent (up to 8×10^{-5} mol l^{-1}), the formation of addition products of the drug to the macrocycle in which the thienoyl chromophore is lost was inferred from absorption and NMR spectra.[12]

3.3 The Photochemical Intermediates

In these molecules photoexcitation either of the S_0-S_1 or of the S_0-S_2 transitions leads to practically quantitative population of the lowest triplet states. This was evidenced by laser flash photolysis.[10-13] This high ISC efficiency is accounted for by the results of the quantum mechanical calculations which show the presence of a π,π^* triplet almost isoenergetic with the lowest n,π^* excited singlet.

In aqueous environment both the formation and the decay of the lowest triplet of the KPF carboxylate anion was observed in the subnanosecond time domain (Figure 3). The T-T absorption has a maximum at 526 nm (a) and is quite similar in shape and intensity to that of benzophenone, taken in the same conditions (c). The close spectroscopic parentage of the lowest n,π^* triplets of the two molecules is patent. However, the two excited states differ in their kinetic evolution: while the benzophenone triplet is very long-lived and rather unreactive (the lifetime is of the order of tens of microseconds), the KPF triplet has a decay time constant of only 250 ps and, with the same time constant, a transient absorption with λ_{max} = 570 nm (b) grows up. This dynamic behaviour was ascribed to the decarboxylation reaction which was suggested to be promoted by the more delocalised and partial π,π^* charge distribution of the KPF lowest triplet state. Alternatively, the presence of an upper state, very close to T_1 and reactive so that it induces fast deactivation of the latter, can be hypothesised.

The study of the intermediates along the pathway to the final product (nanosecond and microsecond domains) led us to propose the mechanism reported in Scheme 1.[10] In the triplet state (1) an intramolecular electron transfer from the carboxyl (donor) to the carbonyl (acceptor) groups occurs with a very low energy barrier and leads to the triplet biradical 2, with λ_{max} = 570 nm and the structure of a benzophenone-like anion. Release of carbon dioxide proceeds adiabatically from **2**, with rate constant 8.3 x 10^6 s^{-1}, and the biradical **3** is formed, undergoing protonation equilibrium with the biradical **4** (pK_a ~ 7.6). This latter at neutral pH has a decay rate of 2.5 x 10^5 s^{-1} and is very sensitive to dissolved oxygen (k_q ~ 2 x 10^9 $M^{-1}s^{-1}$). The conversion to (3-benzoylphenyl)ethane needs ISC and from **4** is rather slow, since it involves a formal intramolecular H-shift and further long-lived intermediates (~10^5 s^{-1}), while from **3** is faster (~ 4 x 10^6 s^{-1} at pH 12.5) since it only requires an intramolecular electron transfer from the oxygen site to the benzylic site, followed by protonation from solvent of the formed carbanion.[6,§]

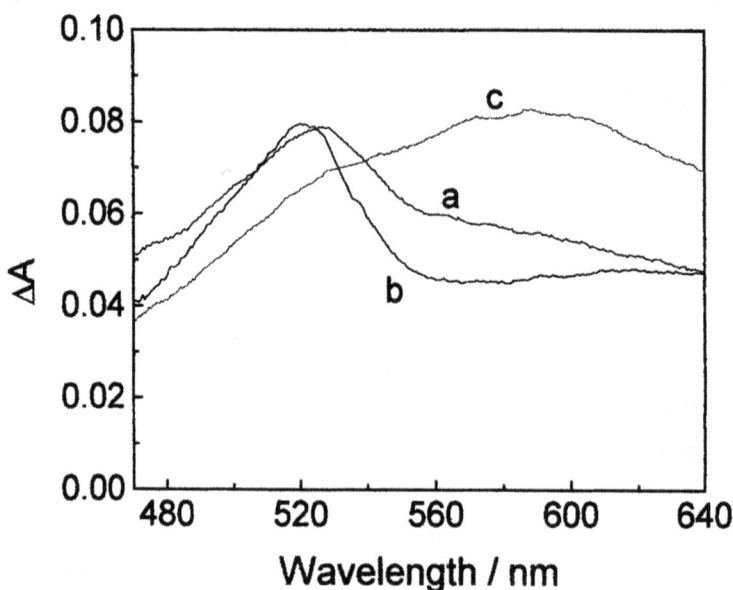

Figure 3 *Absorbance changes observed in KPF, 7.0 x 10^{-5} mol l^{-1} in phosphate buffer 10^{-2} mol l^{-1} at pH 7.4, after a 35 ps laser pulse at 266 nm: (a) 99 ps; (b) 693 ps; (c) benzophenone, 1.1 x 10^{-4} mol l^{-1}, at 99 ps.*

In the KPF-β-CD inclusion complex, where two comparably efficient photochemical pathways are present, one leading to decarboxylation and the other to attack of the cyclodextrin, no subnanosecond triplet decay is observed: the triplet lifetime becomes of the order of 100 ns. This was attributed to the slowing down of the intramolecular electron

---

[§] A minor contribution to the photocleavage of KPF is also provided by photoionisation. This additional mode may become important at high light intensities due to a prevailing biphotonic nature.[10]

transfer process, thermodynamically disfavoured by the location of the C = O in the apolar CD interior. On this time scale, an intrinsically slower process, like H-abstraction from a CD-glucose unit, may compete. A transient with lifetime 250 ns, not observed in the absence of CD, was assigned to the triplet radical pair diarylketyl radical-CD·, recombining upon ISC to form the addition product.[13]

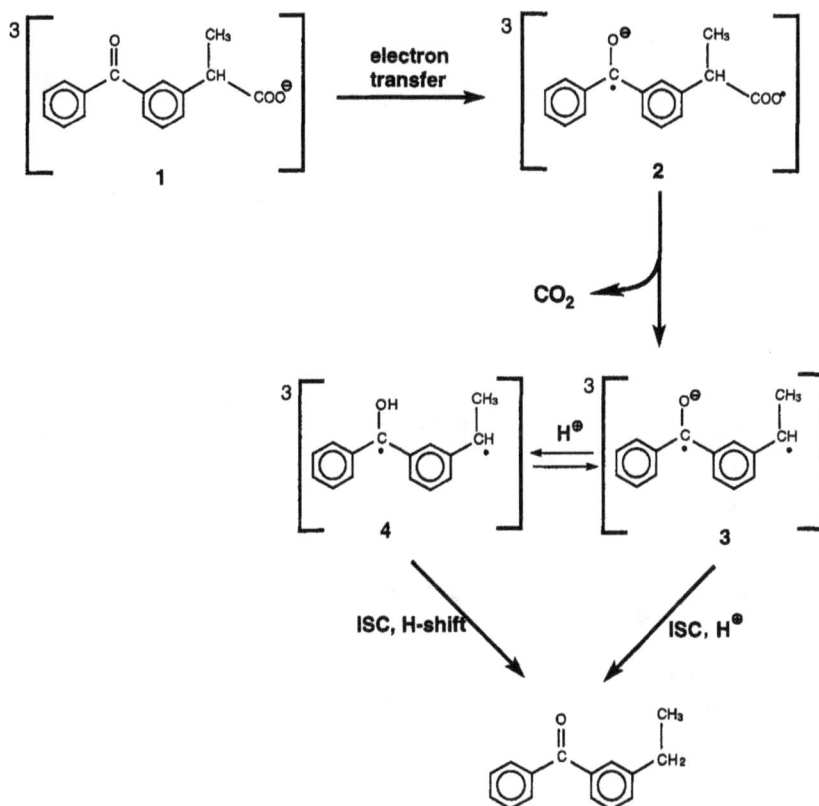

Scheme 1

The features of the SPF photodecarboxylation (oxygen, temperature and concentration dependence) have a direct relation to the properties of the lowest triplet which clearly appears to be the precursor of the reaction. This state, of π,π^* character, is characterised in neutral solution at room temperature by a T-T absorption with λ_{max} = 600 nm and a lifetime in the microsecond domain in deaerated solution. Its decay rate depends on the presence of oxygen ($k_{ox} \sim 10^9$ l mol^{-1} s^{-1}) and on the SPF concentration ($k_{SPF} \sim 2 \times 10^9$ l mol^{-1} s^{-1}) and was estimated to be $\sim 2.5 \times 10^4$ s^{-1} at infinite dilution. Moreover, it increases by increasing temperature following an Arrhenius law with activation parameters $\Delta E^{\neq} \sim$ 8-9 kcal/mol and preexponential factor $\sim 10^{12}$ s^{-1}. These findings indicate that photodecarboxylation actually proceeds from a higher triplet state T**, energetically close to $T_2(n,\pi^*)$. The electronic population of this latter, very similar to that of the n,π^* triplet of KPF, suggests a possible direct involvement of T_2 itself in the reaction.

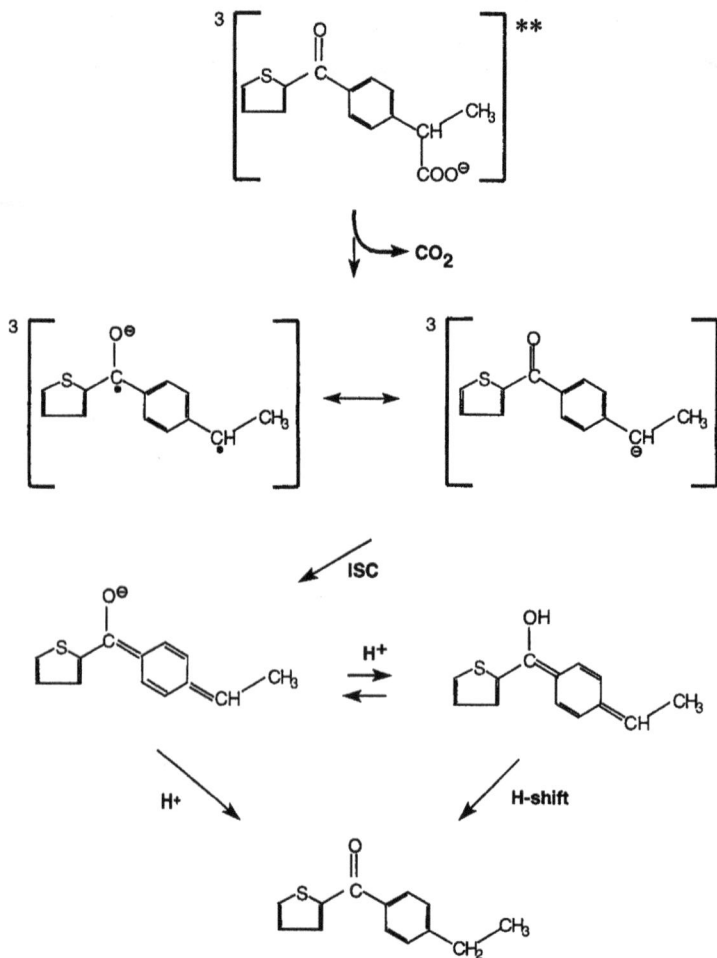

Scheme 2

In Scheme 2 a reaction mechanism consistent with the properties of the observed intermediates is presented. Upon decay of the lowest triplet, a transient with 4 μs lifetime, main absorption at 360 nm and a weak tail at ca. 580 nm, pH insensitive in the range 6-12, was assigned to a decarboxylated triplet anion. This species, in contrast to the KPF case, does not show the acid-base properties of an oxygen centred anion. Its decay rate has a temperature dependence consistent with a spin forbidden process and can be attributed to ISC. The residual absorption has a maximum at 370 nm and intensity which depends on pH in the range 6-10. This suggests that protonation occurs in the ground state at the oxygen site with formation of an enol from which the final product is formed by formal intramolecular H-shift. Accordingly, the rate constant of the last step, ~ 2 x 10^3 s^{-1} at pH 6.5, is much lower in deuterated buffer (~ 60 s^{-1}). This mechanism is further supported by the results of quantum mechanical calculations of the charge distribution in the triplet anion, which indicate that the negative charge is rather delocalised with significant fractions located on the oxygen atom and on the carbon centres of the phenyl substituent.

In the β-CD-SPF inclusion complex a decrease of the ISC efficiency occurs, consistent with a change in the relative position of the ISC-coupled ^{1}n,π* and ^{3}π,π* states, determined by the lower environmental polarity of the carbonyl group. The lowest triplet of the complex exhibits self-quenching and oxygen quenching rate parameters one order of magnitude lower than those of the free molecule, thus making the photodegradation to be rather independent of external agents. Finally, the temperature dependence of the triplet decay rate is consistent with the opening of the photochemical channels (decarboxylation and addition to CD) from the T$_2$ (n,π*) state.[12]

The transients in TPA are similar to those observed in SPF. The lowest triplet (λ_{max} = 600 nm, lifetime ca. 0.8 μs) is of π,π* nature and is characterised by a thermally activated decay. The energy barrier (ca. 7-10 kcal/mol) corresponds to that of decarboxylation and fairly well to the energy separation between T$_1$(π,π*) and T$_2$ (n,π*). As already suggested in the case of SPF, a direct involvement of T$_2$ in the reaction could be hypothesised.[11] With a rate constant equal to that of the decay of the lowest triplet a pH sensitive further transient is formed. This intermediate is characterised by a protonation equilibrium with pKa ~ 8.2 and by a slightly temperature dependent decay, consistent with the occurrence of an activated ISC process. It was identified as an oxygen centred triplet anion, undergoing ISC to a ground state anion. Evolution to the final product involves complex kinetics with submillisecond and millisecond components.

4 CONCLUDING REMARKS

On the basis of the above results some general conclusions can be drawn about the photodecarboxylation in the investigated benzophenone-like propionic acids. The photoreactivity of the carboxylate forms of KPF, SPF and TPA in water is mediated by the lowest triplet state which is formed in almost quantitative yield. The main channel for triplet deactivation seems to be an intramolecular charge transfer, which decreases the negative charge on the carboxylate group and increases it on carbonyl group. As a result, the release of a CO_2 fragment may occur. A triplet anion with prevailing negative charge on the carbonyl oxygen may or may not be distinguishable (via the acid-base equilibrium at this site): it was detected in KPF and TPA, while it was not clearly evidenced in SPF. As already observed in *p*-nitrophenylacetate,[23] the CO_2 detachment appears adiabatic, i.e. the intermediates detected upon this process have the features of triplet species. Before the final decarboxylated products (the ethyl-substituted arylketones) are formed both ISC and a protonation step are required. At neutral pH in SPF the intermediate formation of a ground state enol derivative was clearly evidenced.

Despite the common features, the precursor excited states have a differentiated dynamic behaviour which prefigures peculiar strategies for an effective photoreactivity control to be achieved in each case. The intramolecular electron transfer taking place in the lowest n,π* triplet state (or in a closely lying upper triplet) of KPF occurs practically without any significant energy barrier, i.e. so fast that the process cannot be influenced by external agents like oxygen, concentration, temperature. The only way to control it is to change the environment. In the cyclodextrin cavity the lower polarity lengthens the lifetime of the precursor triplet and allows a reductive process to become competitive. The quantum efficiency of photodecarboxylation is reduced to one half that in water. In SPF (and TPA) the presence of a thienyl ring in the molecular structure makes itself a long-lived π,π* triplet to become the lowest state and the precursor of the photoreaction. Photodecarboxylation is intrinsically less efficient, since it requires thermal activation. Moreover, it has to compete with other deactivation processes of bimolecular character, easily affected by concentration and oxygen. In the case of SPF the change of environment from water to the cyclodextrin cavity further lengthens the triplet lifetime by providing protection from external quenchers. As for KPF, the relative importance of the photodecarboxylation is decreased since photochemical attack of the cyclodextrin macrocycle occurs in competition.

The modified photobehavior of the drugs in β-CD is relevant to the observed protection of biological substrates against phototoxic effects *in vitro*.[$] The formation of radicals by reaction with the saccharide is in fact not expected to cause in turn cell damage because of the fast addition process to the CD itself. Finally, the opening of a reductive photochemical channel in the presence of H-donating groups in hydrophobic, constrained environments could be of toxicological relevance potentially leading to irreversible binding of the drugs to cell components.

[$] The production of singlet oxygen is practically absent in KPF both in aqueous medium and in the presence of β-CD, due to the short triplet lifetime of this molecule in these conditions.[10,13,24] On the contrary, it is rather efficient in SPF[12,24] and TPA[24] and could contribute to the photosensitisation reactions. Owing to the opposite effects of lengthening of the triplet lifetime and decreasing of the bimolecular rate constant for triplet quenching by oxygen in the included drugs, the efficiency of singlet oxygen production is nor affected drastically by CD, so that the protective role of this latter is not expected to be significantly related to singlet oxygen.[12]

References

1. L. L. Costanzo, G. De Guidi, G. Condorelli, A. Cambria and M. Fama, *Photochem. Photobiol.*, 1989, **50**, 359.
2. J. V. Castell, M. J. Gomez-Lechon, C. Grassa, L. A. Martinez, M. A. Miranda and P.Tarrega, *Photochem. Photobiol.*, 1994, **59**, 35; *ibidem*, 1993, **57**, 486.
3. J. V. Castell, M. Gomez-Lechon, D. Hernandez, L. A. Martinez and M. A. Miranda, *Photochem. Photobiol.*, 1994, **60**, 586.
4. G. De Guidi, R. Chillemi, L. L. Costanzo, S. Giuffrida, S. Sortino and G. Condorelli, *J. Photochem. Photobiol. B: Biol.*, 1994, **23**, 125.
5. G. Condorelli, L. L. Costanzo, G. De Guidi, S. Giuffrida and S. Sortino, *Photochem Photobiol.*, 1995, **62**, 155.
6. F. Bosca, M. A. Miranda, G. Carganico and D. Mauleon, *Photochem. Photobiol.*, 1994, **60**, 96.
7. P. Bortolus and S. Monti, *Adv. Photochem.*, 1996, **21**, 1.
8. G. De Guidi, G. Condorelli, S. Giuffrida, G. Puglisi and G. Giammona, *J. Incl. Phen. Mol. Rec. Chem.*, 1993, **15**, 43.
9. G. De Guidi, S. Giuffrida, P. Miano and G. Condorelli, 'Photochemistry and Photobiology', F. Vargas Editor, Trivandrum, India, in press.
10. S. Monti, S. Sortino, G. De Guidi and G. Marconi, *J. Chem. Soc., Faraday Trans.* 1997, **93**, 2269.
11. S. Encinas, M. A. Miranda, G. Marconi and S. Monti, *Photochem. Photobiol.*, in press.
12. S. Monti, S. Sortino and G. De Guidi, 7th Congress of the European Society for Photobiology, Stresa (Italy), 8-13 September 1997, *Abstracts* p.20; S. Sortino, G. De Guidi, G. Marconi and S. Monti, *Photochem. Photobiol.*, in press.
13. S. Monti, S. Sortino, G. De Guidi and G. Marconi, *New J. Chem.*, in press.
14. N. L.Allinger, Y. H. Yuh and J. H. Lii, *J. Am. Chem. Soc.*, 1989, **111**, 8551, N. L. Allinger, *Quantum Chemistry Program Exchange Bulletin*, 1989, **9**, 2.
15. B. Mayer,1994, Mol Doc-V1.2 Program Package.
16. K. S. Kirkpatrick, C. D. Gelasz and M. P. Vecchi, *Science*, 1983, **220**, 671.
17. G. Perrot, B. Cheng, K. D. Gibson, J. Vila, K. A. Palmer, A. Nayeem, B. Maigret and H. A. Scheraga, *J. Comput. Chem.*, 1992, **13**, 1.
18. I. Tinoco, *Adv. Chem. Phys.*,1962, **4**, 113.
19. (a) G. Marconi, S. Monti , B. Mayer and G. Köhler, *J. Phys. Chem.*, 1995, **99**, 3943; (b) B. Mayer, G. Marconi, C. Klein, P. Wohlschann and G. Köhler, *J. Incl. Phen. Mol. Rec. Chem.*, 1997, **29**, 79; (c) G. Grabner, S. Monti, G. Marconi, B. Mayer, C. Klein and G. Köhler, *J. Phys. Chem.*, 1996, **100**, 20068.
20. P. Bortolus, F. Elisei, G. Favaro, S. Monti and F. Ortica, *J. Chem. Soc. Faraday Trans.*, 1996, **92**, 1841.
21. S. Monti, N. Camaioni and P. Bortolus, *Photochem. Photobiol.*, 1991, **54**, 577.
22. I. Jano, *Chem. Phys.Lett.*, 1984, **106**, 60.
23. B. B. Craig, R. G.Weiss and S. J. Atherton, *J. Phys. Chem.*, 1987, **91**, 5906.
24. D. de la Peña, C. Martì, S. Nonell, L. A. Martinez and M. A. Miranda, *Photochem. Photobiol.*, 1997, **65**, 828.

Photostability of Coumarin

Jeffrey M. Lynch and Alicja M. Zobel*
Department of Chemistry
Trent University
Peterborough, Ontario, Canada
K9J 7B8

1 INTRODUCTION

Secondary metabolites are phenolic compounds synthesized by plants which may serve several protectory roles for the plant. Phenolics are widely distributed. Thus, these compounds are very common in fruit and vegetables. In the western diet it is practically impossible to avoid consuming phenolics such as coumarins[1] and flavonoids.[2] These two types of phenolic compounds may possess a variety of biological activities. Weinmann[3] reported that oncologists are interested in the anti-proliferative effects of coumarin and 7-hydroxycoumarin while Stavric and Matula[2] suggested that flavonoids display protective effects against carcinogens not only at a cellular or enzymatic level but also by reducing the bio-availability of carcinogens. Considering that secondary metabolites are present in food as well as in many prescription and non-prescription drugs, an understanding of the photostability of these compounds is of utmost importance. In this paper, we discuss: 1) the localization and function of phenolic compounds in plants with emphasis on coumarin; and, 2) the mechanism of coumarin photostability.

2 LOCALIZATION OF PHENOLIC COMPOUNDS IN PLANTS

Secondary metabolites are located in the vacuole of a plant cell,[4] in the inter-cellular spaces,[5] and on the surfaces of tissues.[6, 7] Every plant cell deposits in its vacuole several different secondary metabolites. In the vacuole, the compounds (glycosides) are relatively inactive, but are activated when released from this compartment.[8] When exocytosis occurs, glucosidase located in the cell wall cleaves the sugar constituent of the secondary metabolites forming aglycones. The removal of the sugar constituent increases the toxicity of the phenolic compounds. Zobel *et al.*[5] proposed that phenolic compounds that are covering the external walls of cells surrounding inter-cellular spaces, will form a defense barrier against microbial attack. Vickery and Vickery[9] found that plant phenolics will accumulate in cells adjacent to those infected by disease so as to protect against micro-organisms. The aglycones can also be extruded to the surface of plant tissues.

Baker[10] reported that phenolic compounds can be embedded in the epicuticular waxes. To date, three large groups of secondary metabolites, the flavonoids,[6] coumarins[7] and glucosinolates[11] have been found to be embedded in the waxes. Wollenweber[12] stated that of the over 400 species known to contain flavonoids, many species had these compounds on their surfaces. On the surface of the plant the phenolic compounds: 1) will kill bacteria and fungal spores;[13] 2) may render an unappealing taste to herbivores;[1, 14, 15] 3) will protect the plant tissues from free radical damage; 4) may be responsible for some dispersion of UV radiation;[16] 5) will form a barrier to protect the plant from UV radiation;[17] and, 6) may transduce UV radiation (absorbed energy) into lower-energy, longer wavelength, usable (i.e. visible) light.[1, 18, 19]

3 SECONDARY METABOLITES

It has been suggested that phenolic compounds have been chosen by natural selection,[20] and may play several important ecological roles including defense and protection from environmental stress. Harborne[21] suggested that phenolic compounds represent one method of defense utilized by plants against herbivores. Rhodes[13] stated that the diversity of the secondary metabolites and their uneven distribution between and within plant families indicates that a single physiological role cannot be attributed to these compounds. Zobel[19] reported that a compound that is an allelochemical in one species does not have to be an allelochemical in another. Therefore, defining the physiological and biochemical role(s) of a secondary metabolite can be very complex; however, Harborne[22] suggested that compounds of the same group share a number of similar properties.

3.1 Chemistry, Distribution and Function of Coumarins in Plants

The biosynthesis of coumarins is complex and involves several different enzymes (Murray *et al.*, 1982).[8] Hydroxycinnamic acids and coumarins are phenylpropanoids[22, 23] as they contain at least one phenylpropane (C_6C_3) structure. All of these compounds are derived from the aromatic amino acid phenylalanine. Phenylalanine is synthesized by the shikimic acid pathway and is converted by PAL to cinnamic acid, a key intermediate in phenylpropanoid biosynthesis.[24, 25] From cinnamic acid, derivatives are synthesized by the substitution of hydroxyl and methoxy groups to the aromatic ring. One such derivative synthesized is *o*-coumaric acid.

Lactonization of *o*-coumaric acid produces coumarin.[23, 25] The simplest coumarin (Figure 1) has no hydroxyl groups (i.e. all R groups are hydrogen) thus, coumarin itself does not absorb UV radiation nor does it possess an anti-oxidant potential. However, several hydroxylated derivatives of coumarin exist (Figure 2) and these compounds can absorb UV radiation,[1] scavenge free radicals,[27] and transduce UV radiation into visible light.[1]

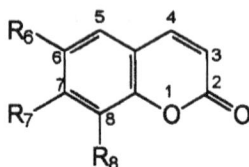

Figure 1 *The typical coumarin ring. Coumarin ($C_9H_6O_2$) has a melting point of 70 °C, a boiling point of 297 °C, and contributes to the scent of many plants.[26] It is a common volatile plant constituent and, depending upon the specific coumarin, the R groups may be H, OH, OCH$_3$, or O-Glucose. Coumarin is freely soluble in water, alcohol, and ether.[26]*

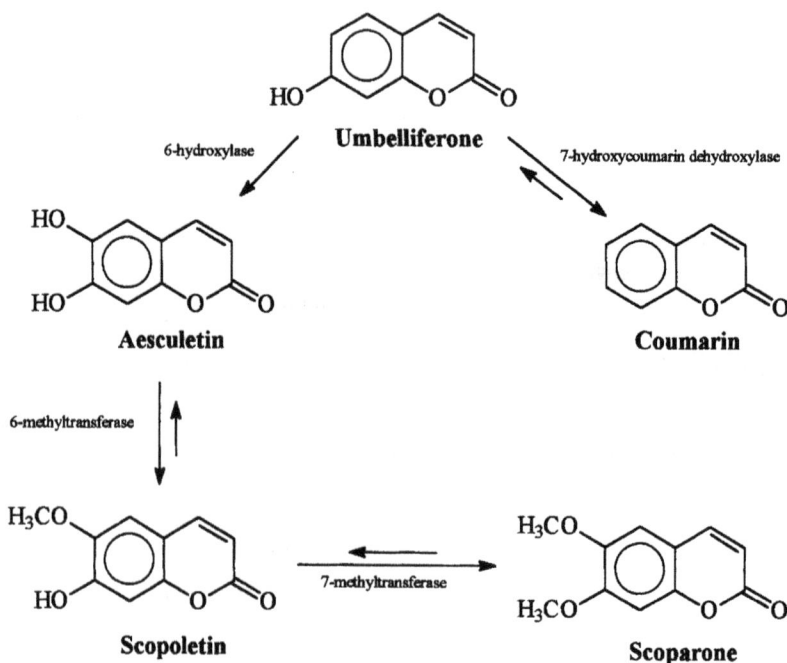

Figure 2   *The biosynthetic pathway of coumarin and its derivatives from 7-hydroxycoumarin.[8] All reactions are reversible.*

Coumarins represent one phytochemical class widely distributed in plants and Harborne[22] stated that the most widespread plant compound is the parent compound, coumarin, which occurs in over 27 plant families. Weinmann[3] reported that over 1000 coumarins have been discovered. Coumarin is synthesized and is metabolically active in shoots,[28] in green tissue, and accumulates in woody tissue during dormancy. Zobel[1] stated that coumarins can occur in all parts of some plants (e.g., *Campanula* plant) yet be restricted to specific organs in other species. Thompson and Brown[29] stated that some species (e.g., *Rutaceae, Umbelliferae*) can have more than 20 different coumarins. On the surface of plant tissues, Zobel *et al.*[30] found that several different coumarins can exist. It is important to understand that the localization of the coumarins is correlated to their role in the plant.[1] The wide distribution and different tissue locations of coumarins make these compounds an ideal class of toxins used by both primitive and higher plants for defensive purposes.

Towers and Yamamoto[31] stated that simple coumarins are phototoxic to viruses, bacteria, and mammalian cells. The coumarins serve to protect the plant against bacteria, fungi and insects.[1, 32] Vickery and Vickery[9] and Zobel[1] suggested that coumarin also acts as a growth inhibitor in plants at high concentrations. Vickery and Vickery[9] stated that at high concentrations, coumarin interferes with plant growth compounds particularly indolylacetic acid (IAA). The inhibition of growth by coumarin protects plants against fungal invasion and Vickery and Vickery[9] suggested that coumarins accumulate in cells adjacent to cells infected by disease.

Plants do not suffer from autotoxicity due to the compartmentalization of secondary metabolites. Produced at the endoplasmic reticulum, coumarin and coumarin derivatives can be transported via vesicles to the cell vacuole for storage as glucosides. Vacuolar storage of *o*-coumaric-acid and of coumarins as glucosides as well as the deposition of coumarin outside of the cell (e.g., on the tissue surface) is advantageous to the plant as it minimizes the risk of autotoxicity. In the vacuole, *o*-coumaric-acid-β-D-glucoside remains until:

- abrasion of the tissue (e.g., by herbivores) occurs and the glucosidases located in the cell wall remove the sugar constituent and the compound spontaneously cyclizes to coumarin; and/or,

- coumarin is released via exocytosis into the inter-cellular spaces (i.e. to kill invading bacteria or viruses) or is extruded onto the surface of the root or leaf to kill bacteria, spores, and render an unappealing taste to animals.

4 FREE RADICALS

A free radical is a species containing one or more unpaired electrons. The highly reactive nature of free radicals makes them potentially dangerous particularly if they are not tightly controlled. Pine[33] (1987) suggested that free radicals can be formed in three general ways:

1. Through homolytic cleavage of $A—B \longrightarrow A^{\bullet} + B^{\bullet}$

2. Molecular reaction with other free $A—B + X^{\bullet} \longrightarrow A^{\bullet} + B—X$

3. Addition of a free radical to an

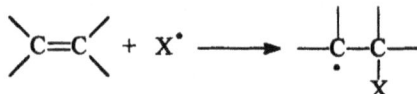

It should be noted that reaction (1) may be induced thermally[30] or photochemically;[31] reaction (2) can be hydrogen ion transfer.

Free radicals are typically short-lived, but one radical can generate other free radicals until a termination reaction occurs (e.g., combination of two free radicals). Targets of free radicals include proteins, DNA,[35-37] carbohydrates and the phospholipid component of cellular membranes.[34, 38, 39] Depending upon the cellular or extracellular target, free radical damage can result in: 1) modified ion transport; 2) modified enzyme activity; 3) mutations and translational errors; and, 4) modification of the structural and functional integrity of the membrane. Most free radicals have a chronic effect on cells.

Antioxidant is a term given to a compound that scavenges free radicals. All radical scavengers are electron donors and possess at least one hydroxyl group. Zobel[19] stated that antioxidants were necessary for plants from the very beginning of land colonization. The large structure and numerous hydroxyl groups of the flavonoids make these compounds excellent antioxidants.[40]

5 UV RADIATION ABSORBED BY PHENOLIC COMPOUNDS

The presence of hydroxyl groups enables a secondary metabolite to absorb UV irradiation[13] as well as scavenge free radicals.[34, 38, 39, 41] From Murray et al.,[8] it was known that different phenolic compounds have different spectral peaks. Zobel[19] suggested that a greater variety of phenolic compounds in any plant would increase the range of its UV absorption capacities.

Secondary metabolites are phenolic compounds that can possess one (phenolic acids), two (coumarins), three (flavonoids and anthocyanins), or more (tannins) benzene rings in the structure. Each carbon atom in a benzene ring has a 2p orbital and these orbitals combine (overlap) to form six B orbitals that are spread uniformly over the ring.[42] When like charges overlap the bonding B orbital results, but when opposite signs overlap the antibonding B* orbital is formed (Figure 3). Solomons[43] stated that the antibonding B* orbital, in comparison to the bonding B orbital, is of a higher energy and is not occupied by electrons when the molecule is in its ground state. However, the B* orbital can become occupied if the molecule absorbs light or radiation of the right frequency and an electron is promoted from the lower energy level to the higher one (i.e. an excited state). After a compound has absorbed UV, which is high-energy radiation, the compound acquires energy, and to revert to the ground state this energy must be released. The amount of energy re-radiated equals the amount absorbed.[44] This can be accomplished by emission of heat or longer wavelength radiation, but if this is not done there is an increase in reactivity, and free radicals may be formed.[1, 33, 34]

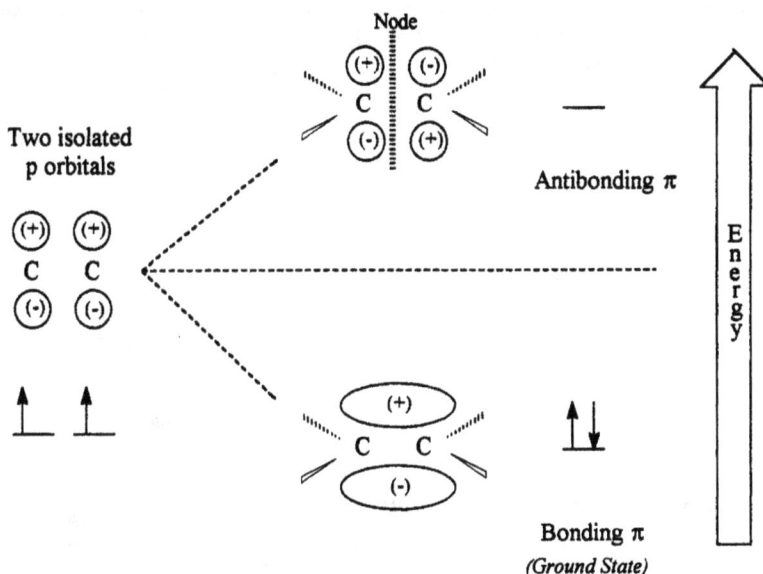

Figure 3 *(as from 43, p. 59) Two p orbitals combine to form two B orbits. The bonding B orbital is of a lower energy and contains both B electrons (with opposite spins) in the ground state. In comparison, the higher energy antibonding B* orbital is not occupied by electrons in the ground state. Excitation of the molecule by light or radiation of a suitable frequency can promote an electron from the lower energy level to the higher energy level.*

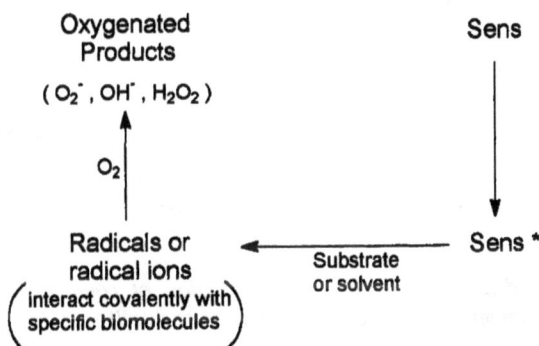

Figure 4 *(as from ref. 37, p. 406). Type I phototoxic mechanism of action that occurs in coumarin. UV-A radiation renders the greatest toxicity of the excited coumarin compound.*

Coumarin is an example of a compound that has an abundance of B electrons and can be excited by UV-A radiation. Downum[37] suggested that the phototoxic mechanism of coumarins is via a Type I mechanism (Figure 4). In the excited state coumarin can react with a suitable substrate to form radicals (Figure 5). Hydroxylated derivatives of coumarin can be scavengers of free radicals *in vitro*.[27]

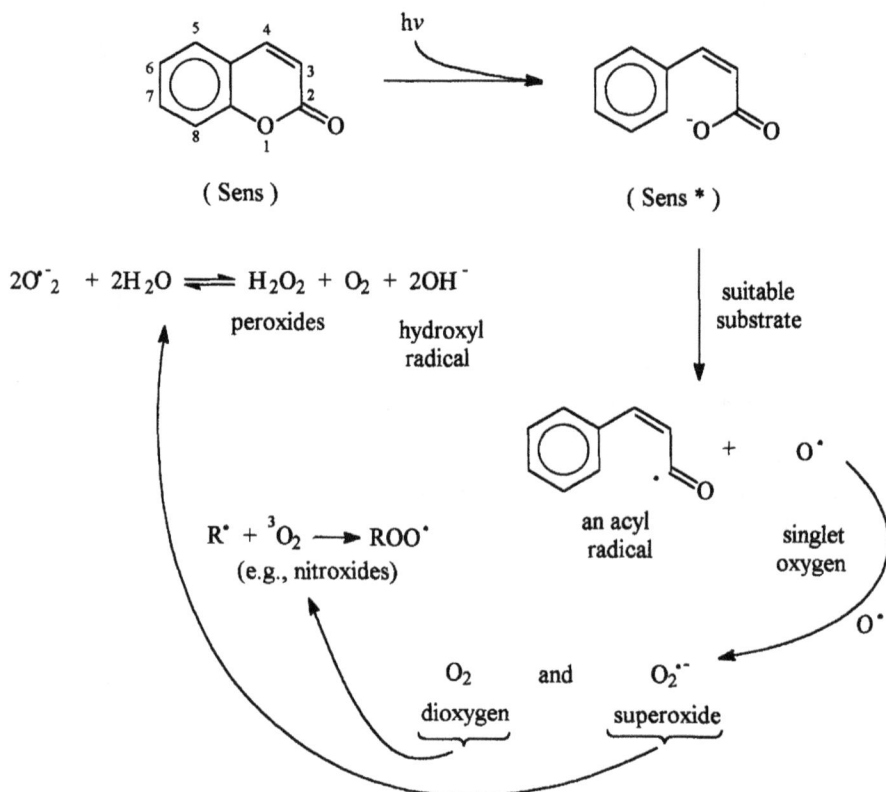

Figure 5. *The radicals produced by the excitation of coumarin by UV-A radiation. Dioxygen (O$_2$) is dangerous in excess (i.e. toxic to the cell) and may combine with other radicals or broaden the effects of other radicals.[34] Rice et al.[34] stated that singlet oxygen reacts as an electrophile whereas superoxide is a weak base and has low reactivity as a radical. The hydroxyl radicals can damage DNA whereas the peroxide radicals damage unsaturated membrane components.*

References

1. A. M. Zobel, in 'Phytochemistry of Fruit and Vegetables', F. Tomas-Barberan and R. Robins, Eds., Clarendon Press, Oxford, 1997, p.173.
2. B. Starvic, and T. Matula, in 'Lipid Soluble Antioxidants: Biochemistry and Clinical Applications', A. Ong and L. Packer, Eds., Birkhauser Verlag, Boston, 1992, p. 274.
3. I. Weinmann, in 'Coumarins: Biology, Applications and Mode of Action', R. O'Kennedy and R. Thornes , Eds., John Wiley and Sons, Toronto, 1997, p. 1.
4. K. Stern, 'Introductory Plant Biology, Fourth Edition', W. C. Brown Publishers, Dubuque, Iowa, 1988.
5. A. M. Zobel, J. Crellin, S. A. Brown and K. Glowniak, *Acta Horticult.*, 1994, **381**, 510.
6. E. Wollenweber and U. Dietz, *Phytochem.*, 1981, **20**, 869.
7. A. M. Zobel and S. A. Brown, *J. Nat. Prod.*, 1988, **51**, 8966.
8. R. Murray, J. Mendez, and S. A. Brown, S.A. 'The Natural Coumarins: Occurrence, Chemistry and Biochemistry', John Wiley and Sons, Toronto, 1982.
9. M. Vickery and B. Vickery, 'Secondary Plant Metabolism', The Macmillan Press, London, 1991.
10. E. Baker, in 'The Plant Cuticle', D. Cutler, K. Alvin, and C. Priece, Eds., Academic Press, Toronto, 1982, p. 139.
11. J. M. Lynch, 'The influence of Enhanced Low-intensity UV Radiation on the Phenolic Content of *Brassica Oleracea* Leaves', Master's Thesis, Trent University, 1998.
12. E. Wollenweber, in 'Studies in Organic Chemistry: Flavonoids, Bioflavonoids', L. Farkas, M. Gabor, and F. Kallay, Eds., Elevier, Amsterdam, 1986, p. 155.
13. M. Rhodes, *Ann. Proc. Phytochem. Soc. Europe*, 1985, **25**, 99.
14. J. Harborne, 'Biochemistry of Phenolic Compounds', Academic Press, New York, 1964.
15. J. B. Harborne, *Annu. Proc. Phytochem.. Soc. Europe*, 1985, **25**, 393.
16. Y. P. Cen and J. F. Bornman, *Physiol. Plant.*, 1993, **87**, 249.
17. A. M. Zobel and J. M. Lynch, *Allelopathy J.*, 1997, **4**, 1.
18. A. M. Zobel and S. A. Brown, S.A.,. *J. Chem. Ecol.*, 1993, **19**, 939.
19. A. M. Zobel, in 'Phenolic Compounds as Bioindicators of Air Pollution', in R. A. Yunes and R. Iqbal, Eds., John Wiley and Sons, Chichester, 1996, p. 100.
20. R. Brouillare, O. Dangles, M. Jay, M., J. Biolley and N. Chirol in 'Polyphenolic Phenomena', A. Scalbert, Ed., INRA Editions, Paris, 1993, p.41.
21. J. Harborne, 'Comparative Biochemistry of the Flavonoids', Academic Press, New York, 1967.
22. J. B. Harborne, 'Phytochemical Methods: A Guide to Modern Techniques of Plant Analysis, Second Edition'. Chapman and Hall, New York, 1984.
23. H. Street and W. Cockburn, 'Plant Metabolism, Second Edition', Pergamon Press, Toronto, 1972.
24. S. A. Brown, S.A., in 'The Biochemistry of Plants: Secondary Plant Products', E.E. Conn, Ed., Academic Press, Toronto, 1981, p. 269.
25. G. Keating, and R. O'Kennedy, in 'Coumarins: Biology, Applications and Mode of Action', R. O'Kennedy and R. Thornes, Eds., John Wiley and Sons, Toronto, 1997, p. 23.
26. S. Budavari, 'The Merck Index, Twelfth Edition', Merck & Co. Inc.; Whitehouse Station, NJ, 1996.

27. M. Paya, B. Halliwell and J. Hoult, *Biochem. Pharmacol.*, 1992, **44**, 205.

28. T. Kosuge, and E. Conn, *J. Biol. Chem.*, 1959, **234**, 2133.

29. H. Thompson and S. A. Brown, *J. Chromatog.*, 1984, **314**, 323.

30. A. M. Zobel, S. A. Brown and J. E. Nighswander, *J. Botany*, 1991, **67**, 213.

31. G. Towers and E. Yamamoto, *Annu. Proc. Phytochem. Soc. Europe*, 1985, **25**, 271.

32. O. Ceska, S. Chaudhary, P. Warrington, G. Poulton and M. Ashwood-Smith, *Experimentia*, 1986, **42**, 1302.

33. S. Pine, 'Organic Chemistry', Fifth Edition. McGraw, Inc., Toronto, 1987.

34. C. Rice-Evans, A. Diplock and M. Symons, 'Techniques in Free Radical Research', Elsevier, New York, 1991.

35. R. Mason, K. Stolze and W. Flitter, in 'Antimutagenesis and Anticarcinogenesis Mechanisms II', Y. Kuroda, D. Shankel, and M. Waters, Eds., Plenum Press, New York, 1990, p. 119.

36. M. Minnunni, U. Wolleb, O. Mueller, A. Pfeifer and H. Aechbacher, *Mutation Res.*, 1992, **269**, 193.

37. K. Downum, *New Phytologist*, 1992, **122**, 401-420.

38. N. Das and L. Ramanathan, in 'Lipid Soluble Antioxidants: Biochemistry   and Clinical Applications', A. Ong and L. Packer, Eds., Birkhauser Verlag, Boston, 1992, p. 295.

39. T. Okuda, in 'Polyphenolic Phenomena', A. Scalbert, Ed., INRA Editions, Paris, 1993, p. 221.

40. N. Salah, N. Miller, G. Paganga, L. Tijburg, G. Bolwell and C. Rice-Evans, *Arch. Biochem. Biophys.*, 1995, **322**, 339.

41. T. Osawa, M. Namiki and S. Kawakishi, S. in 'Antimutagenesis and Anticarcinogenesis Mechanisms II', Y. Kuroda, D. Shankel, and M. Waters, Eds., Plenum Press, New York, 1990, p.139.

42. D. McQuarrie and P. Rock, 'General Chemistry, Second Edition', W.H. Freeman and Company, New York, 1987.

43. T. Solomons, 'Organic Chemistry, Fifth Edition', John Wiley and Sons, Inc., Toronto, 1992.

44. 'Compton's Interactive Encyclopedia', Compton's New Media Inc., Carlsbad, California, 1995.

Photostabilities of Several Chemical Compounds used as Active Ingredients in Sunscreens

John M. Allen, Sandra K. Allen and Brian Lingg
Department of Chemistry
Indiana State University
Terre Haute, IN 47809, USA

1 INTRODUCTION

Sunscreens contain a variety of active ingredients prepared in a cream, oil, or lotion formulation. When applied to skin, these formulations are designed to attenuate potentially harmful solar radiation that reaches the skin. Inorganic active ingredients (*e.g.* titanium dioxide), which are referred to as physical blockers, mainly reflect and scatter solar radiation. Organic active ingredients (*e.g.* 2-ethylhexyl-4-methoxycinnamate) are referred to as organic absorbers because they are designed to absorb solar radiation. We will address issues associated with the photostabilities of organic absorbers in this discussion.

One might well argue that an examination of the photochemical stabilities of organic compounds used as active ingredients in sunscreens is an appropriate starting point for a general discussion of the photostabilities of drugs. After all, sunscreens are specifically designed to absorb UV-A (320 to 400 nm) and UV-B (290 to 320 nm) radiation; the most energetic solar radiation that reaches the Earth's surface. In order to be effective, sunscreens must absorb this radiation and dissipate the acquired energy in a manner that does not lead to destruction of the absorbing molecules (photolysis) nor to the formation of toxic products. In general, sunscreens are remarkably effective at accomplishing this rather daunting task.

Photolysis of organic absorbers is of concern because it may lead to a reduction in efficacy (*i.e.* loss of absorptivity), formation of reactive intermediates (*e.g.* free radicals), or formation of stable photoproducts. The formation of reactive intermediates and stable photoproducts is of concern because these species may be toxic. Indeed, loss of efficacy and formation of toxic photoproducts are two of the central issues associated with drug photostability in general.

Historically, one of the most widely used sunscreen active ingredients has been *p*-aminobenzoic acid (PABA). The use of PABA, however, has declined significantly due to reports of adverse dermatologic effects. Previous investigations into the photochemistry of PABA have centered upon elucidation of stable photoproducts formed when PABA is illuminated in solution.[1-3] Evidence has also been provided for the formation of several reactive intermediates that are produced during photolysis of aqueous PABA solutions.[1,3] These reactive species include a variety of carbon-centered radicals, nitrogen-centered radicals, hydrogen atoms, and aquated electrons (e_{aq}^-). Subsequent reactions of these species with oxygen undoubtedly lead to the formation of a variety of oxidants, many of which are known to produce toxic effects. In two previous studies,[4,5] it was reported that PABA both forms and rapidly reacts with singlet molecular oxygen $^1O_2$, an oxidant that is presumably formed by energy transfer from photoexcited PABA molecules to oxygen.

The photochemistry of 2-ethylhexyl-4-methoxycinnamate (Octylmethoxycinnamate) has been previously investigated.[6] This compound was found to undergo a *trans-cis* isomerization with a fairly high quantum yield ($\phi > 0.5$) and to quench reactive excited state triplets of other molecules. No other photoproducts were reported. In another report of Octylmethoxycinnamate photochemistry,[7] the compound was shown to undergo a [2+2] cycloaddition reaction yielding a dimeric product.

The photochemistry of Parsol-1789, Benzophenone-3, and Padimate-O has been previously investigated using a Hg vapor lamp.[8] It must be noted that an Hg vapor lamp has an emission spectrum that is quite different from natural sunlight. In these experiments, Parsol-1789 and Padimate-O were found to undergo photolysis yielding several stable photoproducts. The authors proposed mechanisms for these photoreactions involving the formation of free radical intermediates. Benzophenone-3 was found to undergo no photolysis.

We report data for the photostabilities of several sunscreen active ingredients in aqueous solutions using simulated sunlight. Chemical actinometry was used to allow normalization of results obtained from simulated sunlight experiments to actual sunlight illumination. Some of the evaluated sunscreen active ingredients were found to be quite stable while others undergo photolysis in simulated sunlight. In addition, photolysis quantum yields were determined at 313 nm for the evaluated sunscreen active ingredients. The following sunscreen active ingredients were evaluated: *p*-aminobenzoic acid (PABA), 2-ethylhexyl-4-(dimethylamino)benzoate (Padimate-O), 2-ethylhexyl-4-methoxycinnamate (Octylmethoxycinnamate), 2-ethylhexyl-2-cyano-3,3-diphenylacrylate (Octocrylene), 2-ethylhexylsalicylate (Octylsalicylate), 2-hydroxy-4-methoxybenzophenone (Benzophenone-3), 2,2′-dihydroxy-4-methoxybenzophenone (Benzo-phenone-8), 1-[4-(1,1-dimethylethyl)phenyl]-3-(4-methoxyphenyl)-1,3-propanedione (Parsol-1789), and 2-aminobenzoic acid, menthyl ester (menthyl anthranilate).

2 EXPERIMENTAL

2.1 Reagents

PABA and Benzophenone-8 were obtained from TCI America, Padimate-O, Benzo-phenone-3, Octocrylene, Octylmethoxycinnamate, and Octylsalicylate were obtained from ISP Van Dyk. Parsol-1789 was obtained from Givaudan. Menthyl anthranilate and valerophenone were obtained from Aldrich. Acetonitrile was obtained from Burdick and Jackson. All reagents used were of the highest purity available and were used as received.

2.2 Instrumentation

Solar simulators are designed and constructed to provide illumination with a spectral energy distribution that closely matches terrestrial sunlight[9]. A solar simulator used for illumination of samples in the experiments reported here incorporating a 1000 W Xe arc lamp, optical bench, and sample illumination chamber was obtained from Spectral Energy Corp. The lamp output was filtered through a water filter with quartz windows to remove most of the IR radiation and optical filters to remove wavelengths below 290 nm. The output of the illumination system was focused onto the face of a 1 cm quartz cuvette (NSG Precision Cells) that was thermally equilibrated with a constant temperature water bath (Haake D1) at 25 °C. A magnetic stirrer was mounted under the cuvette so that the samples could be stirred while being illuminated. An electric shutter was controlled by a darkroom timer (Dimco Gray, Model 900) to provide precise control of illumination times.

A monochromatic illumination system (Spectral Energy Corp.) incorporating a 1000 W Hg/Xe arc lamp, the output of which was passed through a monochromator, was used to make photolysis quantum yield determinations at 313 nm. This instrument was constructed so that samples were illuminated under temperature-controlled conditions with continuous stirring in a similar fashion as with the solar simulator described above.

The high performance liquid chromatograph (HPLC) instrument that was used in all experiments described herein was constructed using a Spectra-Physics Model 200 pump, an Applied Biosystems Model 785 UV detector, and a Rheodyne Model 7125 injector. The HPLC was operated isocratically in reversed-phase using an Alltech 250 mm 5mm C_{18} column and helium-purged acetonitrile/water binary solvent system in all experiments. A C_{18} pre-column and filter were used to protect the analytical column.

UV-Visible spectra were obtained using a Shimadzu UV-2101 spectrophotometer with 1 cm quartz cells (NSG Precision Cells). All spectra for solutions of sunscreen active ingredients were recorded versus the solvent system used to prepare the solutions.

Purified water used for preparing aqueous photolysis solutions was generated by the use of in-house deionized water that was further purified using a Barnstead E-Pure laboratory water purification system with an activated carbon cartridge providing water of >18 MW resistivity.

2.3 Photochemical Experiments

The general procedure that was followed to measure sunscreen active ingredient photostability in the solar simulator was to first prepare an aqueous solution having an initial sunscreen active ingredient concentration of 1×10^{-5} M. In some cases, CH_3CN was added as a co-solvent in order to dissolve the sunscreen active ingredient. The solutions were not de-gassed so that they remained air-saturated. One of the solutions was then placed in a teflon stoppered 1 cm path length quartz cuvette and stirred with a teflon coated magnetic stir bar in the temperature controlled sample chamber of the solar simulator with the shutter closed. An aliquot of the solution was then removed from the cuvette and placed in the dark. Another aliquot was analyzed by HPLC to determine the initial concentration of sunscreen active ingredient prior to any illumination. The sample was then illuminated with the solar simulator under temperature controlled conditions while being stirred continuously. The reaction progress was followed by periodically removing a small aliquot of the solution and determining the concentration of the sunscreen active ingredient. The loss of the active ingredient in the aliquot of the solution that was placed in the dark was also determined after a period of time comparable to the duration of the photochemical experiment. Loss of the sunscreen active ingredient was monitored by HPLC using UV-Visible detection. The detection wavelength used was determined by measurement of UV-Visible spectra for each sunscreen active ingredient compound.

A chemical actinometer was used to measure the irradiance incident upon samples on each day that experiments were conducted so that results of experiments conducted on different days could be compared and so that solar simulator experiments could be normalized to sunlight irradiance. An aqueous solution of valerophenone has been used as a sunlight-range chemical actinometer in previous studies[10,11] and was used for that purpose in all experiments reported herein. A 1×10^{-5} M aqueous valerophenone solution was illuminated under identical conditions as the sunscreen active ingredient solutions described above. The loss of valerophenone was followed with illumination time by HPLC.

A portion of the valerophenone actinometer solution described above was transferred to a stoppered quartz tube. An aliquot of this solution was removed and the initial valerophenone concentration was determined. Another aliquot was set aside in the dark. The quartz tube was then placed in direct, clear-sky sunlight at solar noon during the autumnal equinox in Terre Haute, IN (approximately mid-latitude, U.S.) at 22°C. Aliquots of the solution were removed at regular time intervals and the loss of valerophenone was followed by HPLC.

The procedure used to measure quantum yields for sunscreen active ingredient photolysis was quite similar to that described above for the solar simulator experiments except that the solutions containing sunscreen active ingredients were illuminated using the monochromatic illumination system at 313 nm. The valerophenone actinometer solution was also separately illuminated at 313 nm on each day an experiment was conducted under identical conditions. The loss of the sunscreen active ingredient with illumination time was again followed by HPLC as was loss of valerophenone.

3 RESULTS

3.1 Optical Absorption of Sunscreen Active Ingredients

Chemical compounds that are used as sunscreen active ingredients are typically strong absorbers of either UV-A or UV-B radiation. Absorption spectra of the evaluated sunscreen active ingredients are presented in Figures 1 and 2. Historically, sunscreen active ingredient development has been directed toward those compounds that are effective in the UV-B region. This was primarily due to the belief that only the UV-B region is energetic enough to cause injury to the skin. More recent work has shown that UV-A radiation can also inflict damage. Therefore, a great deal of effort has been expended on the part of sunscreen manufacturers of late to develop effective UV-A absorbers. With the exceptions of Parsol-1789 and Mexoryl (not included in this study), both of which are relatively new sunscreen active ingredients, the UV-A absorbers generally have low molar absorptivities (extinction coefficients).

3.2 Photolysis of Sunscreen Active Ingredients in the Solar Simulator

Each solution containing a sunscreen active ingredient was illuminated in the solar simulator using the procedure described above. Octylsalicylate, Benzophenone-3, and

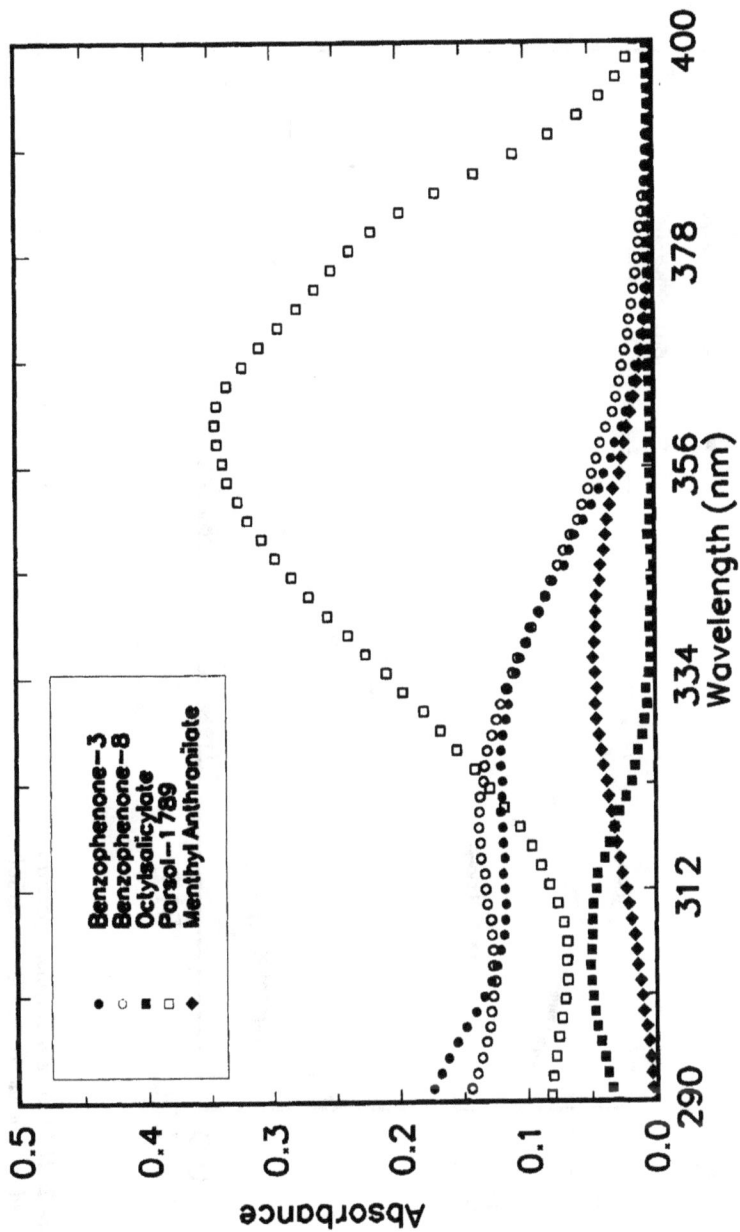

Figure 1 *UV-Visible absorption spectra of 1x10⁻⁵ M aqueous solutions of several UV-A sunscreen active ingredients. Spectra were obtained in 1 cm quartz cuvettes versus solvent.*

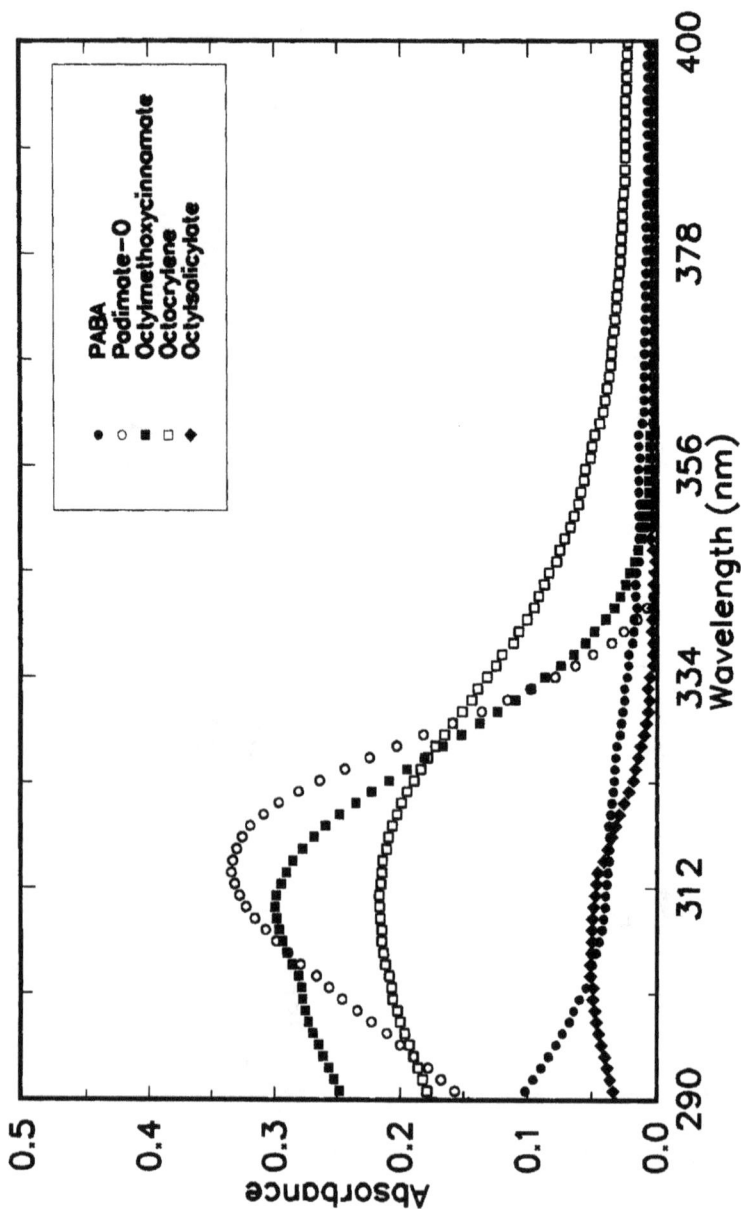

Figure 2 *UV-Visible absorption spectra of 1x10^-5 M aqueous solutions of several UV-B sunscreen active ingredients. Spectra were obtained in 1 cm quartz cuvettes versus solvent.*

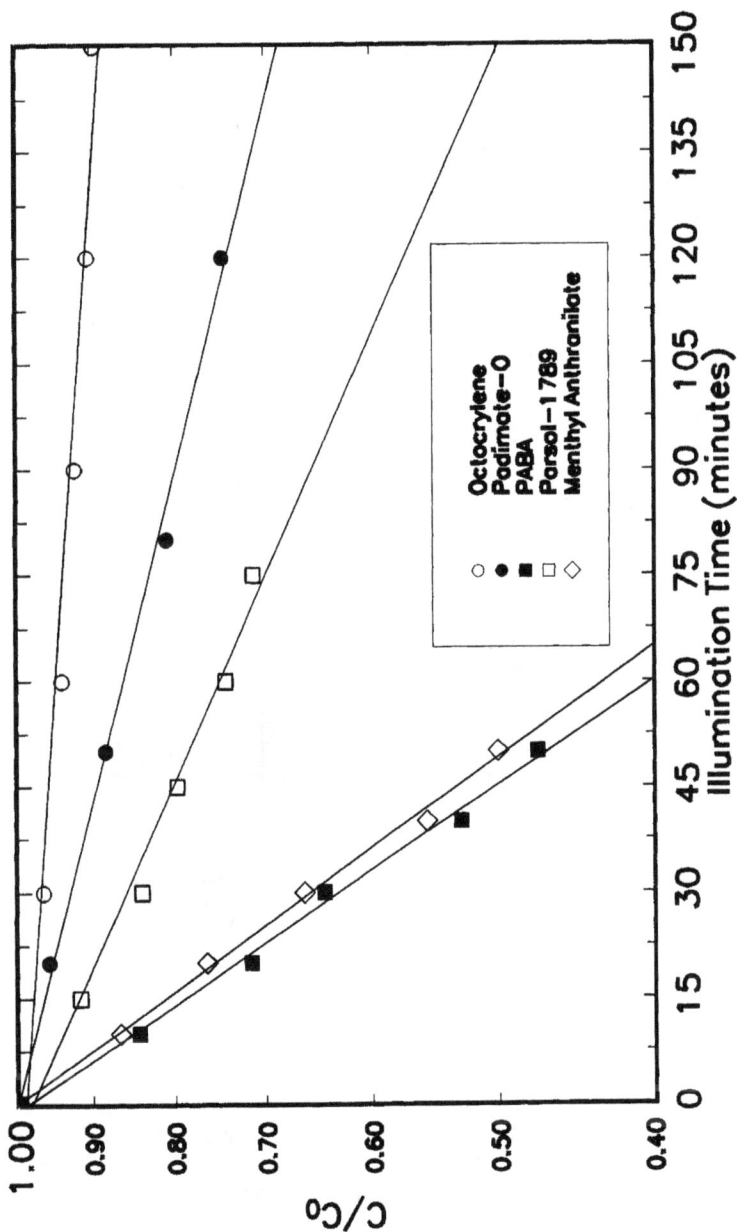

Figure 3 *Photolysis of several individual* 1×10^{-5} *M aqueous solutions of sunscreen active ingredients in the solar simulator. Photolyses were carried out in 1 cm quartz cuvettes at 25 °C with constant stirring.*

Benzophenone-8 did not undergo any measurable photolysis. Photolysis of each of the other sunscreen active ingredients followed first-order kinetics in all cases (Figure 3) with the exception of Octylmethoxycinnamate. In the case of Octylmethoxycinnamate, there was a rapid initial first-order reaction (presumably *trans-cis* isomerization) leading to a photostationary state in which no further reaction was observed. Photostability data are presented in Table 1 for all evaluated compounds except for Octylmethoxycinnamate.

In the Table, $j_{Sol.Sim}$ is the experimentally determined first-order rate constant for photolysis of each of the sunscreen active ingredients in the solar simulator, j_{SUN} is the normalized first-order rate constant for photolysis in sunlight, and $T_{1/2,SUN}$ is the estimated half-life for sunscreen active ingredients when illuminated in sunlight. It should be noted that, since solar irradiance is dependent upon time of year, time of day, latitude, and factors such as cloud cover and haze, values for j_{SUN} and $T_{1/2,SUN}$ are variable.

Table 1 *Photolysis of sunscreen active ingredients in the solar simulator at 25 °C*

Compound	$j_{Sol.Sim}$ $(x10^4 \, sec^{-1})$	j_{SUN} $(x10^5 \, sec^{-1})$	$T_{1/2,SUN}$
PABA	2.50	3.72	5.2 hours
Padimate-O	0.42	0.63	31 hours
Octocrylene	0.13	0.20	96 hours
Octyl salicylate	0	0	No Photolysis
Benzophenone-3	0	0	No Photolysis
Benzophenone-8	0	0	No Photolysis
Parsol-1789	0.75	1.12	17 hours
Menthyl anthranilate	2.2	3.28	5.9 hours

3.3 Determinations of Quantum Yields (ϕ) for Photolysis of Sunscreen Active Ingredients at 313 nm

Determinations of quantum yields for photolysis of sunscreen active ingredient compounds were made using the monochromatic illumination system at 313 nm. This wavelength was chosen because it is present as an intense line in the Hg/Xe arc lamp emission spectrum. The following equation was used to calculate values for ϕ:

$$\phi_{S,313} = \left(\frac{j_{S,313}}{j_{val,313}}\right)\left(\frac{\epsilon_{val,313} \; \phi_{val,313}}{\epsilon_{S,313}}\right)$$

where $\phi_{S,313}$ is the quantum yield for sunscreen photolysis at 313 nm, $j_{S,313}$ is the first-order rate constant for sunscreen photolysis at 313 nm (sec^{-1}), $j_{val,313}$ is the first-order rate constant for valerophenone photolysis at 313 nm (sec^{-1}), $\epsilon_{val,313}$ is the molar absorptivity for valerophenone at 313 nm (101 M^{-1} cm^{-1}),[11] $\phi_{val,313}$ is the quantum yield for valerophenone photolysis at 313 nm (1.00),[11] and $\epsilon_{S,313}$ is the molar absorptivity for the sunscreen at 313 nm (M^{-1} cm^{-1}). Quantum yield data for all evaluated compounds are presented in Table 2. Photolysis was first-order in all cases.

No attempt was made to evaluate the wavelength or temperature dependencies of the photolysis quantum yields for the evaluated sunscreen active ingredients.

Table 2 *Quantum yields for direct photolysis of sunscreen active ingredients at 313 nm*

Compound	$j_{S,313}$ (x10^5 sec^{-1})	ϵ_{313} (M^{-1} cm^{-1})	ϕ_{313}
PABA	2.49	3,512	0.0010
Padimate-O	50.1	32,011	0.0023
Octocrylene	0	21,010	0
Octyl salicylate	0	4,020	0
Benzophenone-3	0	12,009	0
Benzophenone-8	0	13,022	0
Parsol-1789	13.5	8,020	0.0024
Menthyl anthranilate	1.06	2,405	0.0006

4 CONCLUSIONS

One of the important consequences that can arise as a result of sunscreen active ingredient photolysis is a loss of absorptivity. This may, in some cases, be significant if the photoproducts do not absorb radiation in the desired wavelength range. In other cases, absorptivity is retained because the photoproduct or products absorb this radiation. For example, Parsol-1789 undergoes photolysis leading to products that have little or no absorptivity in the wavelength region of interest. As a consequence, photolysis of this active ingredient is likely to lead to a reduction in efficacy. However, as can be seen from an examination of Table 1, photodegradation is quite slow in all cases except for PABA and menthyl anthranilate. On the other hand, Octylmethoxycinnamate undergoes a rapid photochemical reaction presumably involving isomerization from the *trans* to the *cis* isomer. In this case, much of the absorptivity is retained since the *cis* form is a strong absorber of radiation, although the *cis* form has a slightly lower molar absorptivity and there is a shift in λ_{max}. Significant formation of the dimeric product may further reduce Octylmethoxycinnamate absorptivity.

Sunscreen manufacturers have attempted to formulate active ingredients in such a way as to minimize any photodegradation. This strategy involves the use of formulations that can effectively stabilize active ingredients. This is possible because energy from photo-excited active ingredient molecules can be transferred to other molecules present in the formulation, thus removing enough energy so that photolysis does not occur. On the other hand, it is certainly possible that sunscreen active ingredients are effectively *destabilized* by other ingredients in some formulations. Of course, the results presented here do not take into account any stabilization or destabilization of active ingredients that may result from interactions with other components in formulations.

Photolysis of the evaluated sunscreen active ingredients under sunlight illumination conditions appears to be quite slow (except for PABA and menthyl anthranilate), and creative formulations may be effective at reducing rates of photolysis even further. This indicates that any loss of efficacy is likely to be minimal. However, a variety of reactive intermediates and stable photoproducts are formed whenever *any* photolysis occurs. Even though these reactive intermediates and stable products are formed at very low concentrations, it remains to be demonstrated whether or not the formation of these species is capable of producing significant toxic effects during actual sunscreen usage. There is the additional concern (not addressed here) that some sunscreen active ingredient molecules can transfer energy gained from sunlight absorption to other molecules (*e.g.* O_2, DNA, or other added formulation ingredients) without undergoing photolysis themselves. It is possible that these energy transfer reactions may lead to other toxic effects.

The results presented here allow comparison of the photostabilities of several sunscreen active ingredients in the same solvent system under carefully controlled simulated sunlight illumination conditions. The results indicate that PABA, menthyl anthranilate, Parsol-1789, Padimate-O, and Octocrylene undergo sunlight photolysis in aqueous solution with estimated half-lives of roughly 5 to 96 hours. The quantum yields for photolysis at 313 nm are quite small and similar in magnitude. Octyl salicylate, Benzophenone-3, and Benzophenone-8 are photostable and were found to undergo no detectable photolysis.

References

1. C. F. Chignell, B. Kalyanaraman, R. P. Mason, and R. H. Sik, *Photochem. Photobiol.*, 1980, **32**, 563.
2. F. Gasparro, *Photodermatology*, 1985, **2**, 151.

3. A. A. Shaw, L. A. Wainschel, and M. D. Shetlar, *Photochem. Photobiol.*, 1992, **55**, 647.

4. J. M. Allen, C. J. Gossett, and S. K. Allen, *J. Photochem. Photobiol. B: Biol.*, 1996, **32**, 33.

5. J. M. Allen, S. Egenolf, and S. K. Allen, *Biochem. Biophys. Res. Commun.*, 1995, **212**, 1145.

6. P. Morliere, O. Avice, T. Melo, L. Dubertret, M. Giraud, and R. Santus, *Photochem. Photobiol.*, 1982, **36**, 395.

7. J. K. Broadbent, B. S. Martincigh, M. W. Raynor, L. F. Salter, R. Moulder, P. Sjoberg, and M. E. Markides, *J. Chromatog. A*, 1996, **732**, 101.

8. N. M. Roscher, M. K. O. Lindermann, S. B. Kong, C. G. Cho, and P. Jiang, *J. Photochem. Photobiol. A: Chem.*, 1994, **80**, 417.

9. B. C. Faust, *Rev. Sci. Instrum.*, 1993, **64**, 577.

10. B. C. Faust and J. M. Allen, *J. Geophys. Res.*, 1992, **97**, 12913.

11. R. G. Zepp, M. M. Guntz, D. J. Bertino, and W. L. Miller, *203rd ACS Nat. Meeting, Div. Environ. Chem.*, San Francisco, CA, 1992.

An Analytical and Structural Study of the Photostability of some Leukotriene B4 Antagonists

Colin Orford,[a] Michael L. Webb,[a]*, Kaye H. Cattanach,[b] Frank H. Cottee,[b] Richard E. Escott,[b] Ian D. Pitfield,[b] and Jeffrey J. Richards.[b]
a) SmithKline Beecham Pharmaceuticals
New Frontiers Science Park (North)
Third Avenue, Harlow Essex, CM19 5AW, UK.
b) SmithKline Beecham Pharmaceuticals
Old Powder Mills, nr. Leigh
Tonbridge Kent, TN11 9AN, UK.

1 INTRODUCTION

Although many pharmaceutical compounds are know to be unstable to light,[1] few studies have detailed the precise chemical and physical properties that determine the susceptibilities and the resulting degradation products and their levels.[2,3] Although oral pharmaceutical products may be stored in amber bottles and topical products in opaque tubes, topical products are spread onto skin where they may become exposed to sunlight and hence protection for such products by the packaging is not the only consideration and factors around the stability of the product in use should be considered.

It is important that modern analytical assay and impurity profile methods for drug substances and products be stability indicating in that they should be selective for all known degradation products. In this way, a detailed and rational study of the susceptibility of the products to photodegradation may be made in order to provide important information to help design a stable dosage form.

Compounds (1), (2) and (3) shown in Figure 1, are Leukotriene B_4 antagonists which have been in development for use as potential topical drugs for the treatment of psoriasis. Since psoriatic patients often expose their skin to sunlight to improve their condition, the light stability of any topical drug indicated for Psoriasis is very important. The guideline for testing the light-stability of drugs in the solid state is exposure for 1.2 million lux hours.[4] This is usually carried out by exposing a sample of the drug to high intensity light sources such as a xenon discharge lamp for up to 24 hours. The conditions are designed to simulate sunlight through a glass window by using a suitable glass filter giving significant radiation at wavelengths longer than 320.5 nm.[5, 6]

Compounds (1), (2) and (3), were exposed for varying periods in solution in acetonitrile and in the solid state to both xenon light and south light. The latter exposure was carried out by placing the samples of a south facing window ledge. After irradiation the samples were analysed using a number of methods. A relative assay and impurity profile for related compounds was carried out using high performance liquid chromatography (HPLC) and backed up by capillary electrophoresis (CE). Gel permeation chromatography (GPC) and thin layer chromatography(TLC) were also used to search for potential high molecular weight and non-ultraviolet radiation absorbing degradation products. These separation techniques were used in conjunction with a variety of detection systems such as ultraviolet (UV), fluorescence and refractive index (RI). Extensive use of coupled liquid chromatography/mass spectrometry (LC/MS) was also made. Other spectroscopic techniques such as infrared spectroscopy (IR) and nuclear magnetic resonance spectroscopy (NMR) were used to look for any physical and chemical changes.

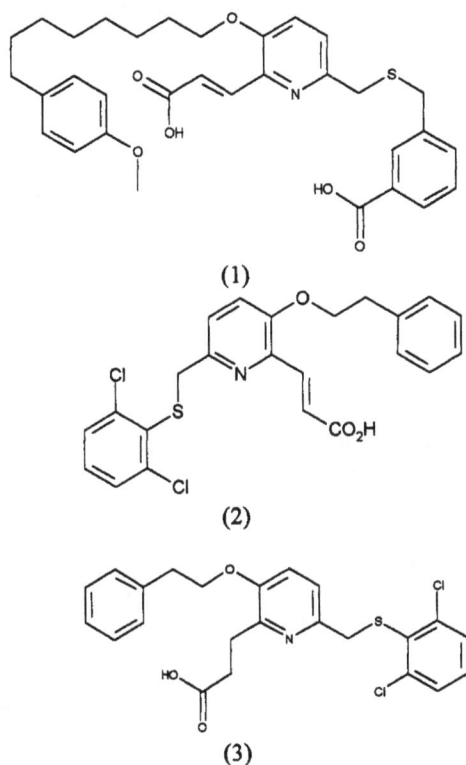

(1)

(2)

(3)

Figure 1 *Structure of (1), (2), and (3).*

The aim of this work was to assess the light stability of the test compounds in solution and in the solid state and to identify and account for all the degradation products. This information was used to try and explain the chemical and physical properties that affect the light stability. A wide range of techniques were used since, even with structurally related compounds, a varying range of degradation products may be formed.

2 LIGHT STRESSING OF (1), (2) AND (3) IN SOLUTION

Table 1 compares the loss of HPLC area of (1), (2) and (3) after irradiation in solution in acetonitrile. In all three cases the loss of assay is accounted for by other peaks in the chromatogram as can be seen from Figure 2. In the cases of (1) and (2) the major degradant is the formation of the E-isomer of the pyridine acrylate moiety. In the case of compound (3) there is no pyridine acrylate system and a large number of minor degradants occur which account for the observed loss in main peak area. The pyridine acrylate moiety is clearly a major factor in the stability of the compounds as its presence gives rise to a UV absorption band with significant absorption at wavelengths greater than 320nm. Light degradation will only follow absorption of light from an electronic transition within the molecule of interest. This strong absorption followed by subsequent isomerisation in the presence of light explains why (1) and (2) are less stable to light in solution than (3). The reverse phase HPLC method is able to account for the all the products of degradation.

Table 1 *The loss in peak area of (1), (2) and (3) in solution in acetonitrile (0.01% w/v)*

	(1)	(2)	(3)
% HPLC main peak area loss after 3 hours in natural light	25%	38%	3%
% HPLC main peak area loss after 5 minutes in xenon light	42%	67%	12%

3 LIGHT STRESSING OF (1), (2) AND (3) IN THE SOLID STATE

Relative stability in the solid state is much more difficult to measure, control and explain than equivalent measurements in solution. It is difficult to measure because any comparative study must ensure identical or near identical exposure for each sample. For this to be controlled the particle size and shape for each compound must be the same, as must be the bulk densities and the area of each sample exposed. Since it is almost impossible to achieve such similarities, any direct comparisons must be interpreted with care and only broad differences and trends taken into account. Table 2 presents data following the exposure of solid samples of (1), (2) and (3) to xenon light. The samples were spread thinly between two glass plates in order to provide a consistent surface for exposure between samples. Only gross differences in stability were compared and therefore no accurate comparisons of particle size and morphology were made. In the case of compound (2), two polymorphic forms of this compound are known and both were exposed to xenon light. The results in terms of loss of HPLC area are summarized in Table 2.

Given the dramatic loss of HPLC peak area for (2) form I using xenon light, the sample was exposed to diffuse light for a period of 24 hours. In this experiment a sample was placed beneath a bank of fluorescent light tubes with a plastic diffuser between light source and sample. Such conditions simulate light stability under normal handling conditions albeit for an extended period. Under these conditions (2) form I degraded by approximately 4% to the same degradant as was formed under xenon light (see discussion in 3.1).

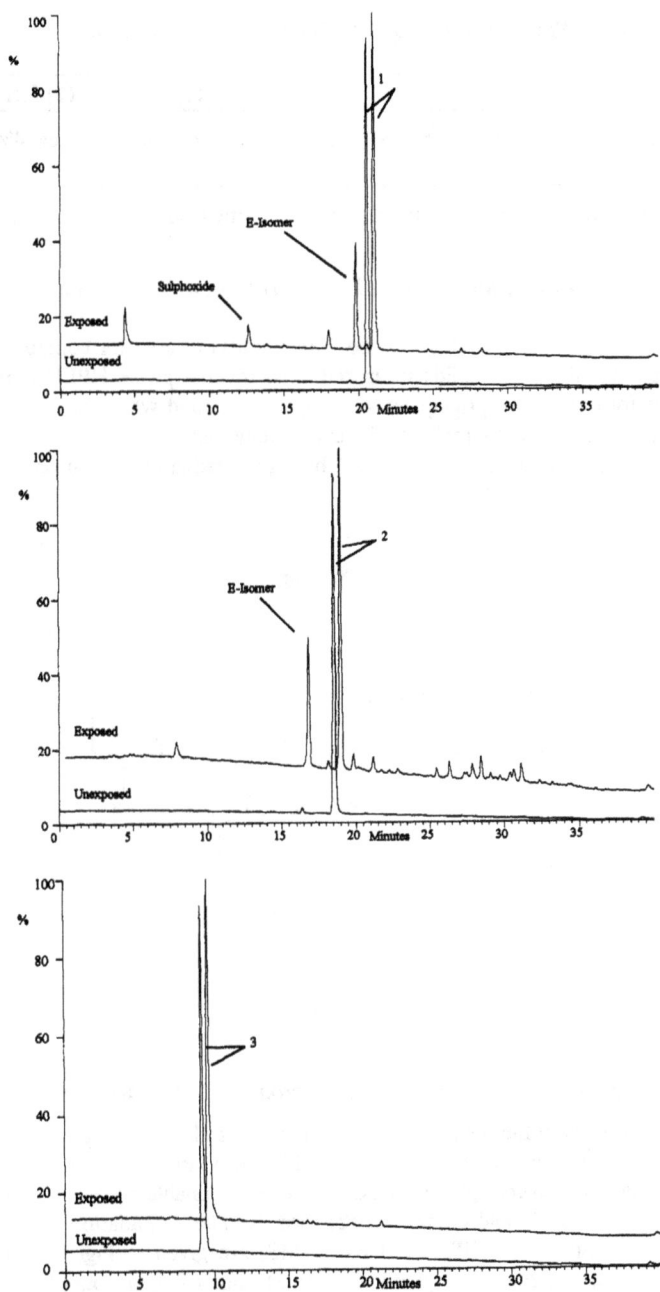

Figure 2 HPLC chromatograms of (1), (2) and (3) in acetonitrile solution.

Table 2 *The loss in HPLC peack area of (1), (2) and (3) in the solid state*

	(1)	(2) form I	(2) form II	(3)
% HPLC main peak area loss after 4 hours in xenon light*	ca.10%	ca.100%	ca. 4%	ca.10%

* Figures are approximate as degradation depends on particle size and area of exposure.

3.1 Analytical Study of Degradants from Light Stressing in Solid-State

The most striking result in Table 2 is the dramatic loss of area for compound (2) form I by comparison with the other solids measured. The reverse phase HPLC impurity profile shows that this form degrades principally to a single compound which has been identified by NMR, MS and X-ray crystallography as the cyclobutane dimer, (4) shown in Figure 3. This dimer is akin to an analogous dimer formed by light irradiation of cinnamic acid and its derivatives.[6]

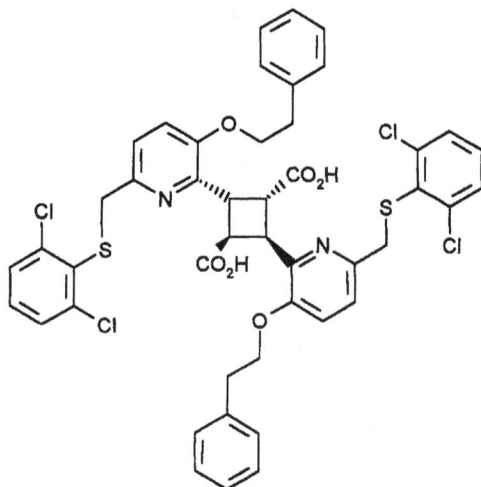

Figure 3 *The structure of (4), the product of irradiated (2) form I.*

The degradation of compound (3) in the solid state is also well explained by the HPLC data. In this case the loss in area may be accounted for by a large envelope of peaks observed in the chromatogram. Reverse phase HPLC, however, is unable to account for the losses recorded for compound (1) and compound (2) form II. The degradation products from the irradiation of 1, identified by HPLC and HPLC/MS, are given in Figure 4. However, the intensity of these peaks in the HPLC profile (Figure 5) cannot fully account for the assay loss. Complementary CE methods were also unable to show any significant products of degradation.

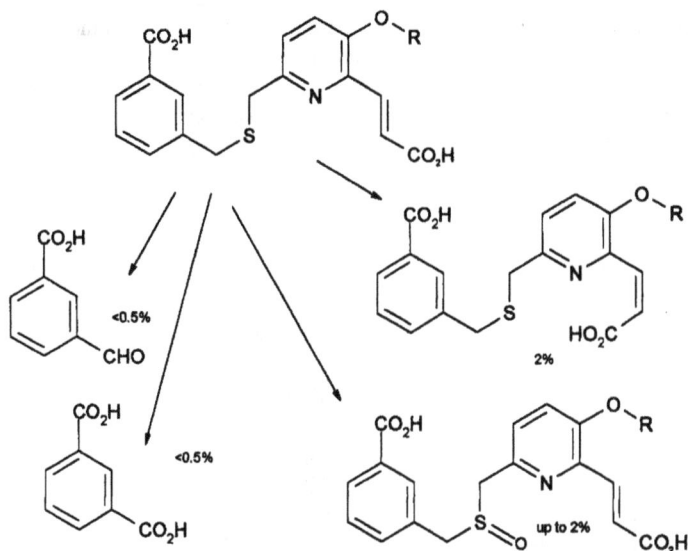

Figure 4   *The degradation products of (1) on exposure to xenon light as deduced by LC/M*

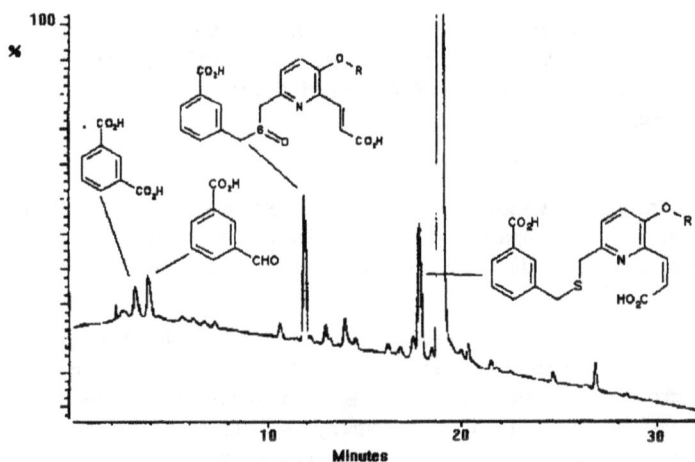

Figure 5   *The HPLC trace for Xenon light degraded (1)*

In order to identify the nature of the major degradation product or products of compound (1) and compound (2) form II, samples of both solids were exposed to xenon light for long periods (several days) to facilitate degradation of over 50%. Examination of the products of this procedure showed significant amounts of material insoluble in acetonitrile. The insoluble material was subsequently isolated for further analysis and was found to be soluble in dimethyl sulphoxide (DMSO) and dimethyl formamide (DMF). Using these solvents in the mobile phase, Gel Permeation Chromatography (GPC) was carried out, the results of which are given in Figures 6 and 7.

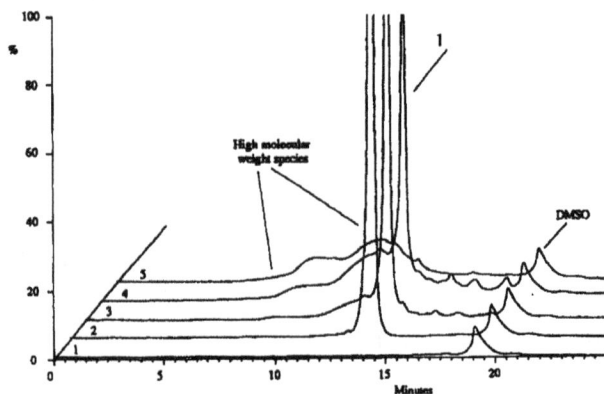

Figure 6 *GPC traces for (1) using UV detection. Column: Styragel HR-1 300x7.8 mm, Sample:80/20THF/DMSO, Eluent: THF (40 deg. C) Detection:UV at 280nm (1=blank, 2=unstressed (1), 3/4 xenon light stressed material, 5=extracted insolubles from xenon light stressed material).*

The GPC traces clearly show significant peaks due to high molecular weight material not present in unirradiated samples. These data suggest that the principal products of light degradation of compound (1) and compound (2) form II may be polymeric in nature. The appearance of the GPC traces suggest the polymeric material is itself a complex mixture.

The insoluble material was analysed using spectroscopic techniques. The proton nuclear magnetic resonance spectrum showed broad signals consistent with polymeric materials. Infrared spectroscopy indicated a loss in the acrylate double bond. Analysis of the insoluble material was attempted using mass spectrometry. A number of ionisation techniques were used, including electron impact, chemical ionisation, electrospray, fast atom bombardment and matrix assisted laser desorption. In all cases no mass spectrometric evidence for high molecular weight material was found. The reason for this may be due to the insoluble and involatile nature of the polymeric materials which may well posses a degree of cross linking making it unamenable to these techniques.

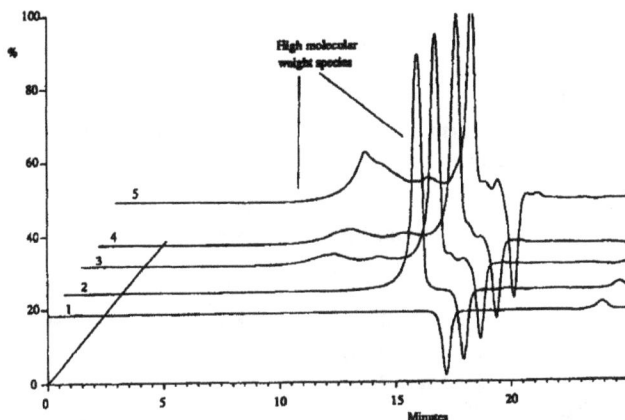

Figure 7 *GPC traces for (1) using refractive index detection. Column: Stryagel HR-1 300 x 7.8 mm, Sample: DMF, Eluent: DMF, Detection: RI (1=blank, 2=unstressed material, 3-5= xenon light stressed material).*

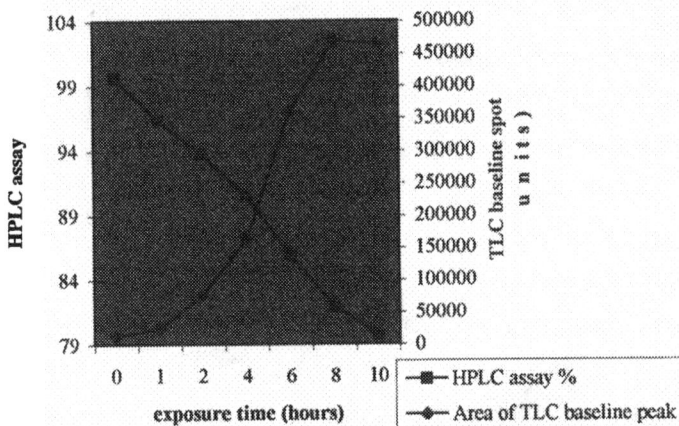

Figure 8 *The correlation between loss of HPLC assay and intensity of baseline spot on TLC plate as measured using a densitometer for light degraded (2) form II.*

The GPC traces of lighted stressed (1) were broad in nature and quantitation of the polymeric species was not possible. The final part of the analytical work was to develop a simple method to indicate the presence of the insoluble materials at low levels and to act as a semi-quantitative or limit test for the polymeric degradants. For this work a Thin Layer

Chromatography (TLC) method was used which exploited the insoluble nature of the degradants. The exposed samples were first dissolved in dimethlyl formamide and applied to a TLC plate and the solvent evaporated. The plate was then developed carefully whereby only the insoluble material was left on the baseline. The insoluble material may be visualized in UV light and may be made to fluoresce. A densitometer scanner was used to record the spot and compare densities of samples irradiated for different lengths of time. The results of this experiment are summarised for (2) form II in Figure 8. In this way a simple method for detecting the insoluble material was developed and routinely used.

3.2 The Influence of Solid State Structure on Light Stability

The most striking feature of Table 2 is the difference in light stability between the two polymorphic forms of (2). Form I degrades almost exclusively to the cyclobutane dimer (4) in the presence of xenon light. Indeed, 24 hours exposure under regular laboratory fluorescent tubes has also been shown to produce around 4% of (4). In the case of form II, however, the light instability is akin to that of (1), where polymeric products are formed.

Clearly the arrangement of atoms in the solid state structure of the two forms of (2) has a profound effect on resultant light stability. In order to examine this effect the x-ray crystal structures of the two forms were measured and examination of the arrangement of the molecules in the crystal lattice permitted a plausible explanation of the observed effects.

Views of the crystal structures are given in Figures 9 (form I) and 10 (form II). In Figure 9 the acrylate double bonds in neighboring molecules in the unit cell are well aligned to undergo a light catalysed cycloaddition reaction in the solid state to (4). In Figure 10, however, the double bonds are too remote for such a reaction to take place in the solid state. Therefore the absorption of light energy in form II promotes another chemical process which initiates principally the formation of polymers in far lower amounts than the formation of (4) in the case of form I.

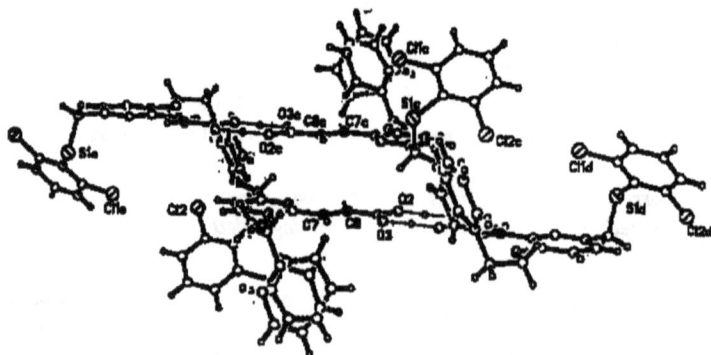

Figure 9 *Crystal structure of (2) form I.*

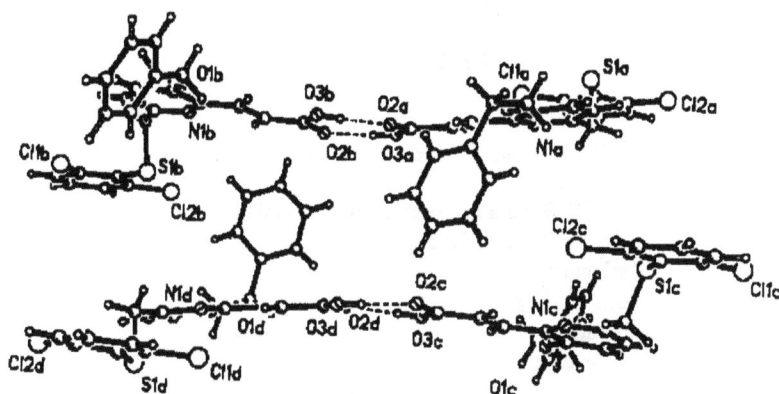

Figure 10 *Crystal structure of (2) form II.*

4 CONCLUSION

The analytical methods used have elucidated the differing nature of the degradation of three structurally related compounds. Indeed, one of the compounds has two crystalline forms which vary dramatically in their degradation indicating the role of solid state structure in degradation pathways. It is important to provide chromatographic methods to monitor such degradation and this work has shown that HPLC, GPC and TLC may be used to ensure the light stability of the three drug substances with respect to exposure to light. The information from these studies, together with information about the wavelengths responsible for the absorption of light permits the formulation scientists to design a dosage form that is stable to light both as a packaged product and in patent use.[7]

5 EXPERIMENTAL

5.1 Xenon Light Source

The samples were all irradiated in a Heraus Suntest CPS Light Cabinet. In order to discriminate against effects due to heating, control samples wrapped in aluminum foil were used to monitor any degradation due to heating. No significant heat degradation was observed.

5.2 HPLC Experimental Conditions

The HPLC experiments were carried out using a Beckman System Gold Liquid Chromatograph This comprising of an 166 single wavelength detector, a 507 autosampler and a 406 pump, all controlled by Gold software version 7.1.

Using the following method conditions:

Column:	Kromasil, 5µm C8, 250 x 4.6 mm
Oven:	40 Degrees
Detection:	By UV at 225 nm
Mobile Phase A:	0.05% TFA(aq) filtered through 0.45 µm filter
Mobile PhaseB:	Acetonitrile
Solvent Programme:	50-90% B via a linear gradient in 30 minutes, hold for 10 minutes. Re-equilibrate to 50-50
Flowrate:	1ml min^{-1}
Injection Volume:	20µl
Analysis Time:	40 minutes
Cycle Time:	55 minutes.

5.3 GPC Experimental Conditions

The GPC Experiments were carried out using a Beckman System Gold Liquid Chromatograph, this comprising of an 166 single wavelength detector, a 507 autosampler and a 406 pump all controlled by Gold software version 7.1.

Using the following method conditions: The refractive index (RI) detector used was an Erma ERC-7515A. The method conditions are given in the Figures 6 and 7.

5.4 TLC Experimental Conditions

The TLC experiments were carried out using the following conditions. The plates were scanned using a Camag TLC Scanner II, controlled by Cats software version 3.

Sample Preparation:	10mg in 0.5 ml DMF
Plate:	10x10cm F_{254nm} Silica Gel plate 200 µm film
Mobile Phase:	10:4:1:1 parts THF: Chloroform: glacial Acetic acid: Formic acid
Application:	2 x5 µl spots. Spot dried with hot air gun between applications. The samples were spotted at 2cm intervals and 1.5 cm in from the edges of the plate.

5.5 Other Analytical Methods

LC/MS experiments were carried out on a PE/Sciex API III Mass Spectrometer equipped with an ionspray source.

The HPLC was delivered using a Hewlett-Packard 1090 Liquid Chromatograph. NMR experiments were carried out using a Jeol GSX400 NMR spectrometer.

Infrared Experiments were carried out using a Nicolet 710 FTIR Spectophtometer.

Acknowledgments. The authors are grateful to Dr. Drake Eggleston of SmithKline Beecham for the x-ray crystal structures on compounds (2) forms I and II and (4) and to Dr. C. Brown and Dr. P. Camilleri for helpful advice and discussions.

References

1. S. R. Byrn, Annu. Rep. Med. Chem., 1985, **20**, 287.
2. Y. Matsuda and R. Masahara, J. Pharm. Sci., 1983, **72**, 1198.
3. A. Albini, Boll. Chim. Farm., 1992, 131, 30.
4. ICH Guideline for the Photostability Testing of New Drug Substances, see p. 66.
5. D. K. Beussink, S Nema and R. J. Washkuhn, Pharm. Technol., 1995, 19, 170.
6. J. Rennert, Photogr. Sci. Eng. 1971, **15**, 60.
7. D. R. Merrifield, P. L. Carter, D. Clapham and F. D. Sanderson, in 'Photostability of Drugs and Drug Formulations'. H. H. Tønnesen, Ed., Taylor and Francis, London, 1996, Chapter 7.

Molecular Mechanisms of Photosensitization Induced by Drugs on Biological Systems and Design of Photoprotective Systems

G. De Guidi,[a] G. Condorelli,[a] L.L. Costanzo,[a] S. Giuffrida, S. Monti[b] and S. Sortino.[a]
[a]Dipartimento di Scienze Chimiche,
Università di Catania,
V.le A. Doria 6, 95125, Catania, Italy
[b]Istituto F.R.A.E. C.N.R.
Via P.Gobetti 101, 40129 Bologna Italy

1 GENERAL SURVEY

This chapter deals with the study of *photosensitization processes induced by drugs* (here called "xenobiotics") on biological systems such as natural and artificial membranes and DNA.

The classes of drugs which will be taken into consideration in the following discussion are the non-steroidal anti-inflammatory drugs and fluoroquinolone antibacterials.

The investigation is carried out through parallel studies of drug photoreactivity in the presence or in the absence of the biosubstrates in order to identify the likely species, either transient or stable, involved in the photosensitization process. This approach involves (i) studies on the drug photolysis mechanism, in particular the determination of the kinetic aspects of the photochemical reaction in various experimental conditions by means of spectroscopic and chromatographic techniques; (ii) studies of the drug phototoxicity *in vitro*, i.e. of human erythrocyte hemolysis, unilamellar liposome peroxidation, and isolated membrane protein crosslink and plasmid DNA photocleavage.

The photosensitization mechanism depends in general on the sensitizer nature and the reaction conditions (aerobic or anaerobic). Drug photodegradation products, free radicals, solvated electrons, singlet oxygen and superoxide anion are the main stable or transient species involved. These species can be responsible for several membrane damages being able to induce lipid peroxidation, protein crosslink or the alteration of the bilayer hydrophobic assembly. Interaction with DNA was studied by various experimental techniques (fluorescence, fluorescence anisotropy, low temperature phosphorescence, circular dichroism, microcalorimetry and topoisomerase assays), through which is possible to assess whether a superficial or an intercalative interaction occurs. The kind of interaction determines the nature of the photoinduced damage; the efficiency of the photosensitization process was determined by measuring supercoiled plasmid linearization rates. It was possible, with the aid of various scavengers or quenchers, to emphasize the different transient species (free radicals, superoxide anion or singlet oxygen) involved in the photocleavage process.

A more detailed investigation of the molecular mechanisms, through which these drugs are able to provoke the above mentioned photoinduced damages, requires time-resolved spectroscopic techniques. These studies joined with steady state techniques are aimed at the elucidation of the factors influencing the molecular reactivity in order to get control of it. Nano and picosecond laser flash photolysis was carried out both in homogeneous media (aqueous or not) and in supramolecular structures like complexes with cyclodextrins or other macromolecules (e.g. proteins or nucleic acids). The application of pulsed techniques provides useful information on the nature of the excited states and of the short-lived intermediates involved in the photosensitization processes. The investigation on the drug photoreactivity in microenvironments with particular characteristics of polarity and in the presence of steric constraints or specific interactions contributes both to the understanding of the photobehaviour in the biological system and to the development of protective strategies against light-induced reactions.

The *protection system design* is particularly addressed to a) inorganic ions; b) drug-cyclodextrin inclusion complexes; c) drug esters with polymeric carriers and polymer networks based on biocompatible macromolecules. Metal complexation (in certain cases with ligands of biological interest) reduces the phototoxicity because of an antioxidant activity and a radical blocking action. The inorganic ions Cu^{II}, Mn^{II}, and Co^{II} act in fact as redox scavengers of photogenerated drug and biomacromolecule radicals, which are the main species responsible for the biological damage. Moreover, copper can scavenge the superoxide anion via its SOD-like activity. β-Cyclodextrin can behave as efficient protective agent because of its ability to associate the drug in a host-guest complex, leading to inhibition of the photosensitization processes. The cyclodextrin can in fact modify the sensitizer photochemistry by decreasing the efficiency of formation of species responsible for toxic effects; moreover, it can limit oxygen diffusion toward the sensitizer, reducing formation of singlet oxygen. Macromolecular conjugates between polyasparthamides and drugs or macrocarriers built via crosslinking by glutaraldehyde or γ-rays, designed to act as prodrug for gradual release of the active species, may in the meantime depress photosensitization processes.

2 PHOTOSENSITIZATION MECHANISMS

Photochemistry in biological systems has been receiving great attention in the last years for several practical applications, such as energy storage, waste degradation, food preservation and photosensitization reactions. As regards the latter aspects we must remember that sun radiation (UVA, UVB, Vis) is able to penetrate the cutaneous layer, even in relatively deep regions, to reach the blood flux (a suitable carrier of "xenobiotic" agents). The presence of a "xenobiotic" able to absorb the incident radiation can start noxious reactions. Among the potential photosensitizers we can find food preservatives, cosmetics, dyes and drugs. These latter can undergo photoreactions, which either may be useful for therapeutic applications or may start undesirable side reactions.[1,2] Side effects such as allergic cutaneous reactions, melanomas and genetic mutations have been reported.[3-5] Thus, a drug is able to induce phototoxicity (when sufficient concentrations are reached in the skin) first of all if it absorbs the sun radiation at wavelengths longer than 310 nm available in the environment (light at these wavelengths can penetrate skin); secondly if photochemical drug decomposition occurs.

This chapter is particularly devoted to the molecular mechanisms of photosensitization at the base of the adverse reactions and to the experimental approach to elucidate them.

Many drugs have been reported to be involved in cases of photosensitization: Table 1 shows some of them classified as pharmacological classes.

Table 1 *List of some drugs (grouped in pharmacological classes) which have been reported in photosensitization reactions.*

Antibacterials	Tetracyclines, Chloramphenicol, Enoxacin, Fluoroquinolones, Nalidixic acid
Tranquillisers	Phenotiazines: Chlorpromazine, Promazines
Diuretics	Chlorotiazides
Antiarhytmics	Amiodarone, Propanolol, Quinidine
Antihypertensives	Methyldopa
Antidepressants	Protriptiline
NSAIDS	Benoxaprofen, Naproxen, Piroxicam, Ketoprofen, Carprofen, Suprofen, Tiaprofenic Acid, Butibufen, Fenbufen, Tolmetin, Benorilate,
Antimicotics	Griseofulvine
Topical bacteriostatics	Bithionol, Phentichlor, Esachlorophen
Antimalarics	Quinine, Chloroquine, Hydroxychloroquine
Antiangine	Nifepidine
Immunosuppressants	Azothiaprine
Antilipemics	Fenofibrate

We are particularly interested in two largely used families of drugs, antibacterial fluoroquinolones and NSAIDs. The former is a modern class of antibiotics, continuously under development with the aim of improving their pharmacological activity and, in the meanwhile, decreasing their (photo) toxicity.[6-15] The latter represents the main therapeutic agent for controlling the pain and the inflammation of rheumatic diseases.[16] Many of them are known to cause both phototoxic and photoallergic reactions. A screening of the *in vivo* and *in vitro* photosensitizing activity of many NSAIDs was achieved by several techniques such as the mouse tail technique,[4] photohemolysis[3] and the photo-basophil-histamine-release test.[17]

As previously outlined, naturally incident light absorption by drugs is directly related to their chemical structure and, if sufficient concentrations are reached in a particular biological compartment, a photodegradation after radiation absorption can occur.[18] This can lead to the formation of noxious transient or stable species, such as free radicals and photoproducts, and/or the promotion of energy transfer to form singlet oxygen (even if this last mechanism is co-operative with the first). Thus, it can be easily explained why ibuprofen, for example, is not phototoxic: although it can form toxic photoproducts when UVB irradiated;[16,19] this happens for short wavelengths which are not able to penetrate skin.[16] On the contrary, indomethacin absorbs above 310 nm, but it is very photostable, so it cannot start a photosensitization process related to its photodegradation.[20]

A molecular mechanism of photosensitization induced by drugs can proceed through several pathways, which can be divided into two different types of mechanisms: photosensitized reactions via radicals (type I) and photosensitized reactions via singlet oxygen ($^1O_2$), (type II) (Scheme 1).

Type I mechanism. This mechanism can be further distinguished whether the reaction occurs in the presence or in the absence of oxygen.

Reactions in the presence of oxygen. In this case there are four possible pathways:

1. An interaction between the excited sensitizer and oxygen occurs with formation of a charge transfer complex. This complex can dissociate in polar solvents with formation of the superoxide anion $^•O_2^-$. This process is in general not very efficient because direct interaction between sensitizer and oxygen more likely leads to energy transfer with formation of singlet oxygen.

2. An interaction of the excited sensitizer (A*) with a substrate (B) can occur, with subsequent electron release or capture (depending on the redox potential of the A*/B system) and formation of anionic radicals which, reacting with oxygen, produce $^•O_2^-$.

3. An interaction between the sensitizer and the substrate can lead to radicals (for example via hydrogen abstraction), which can react with oxygen, thus generating peroxides.

4. The excited sensitizer dissociates generating radicals. These can in turn transform the substrate into a radical or be oxidized by oxygen.

In reactions 1 and 2, superoxide anion formation occurs. This species, because of its instability, is highly reactive and noxious and can decay, promoting oxidation, in four ways:[21]

- Hydrogen abstraction with formation of hydroperoxide $HO_2^•$
- Electron capture with formation of O_2^{2-} and hence H_2O_2
- Electron release in redox reaction e.g. with Fe^{2+} in the Haber-Weiss cycle[22]
- Dismutation with formation of $^1O_2$ and H_2O_2

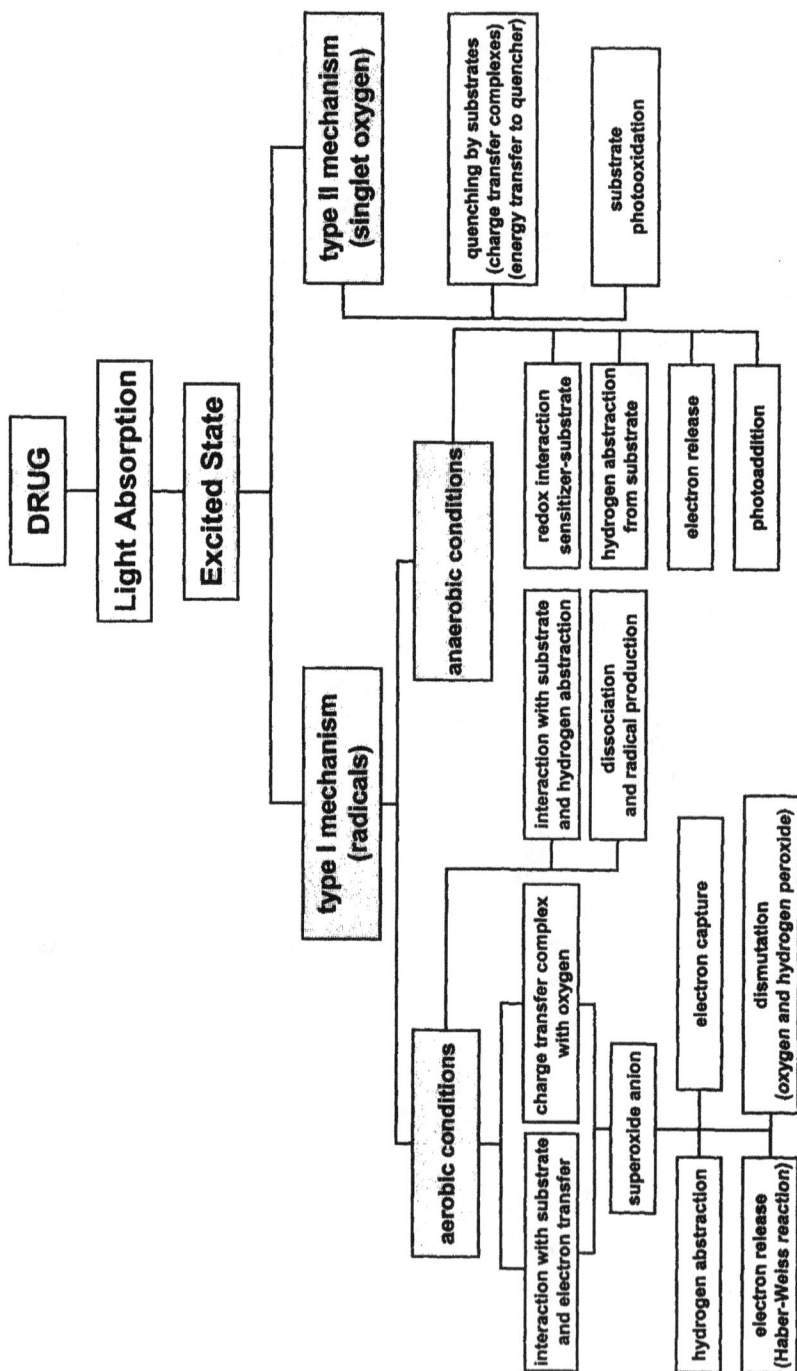

Scheme 1 *Molecular mechanism of photosensitization.*

Reactions in the absence of oxygen. By using deaerated media in order to avoid the oxygen action, thus mimicking processes occurring in anaerobiosis, we can observe four typical pathways:

1.   Redox interaction between sensitizer and substrate with subsequent formation of radical anions.

2.   Hydrogen abstraction from the substrate with formation of radicals.

3.   Electron release in an aqueous medium.

4.   Photoaddition (as in the case of furocumarines with DNA).[23]

Type II mechanism. Once formed, singlet oxygen can decay by:

1. Emission of phosphorescence (λ_{em} = 1.27 μm): however, this process has a low quantum yield.

2. Non-radiative deactivation by collisions with solvent.

3. Physical quenching by quenchers (Q) which can occur by formation of charge transfer complexes or by energy transfer to Q; in the latter case, an efficient intersystem crossing process is necessary, promoting decay of $^3Q^*$ before this one is oxidized by $^1O_2$.

4. Photooxidation of a substrate. This includes a) 1-2 and 1-4 addition to unsaturated compounds with formation of peroxides[24,25] and b) reactions with heteroatom-containing compounds.

3 MOLECULAR TARGETS OF PHOTOSENSITIZATION

3.1 Membrane Targets

Whatever the photodamaging species (stable or transient) involved in the sensitization is, cell membrane is primarily involved as regards the chemical alterations of its principal constituents: lipid and proteins.[26]

3.1.1 Erythrocyte hemolysis. A first check of the photodamaging activity can be obtained through irradiation of aerated, deaerated and oxygen saturated samples of red blood cell (RBC) suspension containing the sensitizer and measure of the absorbance decrease at 650 nm, where the optical density is proportional to the number of intact RBCs[27] (Figure 1). The study of the rate of delayed hemolysis as a function of the time of irradiation, of the sensitizer and oxygen concentrations and of the presence of antioxidants, free radical scavengers, singlet oxygen quenchers, and D_2O (causing an increase of the lifetime of $^1O_2$), provides information about the mechanism.[28,29] These data, together with the elucidation of the photodegradation pathway of the drug (with the isolation and characterization of potentially toxic photoproducts), give a first picture of the photosensitization mechanism in membrane.

3.1.2 Lipid peroxidation. A simplified system used to discriminate target molecules on membrane is given by artificial bilayers (unilamellar liposomes of phosphatidylcholine from egg yolk prepared with the solvent injection technique)[30] which permit the evaluation of the photosensitized damage on the lipid portion. The integrity of the liposome membrane can be measured by efflux of trapped markers, such as glucose-6-phosphate (determined spectrophotometrically at 340 nm), after addition of glucose-6-phosphate dehydrogenase and NADP.[31] Photodynamic lipid peroxidation has been extensively studied and an excellent source of information can be found in a work of Girotti.[32] The formation of peroxides in the irradiated samples can be followed through the reaction with thiobarbituric acid (TBA),[33,34] also if this method is subjected to limits, particularly when cells and tissues are analysed. The iodometric approach, also if more complicated and subjected to interference with O_2, allows determination of total peroxides, including those derived from cholesterol.[35] Moreover, the oxidation products of cholesterol by $^1O_2$ and radical attack are different[29] and their characterization provides evidence for the participation of these two species in the photosensitization mechanism.

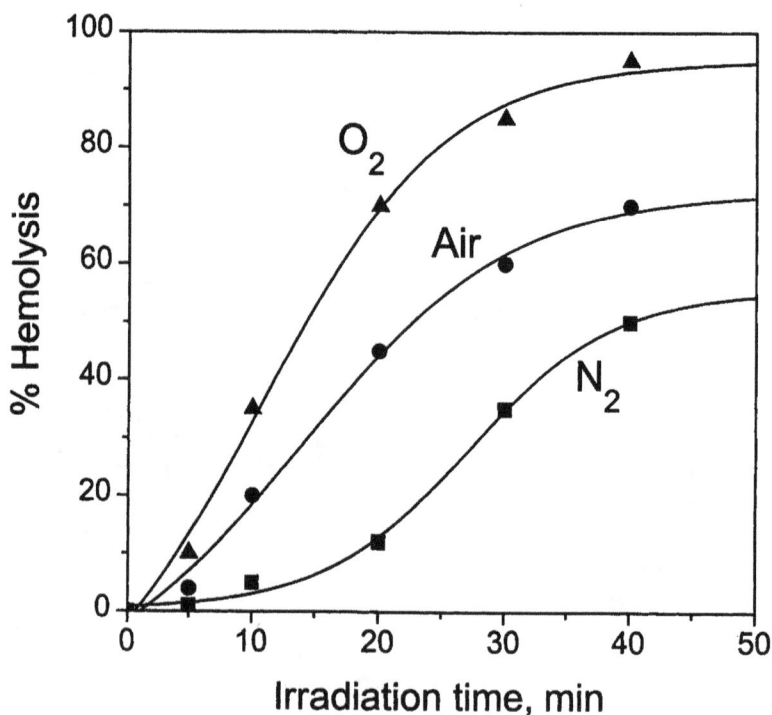

Figure 1 *Typical photohemolysis plot in nitrogen, air and oxygen saturated solutions of the NSAID Carprofen.*

In the presence of oxygen, Type I mechanism involves the superoxide anion. Even if it is known that this species is not directly responsible for lipid peroxidation, it can promote, via Haber-Weiss reaction in the presence of traces of metal ions, formation of the hydroxyl radical, which is one of the most harmful species as regards the photoinduced damage. This is due to its ability in abstracting hydrogen from the bio-substrate. Reduction of photohemolysis, lipid peroxidation and marker efflux operated by SOD suggests the presence of $O_2^-\cdot$, which can be evidenced also by spectroscopic determination of the reduction of cytochrome III.[36] In anaerobic conditions, the same method can be useful for detecting solvated electrons produced during the irradiation.[37,38]

3.1.3 Protein photoinduced crosslinking. Protein and lipid components are closely related in the membrane structure, but treatment of erythrocyte ghost membranes with a surfactant permits the isolation of the protein fraction, which can be used for the study of photosensitized protein crosslinking. This process can be studied by detecting high molecular weight protein components by SDS polyacrylamide gel electrophoresis.[39]

3.2 DNA Targets

3.2.1 DNA-drug interaction. The study of photosensitized DNA damage is needed for a deep understanding of both phototoxicity and phototherapy. It contributes to clarify the mechanisms of degenerative skin diseases and of the cell toxicity of potential anticancer drugs. Under this respect it is very important to consider the biodistribution of the photosensitizing agent and its binding mode to DNA.[40-43] In fact, the photosensitizer can associate to DNA either through a surface interaction such as hydrogen-bonding or van der Waals forces along the grooves of the helix or through non-covalent intercalation via π-stacking of aromatic heterocyclic groups between base pairs. In the latter case a site-selective photocleavage can be promoted. The study of the influence of parameters like hydrophobicity, geometry, size, shape and ability in the formation of H-bonds with the base pairs permits the assessment of the affinity between the photosensitizer and DNA. Useful information about modes of interaction is provided from induced linear and circular dichroism, fluorescence anisotropy measurements and low temperature phosphorescence[44] as well as by other techniques such as microcalorimetry and topoisomerase assays.

Interesting information can be obtained by studying the emission of nucleic acid bases in the absence and in the presence of the drug at 77 K. In the latter case, the appearance of the typical phosphorescence maximum of the adduct drug-DNA indicates the occurrence of short distance energy transfer suggesting close contact between DNA bases and sensitizer.[43, 45] The changes in fluorescence intensity and/or UV spectra due to drug-DNA association can be used to estimate binding constant through non-linear regression analysis based on mathematic models such as that proposed by McGhee and von Hippel.[46]

The presence of circular dichroism signals from a non optically asymmetric drug molecule is also indicative of drug-DNA interaction. The positive band of calf thymus DNA, centered at 275 nm decreases with the addition of the increasing concentrations of the drug (such as Suprofen),[43] whereas a new band with maximum corresponding to the maximum of the UV spectrum of the drug grows up. The circular dichroism signal originates mainly from the dipole-dipole interaction between the electric dipole moment of the drug transitions with the dipoles represented by the bonds of the chiral double helix of DNA. The CD spectrum is expected in this case to have the same shape as the absorption spectrum.[43,44]

3.2.2 DNA photosensitization. The mechanism of photoinduced DNA damage, leading to nucleic acid oxidation and single strand break, involves three main pathways: i) participation of hydroxyl radicals, known as one of the most noxious species in promoting DNA damage,[47] ii) electron transfer[48] and iii) oxidation via singlet oxygen.[49] Analysis of nucleoside photoproducts permits a preliminary discrimination between type I and type II mechanisms. The photosensitization process efficiency is determined by measuring supercoiled plasmid linearization rates. This is obtained by monitoring the changes in conformation of supercoiled form I of DNA, which is at first converted in relaxed open circular form II via single strand break and finally to linear form III through double strand break.[43,50] DNA forms are isolated via agarose gel electrophoresis.

Moreover, it is possible to follow the kinetics of photoinduced DNA cleavage in the presence of various scavengers or quenchers (also in an oxygen-modified atmosphere), with the aim of emphasizing the role of the different transient species (free radicals, superoxide anion or singlet oxygen) involved in the photocleavage process.[49] If DNA photoinduced damage proceeds via a type I mechanism, a hydrogen abstraction by the photogenerated drug radicals more likely involves DNA than solvent or scavengers, if an efficient drug-DNA interaction occurs, (nevertheless, a less efficient production of hydroxyl radicals from solvent cannot be excluded). The general type I-mediated pathway can involve hydrogen abstraction from a sugar of DNA and this process leads directly to DNA breakage.[51]

In the case a type II mechanism is operative, the presence of singlet oxygen can be shown in experiments carried out with quenchers like sodium azide and 1.4-diazabicyclo[2.2.2]octane. If the quencher is able to reduce ssb with high efficiency in aerobic conditions, this is a strong indication of the involvement of singlet oxygen in ssb (also if it is a controversial matter). An excellent review by Piette gives a wide discussion on this point.[52] Further evidence for participation of singlet oxygen can be provided by photocleavage experiments carried out in D_2O (to increase the lifetime of singlet oxygen): in this case the cleavage efficiency is increased.[43]

The presence of efficient type II processes can give indication of the predominance of a surface binding mode of the sensitizer to DNA compared to the intercalative one because in the latter case the sensitizer cannot be available for bimolecular energy transfer with oxygen.[53,54]

A further contribution to the elucidation of the mechanism of drug induced DNA photodamage can be provided by using 2'-deoxyguanosine as a DNA model compound. The major photo-oxidation products of this nucleoside can be identified and classified according to the formation mechanism (mediated by a radical or singlet oxygen).[55]

Moreover, DNA sequencing through acrylamide gel electrophoresis gives detailed information about site selectivity of sensitizer binding to specific DNA base sequences. DNA sequencing can be performed using either restriction fragments of plasmides or oligonucleotides, which can be synthesized according to a specific sequence design.[56]

Finally, interesting data can be provided from the photophysical and photochemical behaviour of the sensitizer in the absence and in the presence of DNA. Time resolved photochemical experiments could enlighten the role played by transient species in DNA sensitization.[57]

4 DRUG PHOTOCHEMISTRY AND PHOTOPHYSICS

The investigation is performed through parallel studies of drug photoreactivity in the presence or in the absence of the biosubstrates in order to identify the likely species, either transient or stable, involved in the photosensitization process. The experimental approach involves studies of direct photolysis in various experimental conditions. Isolation and characterization of the photoproducts with the aid of spectroscopic and chromatographic techniques is also directed to assess a possible role of these molecules in the photosensitization process.[58,59]

For example, one of the most frequently cited classes of drugs as regards phototoxic side effects is that of the NSAID's arylpropionic acids. Under UVA irradiation, these drugs undergo decarboxylation as the main photochemical process in aqueous or methanolic solutions.[60] Subsequent thermal steps lead to production of several photoproducts, some of which display toxic and/or phototoxic properties. Transients like free radicals, oxygen activated species and solvated electrons have been indicated to be potentially responsible for the observed photosensitizing effects. Hydrogen abstraction and oxidative steps are involved in these processes.

Detailed investigation of the molecular mechanisms, through which the drugs are able to provoke the photoinduced damages, requires pulsed spectroscopic techniques, in particular the use of nano- and picosecond laser flash photolysis. These studies, carried out in homogeneous media (aqueous or non aqueous)[61-63] and in supramolecular structures like complexes with cyclodextrins[64,65] or other macromolecules (e.g. proteins or nucleic acids), allow obtaining useful information on the nature of the excited states and of the short-lived intermediates involved in the photosensitization processes. For example, time resolved studies indicated that an intramolecular electron transfer in the triplet state is at the basis of the decarboxylation process of diaryl-ketone-propionic acids. Inclusion of these molecules in β-cyclodextrin reduces the photodecarboxylation quantum yield by slowing down the rate of this intermediate step. This makes possible competition by another deactivation process, consisting of hydrogen abstraction from the cyclodextrin (Scheme 2).[64-65]

Thus the investigation on the drug photoreactivity in microenvironments with particular polarity characteristics and in the presence of steric constraints or specific interactions will contribute to the understanding of the photoreactivity in the biological system and to the development of protective strategies against light-induced reactions.[66]

To provide further information useful to the understanding of the photochemistry and the photosensitizing properties of the drugs, it is important to investigate the effect of pH on the spectroscopic and photochemical behaviour. For example, the absorption and emission properties of enoxacin, as well as the photodegradation quantum yield, can be strongly affected by pH.[62,67] Nanosecond flash photolysis experiments confirm that the yield of absorbing transients reflects the pH-dependence and the overall results represent a key step for the understanding of the phototoxic reactions induced by this drug in physiological conditions (Fig. 2).

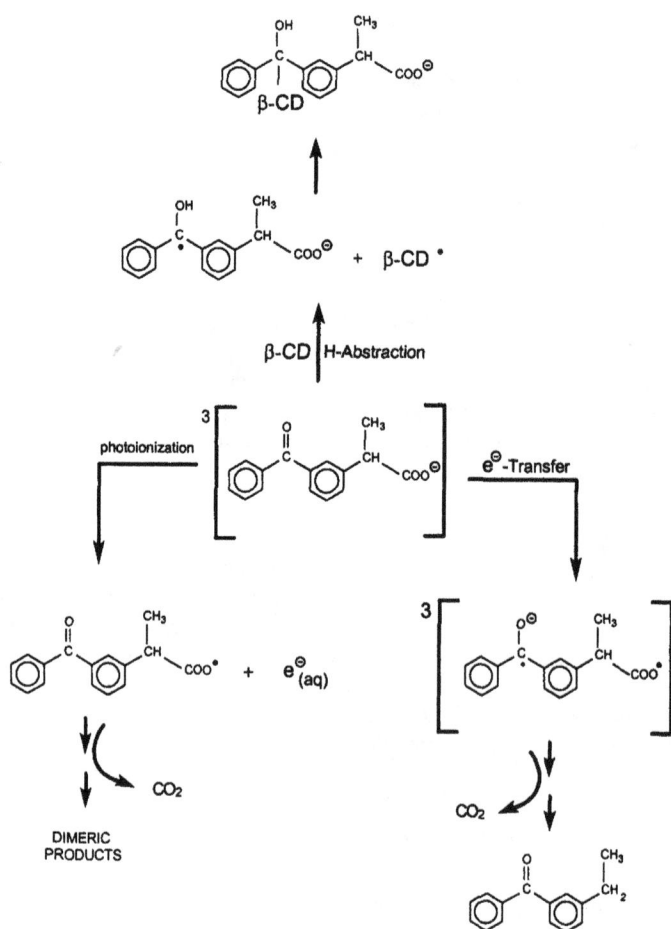

Scheme 2 *Photodegradation scheme of Ketoprofen in the absence and in the presence of β-cyclodextrin.*

In Scheme 3 a general picture of the various aspects of the photosensitization processes is reported, with particular attention to the target biological components proteins, lipids, cholesterol, and DNA, which are more frequently reported to be targets for damaging species originating from light absorption and/or degradation of drugs. The scheme illustrates the pathways which can initiate a damaging process from either the starting drug (as well as a related metabolite) or a photoproduct with sensitizing properties.

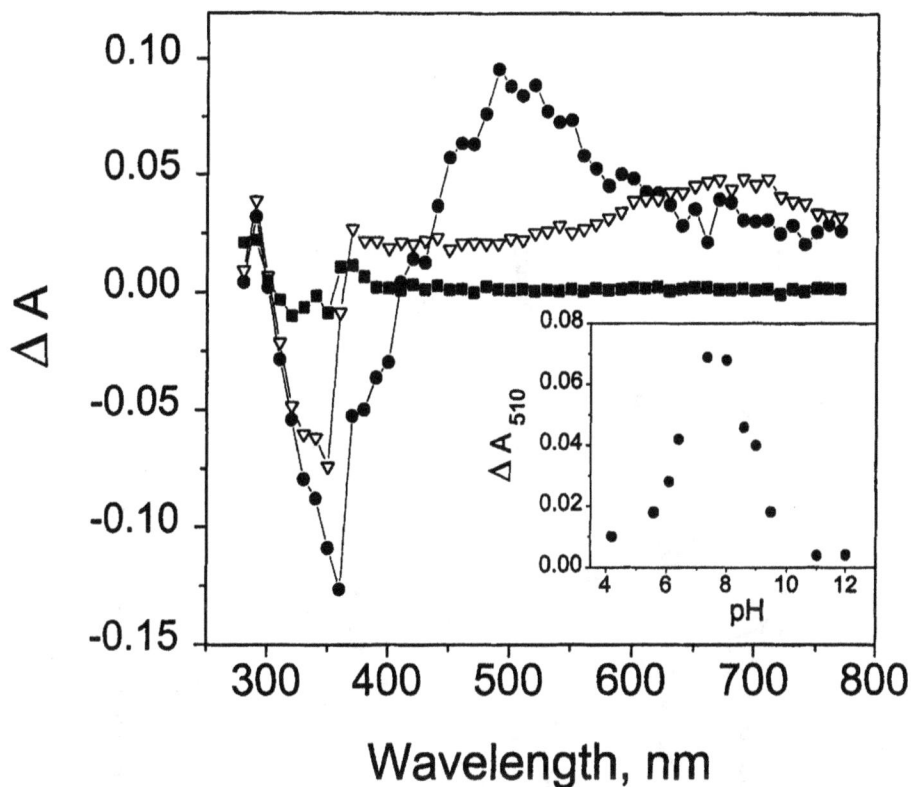

Figure 2 *Transient absorption spectra of Enoxacin at various delay times after the excitation pulse; inset: pH effect on the yield of transients.*

5 PHOTOPROTECTION SYSTEMS

The design of a photoprotective system presents the double advantage of reducing the phototoxic effect of the drug and, in the meantime, of shedding more light on the molecular mechanism of photosensitization.

5.1 Inorganic Ions and their Complexes

The inorganic ions Cu^{II}, Mn^{II}, Co^{II}, are able to reduce the photoinduced damage. These ions protect only within a range of concentrations, since for higher amounts toxic effects are observed. These species can act by redox scavenging of the photogenerated drug and biomacromolecule radicals,[68,69] which are the main responsible species for the biological

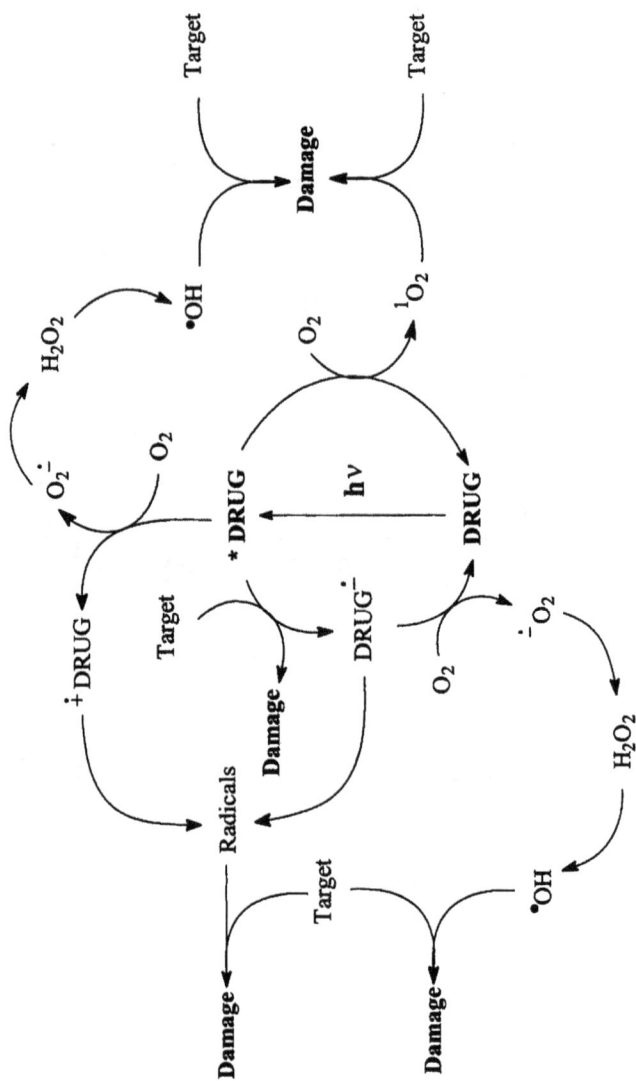

Scheme 3 *General scheme of photosensitization induced by drugs in biological systems.*

damage. Copper acts at a lower concentration range and its activity is enhanced due to its superoxide dismutase-like activity leading to the scavenging of the superoxide anion. Thus, metal complexation (in certain cases with ligands of biological interest, such as linear and cyclic dipeptides as well as with mono- and di-functionalysed cyclodextrins) exerts an antioxidant activity and a radical blocking action protecting from drug induced photohemolysis,[36,70] and lipid peroxiation.[59,70,71] The particular effect of copper confirms the involvement of the superoxide anion (and of its decay product, the hydroxyl radical) in the photoinduced sensitization mechanisms. Inorganic ions also decrease drug-photoinduced DNA cleavage.[72] The protective efficiency decreases with the increase of concentration of the free drug in the bulk of the solution, due probably to a scavenging action of radicals generated in the photolysis. The overall results further confirm the key role played by oxygen-mediated type I mechanism in drug photosensitization in cell membranes.

5.2 Cyclodextrins

β-cyclodextrins can constitute an efficient photoprotective agent as regards drug photoinduced damage on biological targets.[73,74] As reported above (paragraph 4), the inhibitory action detected in photosensitization processes is related to cyclodextrin ability in including the drug through a host-guest process. This process a) can modify the sensitizer photochemistry by decreasing the quantum yield of formation of the species responsible for phototoxicity; b) opposes to oxygen diffusion toward the sensitizer, and if a type II mechanism is operative, a reduction of singlet oxygen, a particularly efficient agent in inducing damages on various biological targets, occurs.[64,65] On the other hand, it is to be considered that the use of this additive is valid only in a restricted range of concentrations, due to the fact high levels of cyclodextrin are toxic towards cells.[73] Moreover, in some cases the cyclodextrin complexation determines an overall increase of the drug photodegradation quantum yield,[65] due to an increase of the efficiency of other deactivation processes such as hydrogen abstraction from the cyclodextrin. Consequently, care has to be taken in therapeutic administration of drug-cyclodextrin associates.

5.3 Macromolecular Conjugates and Polymeric Carriers

As regards macromolecular conjugates between polyasparthamides and drugs or macrocarriers built through crosslinking by glutaraldehyde or γ-rays, the design of a prodrug which can allow a gradual release of the active species and, meanwhile, decrease photosensitization damages, is taken into account. In the first case, a macromolecular conjugate between α,β-poly(N-hydroxyethyl)-DL-aspartamide and a non steroidal anti-inflammatory drug, such as diflunisal, ketoprofen or naproxen leads to a strong reduction of photosensitized cell damage and this effect seems to be independent on the molecular mechanism of the photochemical reaction.[75]

In the second case, an anti-inflammatory drug such as suprofen is incorporated in polymer networks based on biocompatible macromolecules, such as α,β-polyasparthydrazide (PAHy) and α,β-poly(N-hydroxyethyl)-DL-aspartamide (PHEA) crosslinked by glutaraldehyde or γ-rays, respectively. Swelling tests and photosensitization experiments in erythrocyte membranes carried out in aqueous medium indicate that the drug is released in a sustained way both from PAHy and PHEA microparticles. Moreover, incorporation of suprofen in PAHy and PHEA networks provides a significant reduction of the drug photosensitizing activity.[76] The more likely hypothesis is that the photogenerated radicals entrapped in the hydrogel are scavenged by the host organic functional groups, which can act as efficient hydrogen donors. A similar behaviour was observed in the host-guest photochemistry in the presence of cyclodextrin.[65]

Acknowledgments. We thank CNR and Istituto Superiore della Sanità - Progetto: "Proprietà chimico-fisiche dei medicamenti e loro sicurezza d'uso"- grants.

References

1. D. V. Ash and S. B. Brown, *Eur. J. Cancer*, 1993, **29A**, 1781.
2. I. E. Kochevar, *Photochem. Photobiol.*, 1987, **45**, 891.
3. B. Ljunggren, *Photodermatology,* 1985, **2**, 3.
4. B. Ljunggren and K. Lundberg, *Photodermatology*, 1985, **2**, 377.
5. M. A. Miranda, 'Alternative In Vitro Models To Animal Experimentation in Pharmacotoxicology', Farmaindustria, Barcelona, 1992, p. 239.
6. E. D. Shelley and W.B. Shelley, *Cutis*, 1988, **42**, 24.
7. D. Peyramond and F. Biron, S. Tigaud et al., *Rev. Infect. Dis.*, 1989, **11**, S1269.
8. R. T. Schefe, *Int. J. Dermatol.*, 1993, **32**, 413.
9. S. T. Nedorost and J.W.E. Dijkstra, *Arch. Dermatol.*, 1989, **125**, 433.
10. E. V. Granowitz, *J. Infect. Dis.*, 1989, **160**, 910.
11. J. Ferguson and B. E. Johnson, *Br. J. Dermatol.*, 1990, **123**, 9.
12. Y. Kurimaji and M. Shono, *Contact dermatitis*, 1992, **26**, 5.
13. N. J. Lowe, T. D. Fakouki, R. S. Stern, T. Bourget, B. Roniker and E. A. Swabb, *Clin. Pharmacol. & Ther.*, 1994, **56**, 587.
14. W. Christ, T. Lehnert and B. Ulbrich. *Rev. Infect. Dis.*, 1988, **10**, 141.
15. T. Horio, H. Miyauchi, Y. Asada, Y. Aoki and M.Harada, *J. Dermatol. Sci.*, 1994, **7**, 130.
16. J. G. Lombardino, *Chem. Pharmacol. Drugs*, 1985, **5**, 253.
17. B. Przybilla, U. Schwab-Przybilla, T. Ruzicka and J. Ring, *Photodermatology*, 1987, **4**, 73.
18. I. E. Kochevar, *Arch. Dermatol.*, 1989, **125**, 824.
19. J. V. Castell, J. Gomez-Lechon, M. A. Miranda and I. M. Morera, *Photochem. Photobiol.*, 1987, **46**, 991.
20. D. E. Moore and P. P. A. Chappuis, *Photochem. Photobiol*, 1988, **47**, 173.
21. C. Hasselman, C. Pigault, R. Santus and G. Laustriat, *Photochem. Photobiol.*, 1978, **27**, 13.
22. M. Lazàr, J. Rychly, V. P. KlimoPelikàn and L. Valko, in 'Free Radicals in Chemistry and Biology', CRC Press. Inc., Boca Raton, Florida, 1989, p. 5.
23. P. Vigny, F. Gaboriau, L. Voituriez and J. Cadet, *Biochimie*, 1985, **67**, 317.
24. C. S. Foote, in 'Porphyrin Localization and Treatment of Tumors ' (D. R. Doiron, and C. J. Gomer, Eds.), AR Liss, New York, Vol. 3.

25. C. Hasselman, C. Pigault and G. Laustriat, *J. Phys. Chem.*, 1982, **86**, 1755.

26. R. Santus and J. P. Reyftmann, *Biochimie*, 1986, **68**, 843.

27. P. Valenzeno and J. W. Trank, *Photochem. Photobiol.*, 1985, **42**, 335.

28. G. De Guidi, R. Chillemi, L. L. Costanzo, S. Giuffrida and G. Condorelli, *J. Photochem. Photobiol. B:Biol.*, 1993, **17**, 239 and literature there cited.

29. A. Frolov and G. P. Gurinovich, *J. Photochem. Photobiol. B:Biol.*, 1992, **13**, 39.

30. M. H. Kremer, W. M. J. van der Esker, C. Pathmamanoharan and P. H. Wiersema, *Biochemistry*, 1977, **16**, 3932.

31. A. W. Girotti and J. P. Thomas, *J. Biol. Chem.*, 1984, **259**, 1744.

32. A. W. Girotti, *Photochem. Photobiol.*, 1990, **51**, 497.

33. R. O. Recknagel,. and E. A. Glende, in 'Method in Enzymology', L.Packer, Ed., Academic Press, New York, 1984, Vol. 105, p. 331.

34. A. W. Girotti, J. P. Thomas and J. E. Jordan, *J. Free Rad. Biol. Med.*, 1985, **1**, 395.

35. M. Hicks and J. M. Gebicki, *Anal. Biochem.*, 1979, **99**, 249.

36. L. L. Costanzo, G. De Guidi, S. Giuffrida, E. Rizzarelli and G. Vecchio, *J. Inorg. Biochem*, 1993, **50**, 273.

37. J. Kagan, T. P. Wang. I. A. Kagan, R. W. Tuveson, G. R. Wang and J. Lam, *Photochem. Photobiol.*, 1992, **55**, 63.

38. G. De Guidi, R. Chillemi, L.L. Costanzo, S. Giuffrida, S. Sortino and G. Condorelli, *J. Photochem. Photobiol., B: Biol.*, 1994, **23**, 125.

39. G. Condorelli, G. De Guidi, S. Giuffrida and L. L. Costanzo, *Coord. Chem. Rev.*, 1993, **125**, 115.

40. T. R. Krugh, *Curr. Op. Struct. Biol.*, 1994, **4**, 351.

41. D. Kessel and K. Woodburn, *Int. J. Biochem.*, 1993, **25**, 1377.

42. A. M. Pyle, J. P. Rehmann, R. Meshoyer, C.V. Kumar, N. J. Turro and J. K. Barton, *J. Am. Chem. Soc.*, 1989, **111**, 3051.

43. G. Condorelli, L. L. Costanzo, G. De Guidi, S. Giuffrida and S. Sortino, *Photochem. Photobiol.*, 1995, **62**, 155.

44. K. Jansen, P. Lincoln and B. Nondén, *Biochemistry*, 1993, **33**, 417.

45. U. Sehlstedt, S. K. Kim, P. Carter, J. Goodisman, J. F. Vollano, B. Nordén and J. C. Dabrowiak, *Biochem.*, 1994, **33**, 417.

46. J. McGhee and P. von Hippel *J. Mol. Biol.*, 1974, **86**, 469.

47. N. Paillous and P. Vicendo, *J. Photochem. Photobiol. B: Biol.*, 1993, **20**, 203.

48. M. Kelly, D. J. McConnell, C. OhUigin, A. B. Tossi, A. Kirsch-De Mesmaeker, A. Masschelein and J. Nasielski, *J. Chem. Soc., Chem Commun.*, 1987, 1821.

49. J. Piette, *J. Photochem. Photobiol. B: Biol.*, 1991, **11**, 241.

50. T. Artuso, J. Bernadou, B. Meunier, J. Piette and N. Paillous, *Photochem. Photobiol.*, 1991, **54**, 205.

51. I. E. Kochevar and D. A. Dun, 'Bioorganic Photochemistry', H. Morrison, Ed., J.Wiley, New York, 1989, Vol. 1, p. 273.

52. J. Piette, *J. Photochem. Photobiol. B: Biol.*, 1991, **11**, 241.

53. J. A. Hartley, K. Reszka and J. W. Lown., *Free Rad. Biol. Med.*, 1988, **4**, 337.

54. K. Kobayashi and T. Ito, *Photochem. Photobiol.*, 1972, **25**, 385.

55. J. L. Ravanat, M. Berger, F. Benard, R. Langlois, R. Ouellet, J. E. Van Lierr and J. Cadet, *Photochem. Photobiol.*, 1992, **55**, 809.

56. J. Sambrook, E. F. Fritsch and T. Maniatis, 'Molecular Cloning', C.S.H. Press, New York, 1989, Vol. 1-3.

57. J. Kelly, W. J. M. Van Der Putten and M. D. J. Mc Connell, *Photochem. Photobiol.*, 1987, **45**, 167.

58. J. V. Castell, M. J. Gomez-Lechon, M. A. Miranda and I. M. Morera, *J. Photochem, Photobiol. B: Biol.*, 1992, **13**, 71.

59. S. Giuffrida, G. De Guidi, S. Sortino, R. Chillemi, L. L. Costanzo and G. Condorelli, *J. Photochem. Photobiol. B: Biol.*, 1995, **29**, 125.

60. G. Condorelli, L. L. Costanzo, G. De Guidi, S. Giuffrida, P. Miano, S. Sortino and A. Velardita, *EPA Newsletter,* 1996, **58**, 60.

61. S. Monti, S. Sortino, G. De Guidi and G. Marconi, *J. Chem. Soc., Faraday Trans.*, 1997, **93**, 2269.

62. S. Sortino, G. De Guidi, S. Giuffrida, S. Monti and A. Velardita, *Photochem. Photobiol.*, 1997, accepted for publication.

63. E. Fasani, M. Mella, S. Monti, S. Sortino and A. Albini, *J. Chem. Soc., Perkin Trans. 2*, 1996, 1889.

64. S. Monti, S. Sortino, G. De Guidi and G. Marconi, *New J. Chem.*, 1997, accepted for publication.

65. S. Sortino, G. De Guidi, G. Marconi and S. Monti, 1997, submitted; S.Monti, S.Sortino and G. De Guidi, Proceedings of the 7th Congress of the European Society for Photobiology, Stresa, Italy, 1997, S30.

66. P. Bortolus and S. Monti, *Adv. Photochem.*, 1996, **21**, 1.

67. P. Bilski, L. J. Martinez, E. B. Koker and C. F. Chignell, *Photochem.Photobiol.*, 1996, **64**, 496.

68. L. L. Costanzo, G. De Guidi, S. Giuffrida, S. Sortino and G. Condorelli, *J. Inorg. Biochem.*, 1995, **59**, 1.

69. L. L. Costanzo, S. Giuffrida, G. De Guidi, S. Sortino, G. Condorelli and G. Pappalardo. *J. Inorg. Biochem.*, 1995, **57**, 115.

70. G. Condorelli, L. L. Costanzo, G. De Guidi, S. Giuffrida, E. Rizzarelli and G. Vecchio, *J. Inorg. Biochem.*, 1994, **54**, 257.

71. P. Bonomo, E. Conte, G. De Guidi, G. Maccarrone, E. Rizzarelli and G. Vecchio, *J. Chem. Soc., Dalton Trans.,* 1996, 4351.

72. S. Giuffrida, G. De Guidi, P. Miano, S. Sortino, G. Condorelli and L. L. Costanzo, *Photochem. Photobiol.*, 1996, **63**, 455.

73. G. De Guidi, G. Condorelli, S. Giuffrida, G. Puglisi and G. Giammona, *J. Inclusion Phenom.*, 1993, **15**, 43.

74. T. Hoshino, K. Ishida, T. Irie, F. Hirayama, K. Uekama and M. Yamasaki, *J. Incl. Phenom.*, 1988, **6**, 415.

75. G. Giammona, G. Cavallaro, G. Fontana, G. De Guidi and S. Giuffrida, *J. Pharm. Sci.* 1996, **4**, 273.

76. G. Giammona, G. Pitarresi, V. Tomarchio, G. De Guidi and S. Giuffrida, *J. Control. Release,* 1998, in press.

A Comparison between the Photochemical and Photosensitising Properties of Different Drugs

M. Tronchin,[1*] F. Callegarin,[1] F. Elisei,[2] U. Mazzucato,[2] E. Reddi[1] and G. Jori[1]

[1] Department of Biology, University of Padova
via Trieste 75, I-35120 Padova, Italy
[2] Department of Chemistry, University of Perugia,
via Elce di Sotto 8, I-06100 Perugia, Italy

1 INTRODUCTION

The study of the stability of commonly used drugs to the UV or visible radiation, as well as the phototoxicity of the original drug or its photoproducts represents a relatively new field of investigation. A database search on Medline® shows no article on this topic before 1973. Most of the scientific papers (74%) were produced after 1983.

In order to start a research line about the photostability and the photosensitising properties of drugs we selected four compounds, belonging to different classes of therapeutic molecules, and characterised by absorption bands in the near-UV and visible spectral range.

13-cis-Retinoic acid (Isotretinoin) is used against several skin diseases,[1] especially for the treatment of severe, recalcitrant nodulistic acne,[2] as well as for severe Gram-negative folliculitis and rosacea[3] not responding to traditional therapy. Isotretinoin therapy helps in healing wounds, UV-damaged and glucorticoid-damaged skin by enhancing keratinocyte proliferation and generation of extracellular matrix.[4] It has also been used as an anticancer and chemopreventer drug[5,6] even though some toxic effects, such as teratogenic action in pregnant women have been reported.[2] 13-cis-Retinoic acid, therefore, deserves further investigations in order to identify other possible side effects.

Nifedipine is a calcium channel blocker which is very useful in hypertension therapy, although vasodilatation and reflex tachycardia often represent undesired effects.[7] Myocardial ischaemia is often prevented by this drug.[8] Nifedipine is also used as a prophylactic treatment to prevent post-partum preeclampsia-eclampsia.[9] High altitude pulmonary oedema may be cured with nifedipine[10] and some authors have proposed this drug also for atherosclerosis[11] and cocaine-addition[12] therapy.

Chloramphenicol was the first broad-spectrum antibiotic drug to be isolated and represents a milestone in therapy: using this name as a keyword more than 23,000 articles are quoted in Medline®. Its main drawback is the possibility of aplastic anaemia,[13] which is particularly frequent in pediatric patients.

Chlorpromazine is a phenothiazine-type psychotropic drug, which has been used for more than 40 years in several psychiatric disorders that can be gathered under the nonspecific definition of 'agitation'.[14] This drug is also used against acute migraine[15] and it is the standard treatment for intractable hiccups.[16] More recently, some researchers proposed it as an antitumor agent[17] and against *V. cholerae* and ETEC-induced diarrhoea.[18] Photosensitivity is documented with this drug[19,20] although some researchers report that low doses may have a photoprotective action.[21]

2 MATERIALS AND METHODS

13-cis-Retinoic Acid (RET), Nifedipine (NIF), Chloramphenicol (CAF) and Chlorpromazine (CLP) were purchased from Sigma and used without further purification. All other reagents were spectroscopic-grade commercially available material.

Absorption measurements were normally performed by using a Lambda-2 Perkin-Elmer spectrophotometer connected to a computer for data digitalisation. Raw data were stored on a floppy disk and elaborated after the experiment. Occasionally, a Lambda-5 Perkin-Elmer spectrophotometer was used for single-wavelength readings.

Fluorescence data were normally obtained using an MPF-4 Perkin-Elmer spectro-fluorimeter connected to a chart recorder. Occasionally data were collected with an LS-50 Perkin-Elmer spectrofluorimeter connected to a computer for instrument control and data digitalisation.

The experimental setup used for the determination of triplet spectra (absorption maxima, λ_{max}, and extinction coefficients $\Delta\varepsilon_T$) and yields (ϕ_T) was that described elsewhere.[22,23] The excitation wavelength of 347 nm from a ruby laser (J.K., second harmonics) was used in nanosecond flash photolysis experiments (pulse width *ca.* 20 ns and energy < 5 mJ pulse^{-1}). Absorption spectra were recorded every 10 nm over the 300-800 nm spectral range averaging at least 10 shots per wavelength recorded. The triplet lifetimes (τ_T) were measured with laser fluence $\leq$ 1 mJ pulse^{-1}; at higher excitation energies triplet-triplet annihilation affects the decay kinetics of the triplet state of these compounds.

Singlet oxygen quantum yields (ϕ_Δ) were determined using the kinetic 1270 nm luminescence comparative techniques. The excitation source was the same laser apparatus as described above for flash photolysis. Luminescence was collected a few mm far from samples (in order to increase the signal) by a Germanium diode (Judson J16 8 Sp). To avoid interference with excitation and possible fluorescence light the direction of the Germanium diode was 90° rotated with respect to excitation light direction and two cut-off filters (Kodak Wratten 87C, 870nm, and Oriel 51362, 1050 nm, respectively) were interposed between the samples and the diode. The electric signal from the diode was amplified by a digital analyser Tektronix DSA 602, and digitalised to a computer (Tektronik PEP 301) for data storage and analysis. Logarithmic decay curves were extrapolated to zero time (I_0) and a calibration curve of I_0 *vs.* laser energy (L_E) was built. In the linearity range (laser fluence $\leq$ 2 mJ pulse^{-1}) the slope of the calibration curve M was used to calculate Φ_Δ using the formula

$$\Phi_{\Delta(x)} = \Phi_{\Delta(ref)} * M_{(x)} / M_{(ref)} \tag{1}$$

with phenalenone ($\Phi_\Delta = 0.97$) as reference.[24]

Irradiation of samples for photostability and photosensitsation experiments was performed with a Laser Point xenon lamp (mod. LX125/UV). Cuvettes were magnetically stirred and kept at 25 °C during irradiation by a 3-wall water-thermostatted cell. A water filter was present between the lamp and the samples in order to cut IR-radiation and to protect the molecules from thermal reactions. The fluence rate was modulated by varying the distance between the lamp and the sample. A light intensity *vs.* distance calibration curve was routinely performed with a LI-COR field photometer (mod. LI-185B) in order to ensure data reproducibility. For lower fluence rates neutral grey quartz filters were used.

Photostability of drugs was followed as a function of the irradiation time by measuring spectral changes. All samples were irradiated in air-saturated or nitrogen-flushed solution, respectively, in order assess the role of oxygen in drug degradation pathways.

The decay of tryptophan ethyl ester was followed by measuring the fluorescence emission spectrum of the molecule upon excitation at 290 nm with the MPF-5 spectophotofluorimeter used above. Peak intensities were corrected for trivial reabsorption. TrpEE concentration at each irradiation time was calculated by

$$[TrpEE]_x = [TrpEE]_0 * Fluor_x / Fluor_0 \tag{2}$$

where subctripts 0 and x refer to time 0 and time x, respectively.

Histidine photosensitisation studies were carried out in phosphate buffer solution (pH = 7.40) using the procedure described by Horinishi *et al.*[25] This method enables one to quantify histidine using a specific reaction between the imidazole side chain and a diazonium salt. The reaction product was detected using the absorption maximum at 480 nm. For each experiment a calibration curve with known amounts of histidine was built. Water-insoluble molecules (NIF) were studied adding Triton X-100 to a final 2% concentration in order to avoid drug aggregation.

Albumin and lysozyme photosensitisation was followed by measuring the decrease in the content of their tryptophan and histidine residues. Lysozyme photosensitisation was also followed by using its ability to lyse *Micrococcus lysodekticus* cells in pure water. Bacterial cell lysis was measured following the sample turbidity at 540 nm. In the saturation range (enzyme activity directly proportional to enzyme concentration) the activity was quantified and used to estimate the percentage of functional molecules.

Differential absorption spectra were measured by using 2-compartment cuvettes. Reference cuvettes contained albumin and a drug in two separated compartments. Sample cuvettes had albumin and drug in the same compartments.

Cholesterol photoproducts were studied by thin layer chromatography (TLC) following the NaBH₄-reduction method described by Girotti *et al.*[26] This procedure transforms all unstable photoproducts into a stable alcohol that may run a TLC without undergoing further transformation. The elution mixture was composed of heptane/ethylacetate (1:1, v/v). The spots were identified by spraying 50% H_2SO_4 over the layer and warming it (100 °C) until a blue/violet colour appeared. For each experiment a standard spot with cholesterol and its photoproducts were chromatographed together with the analysed samples. 7-Chetocholesterol (7α-OH and 7β-OH, respectively, after $NaBH_4$ reduction) was purchased while 5α-hydroperoxy-cholesterol (5α-OH after reduction) was obtained from cholesterol by photosensitisation with rose bengal[27] in the same apparatus which was used for drug irradiation.

3 RESULTS AND DISCUSSION

13-*cis*-Retinoic acid in ethanol (Fig. 1), showed a 344.5 nm absorption peak (molar extinction coefficient 3.5 x 10^4 M^{-1} cm^{-1}) and a 487.5 nm fluorescence emission peak. The absorbance was negligible at wavelengths longer than 450 nm. Nifedipine (Fig. 2) had a relatively sharp absorption peak at 237 nm ($\varepsilon = 1.7$ x 10^5 M^{-1} cm^{-1}) and a much broader absorption at ca. 350 nm ($\varepsilon_{357} = 4.0$ x 10^3 M^{-1} cm^{-1}). Again, the absorbance was negligible at $\lambda > 450$ nm, while no fluorescence was detected. Chloramphenicol (Fig. 3) showed a UV spectrum with a peak at 274 nm ($\varepsilon = 8.5$ x 10^2 M^{-1} cm^{-1}) and no fluorescence emission, while chlorpromazine (Fig. 4) displayed a main, sharp peak, at 258 nm ($\varepsilon = 1.1$ x 10^3 M^{-1} cm^{-1}) and a minor peak at 311 nm ($\varepsilon = 4.9$ x 10^2 M^{-1} cm^{-1}). Its fluorescence emission spectrum had a single peak at 450 nm.

Table 1 *Triplet state properties and singlet oxygen yields in ethanol*

Drug	λ_{max} (nm)	Lifetime (µs)	Triplet Quantum Yield	Singlet Oxygen Quantum Yield
Chlorpromazine	460	10	0.52	0.48
Nifedipine	-	< 0.1	-	< 0.01
Retinoic Acid	420	40	< 0.02	< 0.01
Chloramphenicol	-	< 0.1	-	< 0.01

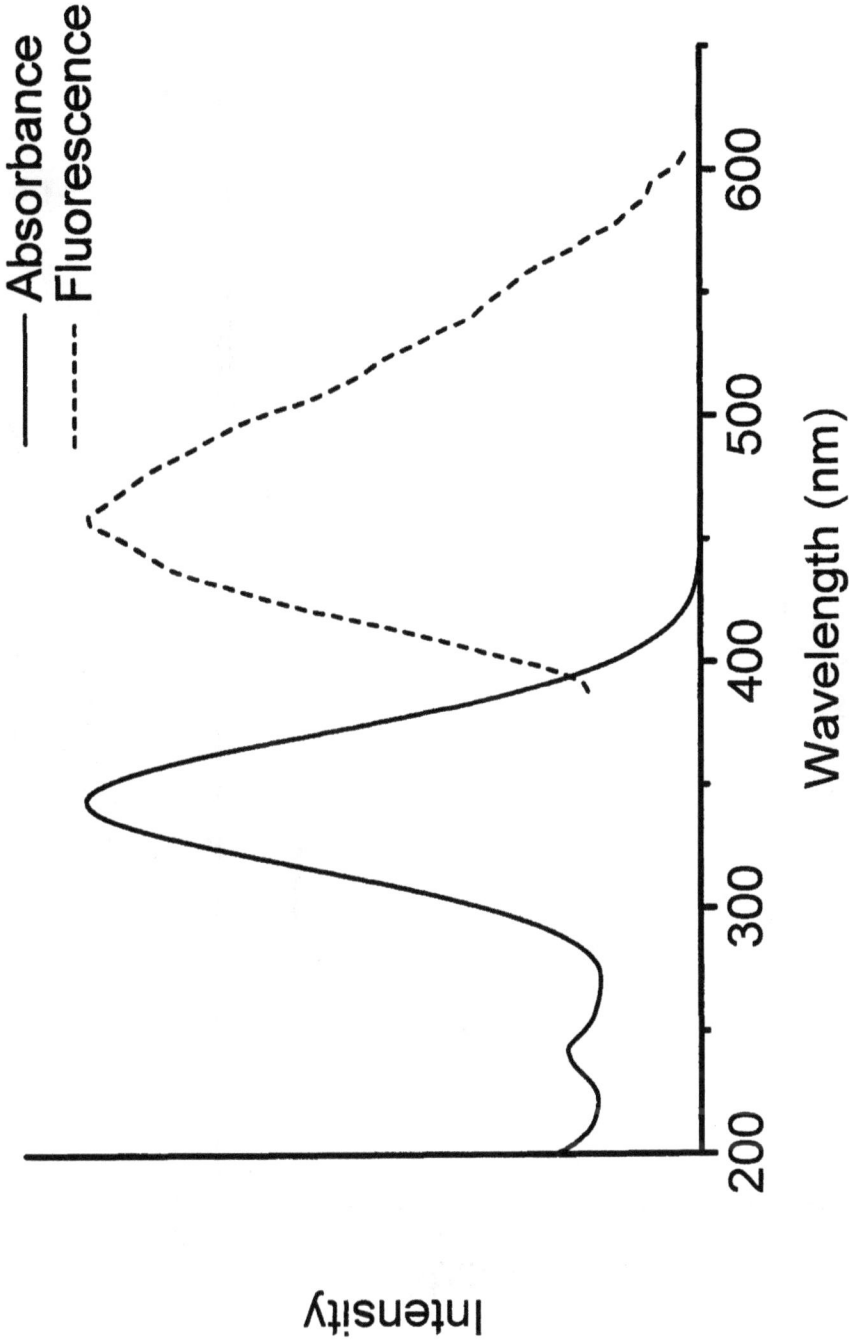

Figure 1 *Absorption and fluorescence emission (exc. = 344 nm) spectra of 13-cis-retinoic acid in ethanol.*

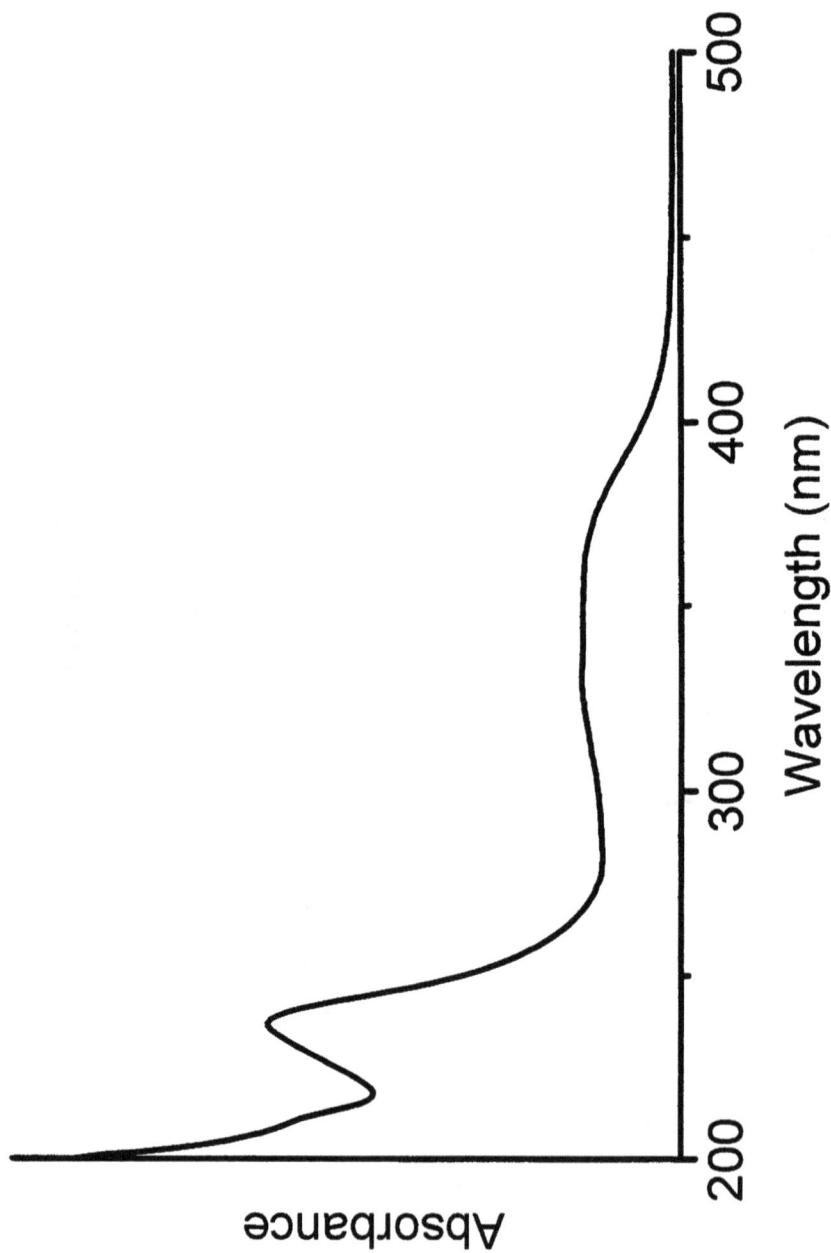

Figure 2 *Absorption spectrum of nifedipine in ethanol.*

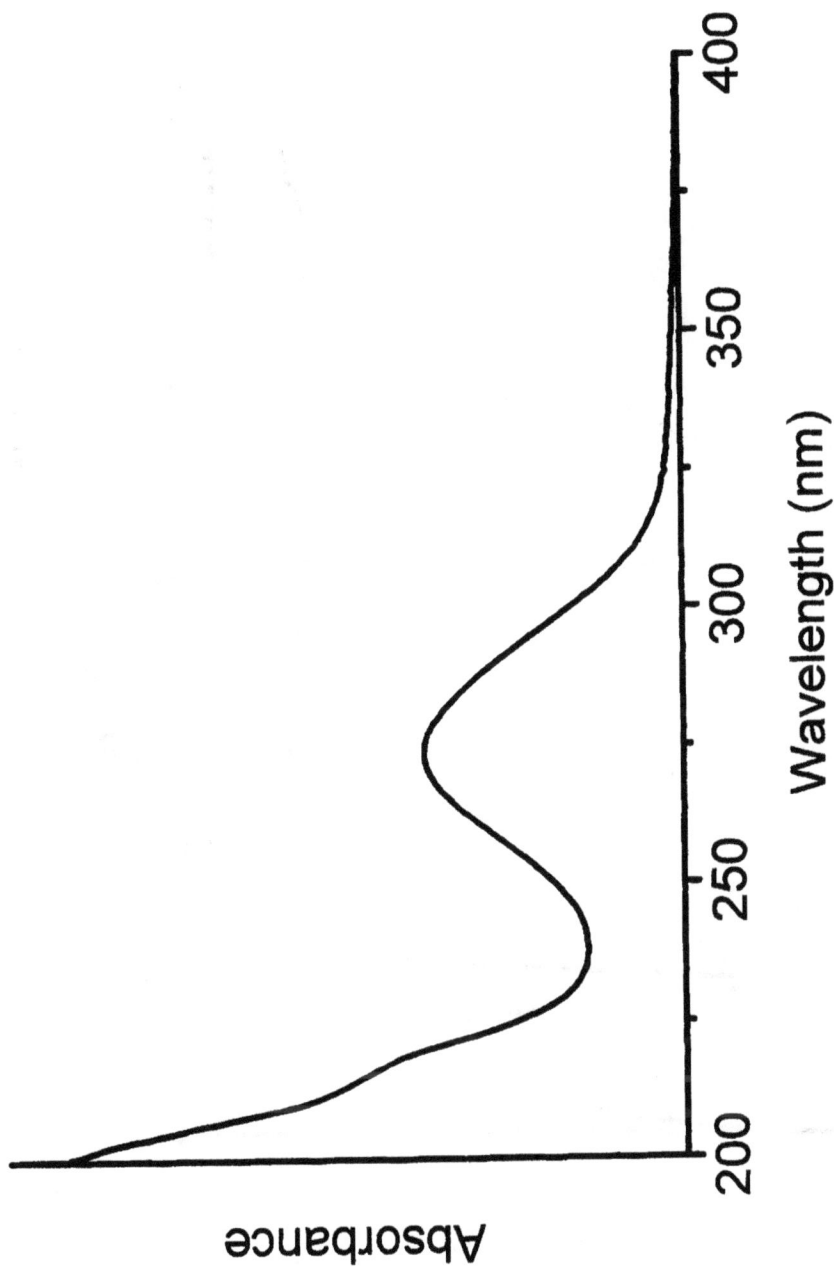

Figure 3 *Absorption spectrum of chloramphenicol in ethanol.*

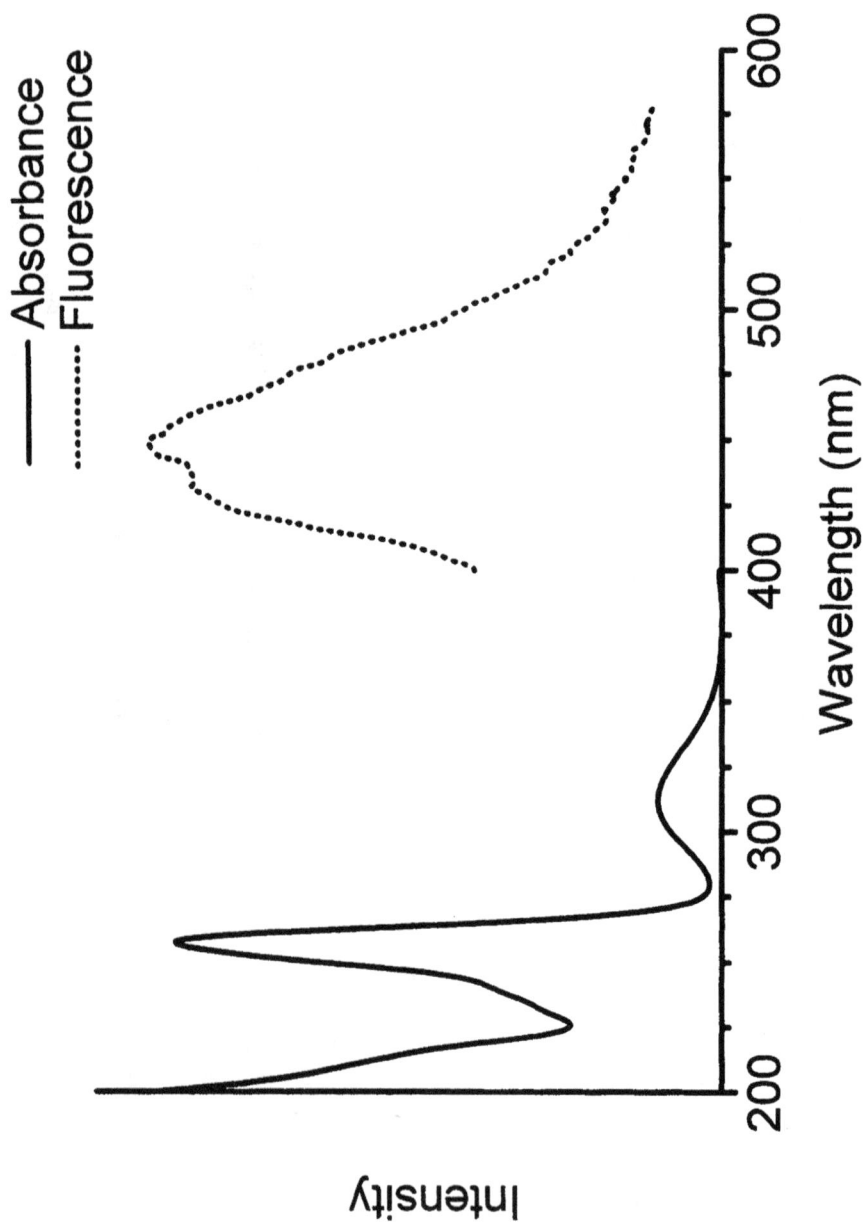

Figure 4 *Absorption spectrum of chlorpromazine in ethanol.*

In laser flash photolysis experiments (Table 1) direct excitation of CLP produced a T_1 state with high efficiency ($\Phi_T = 0.52$) and a triplet-triplet absorption maximum at 460 nm. The lifetime of the transient was 10 μs. No detectable transient was generated from the other compounds under the same conditions. On the other hand, all compounds were able to quench efficiently the lowest triplet state of phenanthrene in ethanol (triplet energy of 62 kcal mol^{-1}) and xanthone in CH_3CN (triplet energy of 74 kcal mol^{-1}). This evidence strongly supports the hypothesis that the triplet state properties of the other molecules (quantum yield and/or lifetime) generated by direct irradiation are below the detection limit of our instrument. In sensitised conditions (i.e. in presence of phenanthrene or xantone) 13-*cis*-retinoic acid showed a detectable triplet-triplet absorption at 420 nm ($\tau = 40$ μs). Such a sensitised triplet formation was not observed for nifedipine and chloramphenicol.

A parallel trend was observed between the triplet and the singlet oxygen formation quantum yields. A well-detectable singlet oxygen formation was observed for chlorpromazine ($\Phi_\Delta = 0.48$) and no signal was generated from the other compounds (Table 1).

As regards photostability 13-*cis*-retinoic acid (Fig. 5) shows a disappearance in its 344 nm peak and a new peak appeared at 261 nm. No isosbestic point was detected, indicating that more than one intermediate was involved in the reaction. No significant difference was observed between air-saturated and nitrogen-flushed solutions during irradiation.

Nifedipine, on the other hand, exhibited three isosbestic points (Fig. 6) indicating the presence of a single intermediate in the decay process. The 357 nm broad peak decreased during irradiation while a new peak developed at 282 nm. Again the presence of oxygen did not modify the degradation spectra and the decay kinetics (Fig. 7).

On the contrary, chlorpromazine, showed a strong influence of oxygen upon its decay mechanism: oxygen strongly reduced the degradation rate (Fig. 8) converting the reaction from an apparent first-order to an apparent zero-order (no concentration dependence was studied to date in order to test this hypothesis). Since the first triplet state of this molecule is strongly efficient (Table 1) in producing singlet oxygen (92% of triplets generated singlet oxygen under our conditions), an alternative decay pathway of this excited state could be the limiting step in the degradation pathway.

As regards the photosensitising properties (Table 2), 13-cis-retinoic acid appeared to be quite inefficient toward all the substrates studied by us.

Nifedipine, on the other hand, was a fairly efficient photosensitiser toward TrpEE. The sensitisation process competed with the photodegradation but also more efficient ($Kv_{(sens)} = 4 \times 10^{-8}$ M s^{-1} and $\tau_{(TrpEE)} = 108$ s, while $Kv_{(degr)} = 1 \times 10^{-6}$ M s^{-1} and $\tau_{(NIF)} = 52$ s), although large excess of nifedipine was needed in order to achieve the complete modification of TrpEE ($Kv_{(sens)} = 4 \times 10^{-8}$ M s^{-1} and $\tau_{(TrpEE)} = 108$ s, while $Kv_{(degr)} = 1 \times 10^{-6}$ M s^{-1} and $\tau_{(NIF)} = 52$ s). No degradation of histidine, albumin or cholesterol was observed with nifedipine, probably because the rate of photosensitised modification of these targets was substantially lower than that typical of nifedipine photodegradation. Moreover, differential spectra and preliminary analysis with elution chromatography, showed no dark interaction between nifedipine and albumin. Analogously, lysozyme activity was not modified after irradiation in the presence of nifedipine.

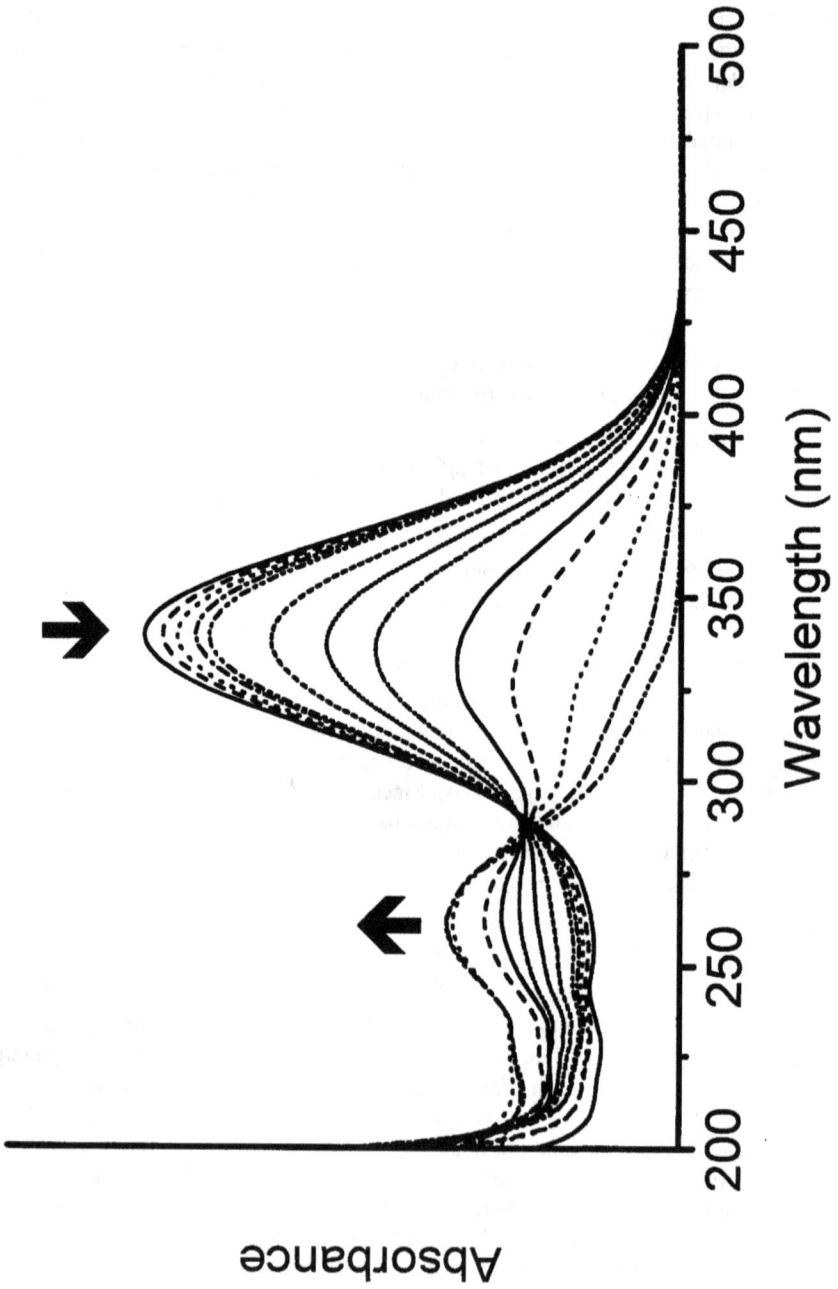

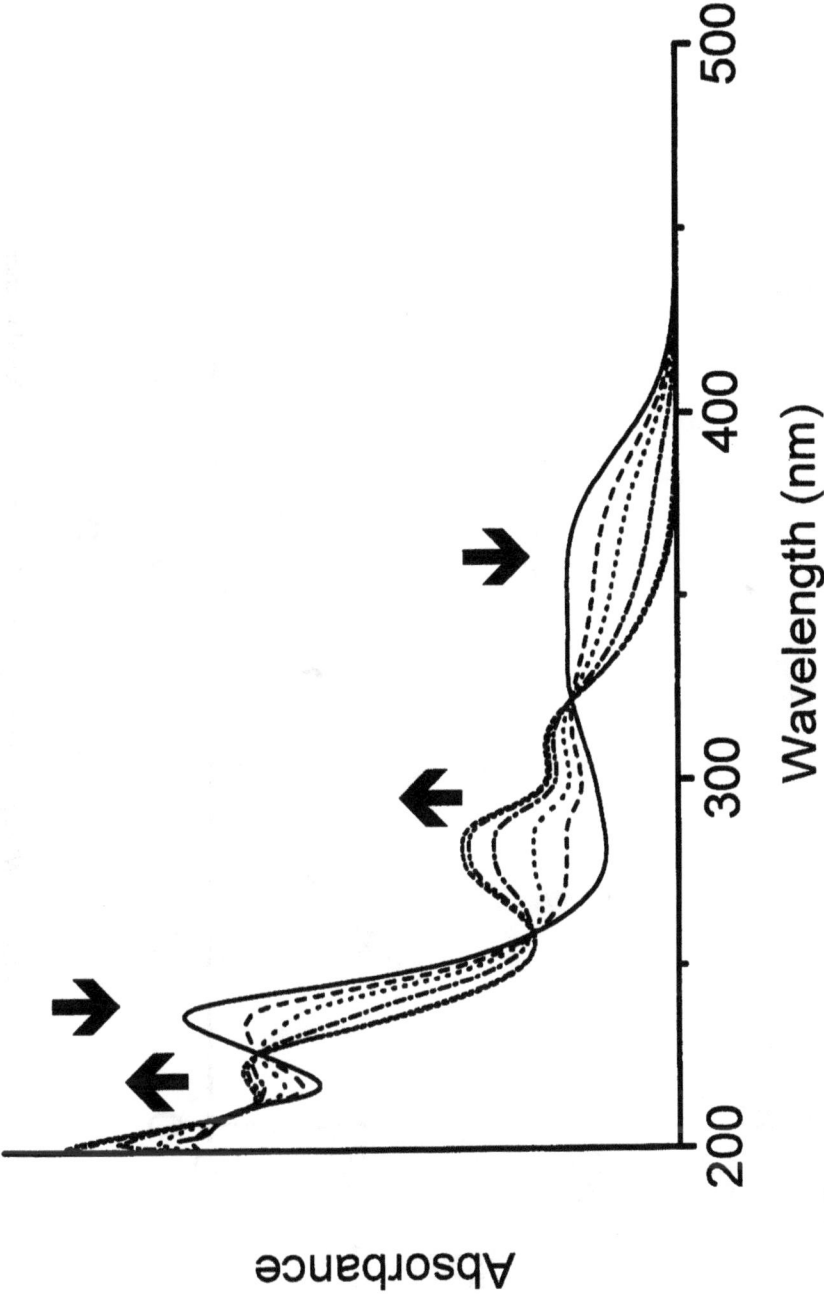

Figure 6 *Time-dependent absorption spectra of nifedipine irradiated in air-saturated ethanol solution.*

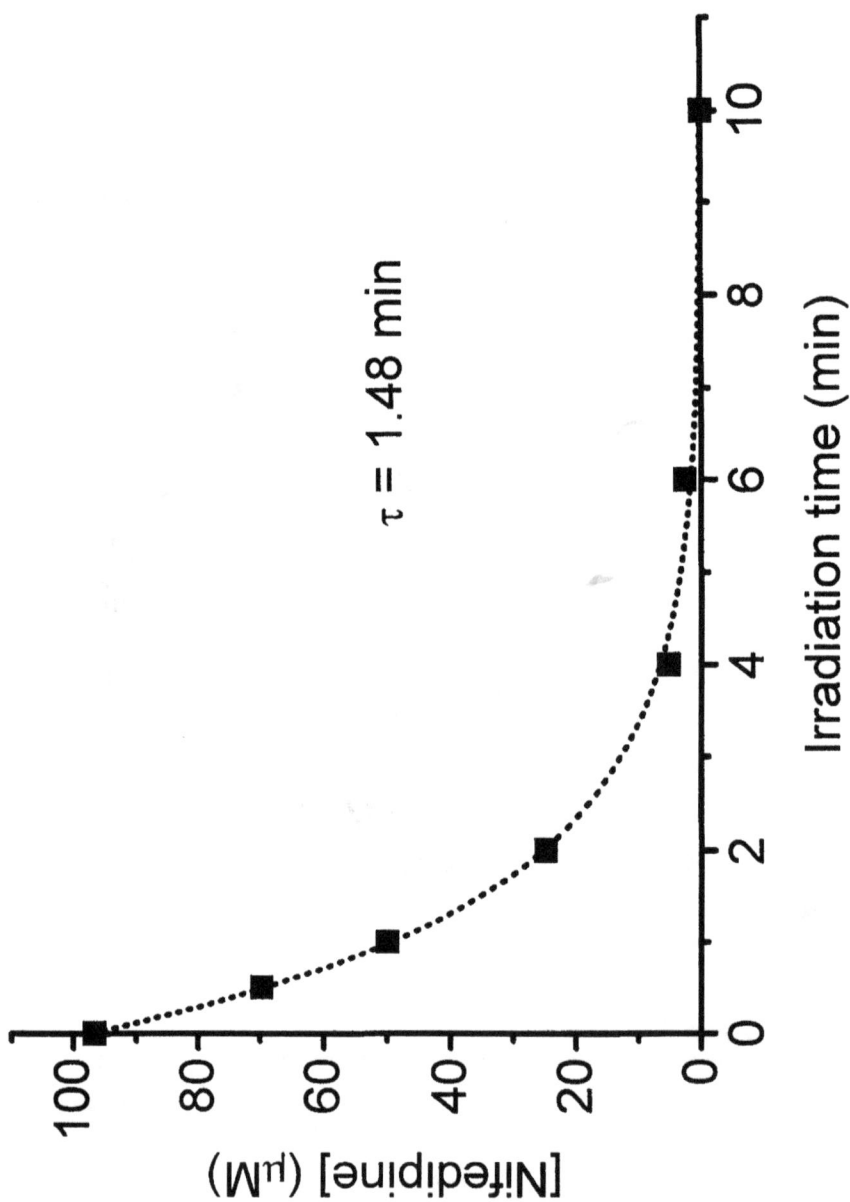

Figure 7 *Nifedipine photodegradation in air-saturated ethanol solution.*

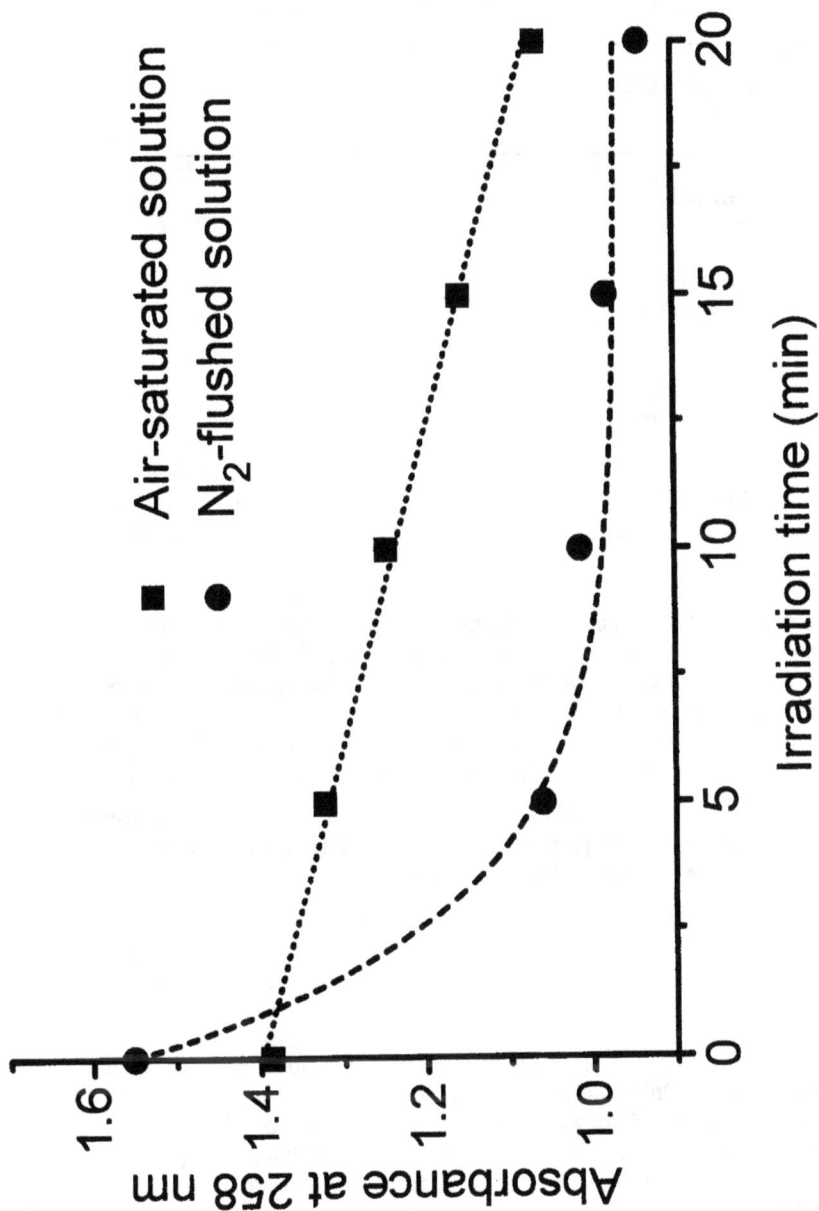

Figure 8 *Chlorpromazine photodegradation in air-saturated and nitrogen-flushed ethanol solutions.*

Table 2 *Summary of photosensitising properties of selected drugs*

Substrate	Chlorpromazine	Nifedipine	Retinoic Acid
Tryptophan Ethyl Ester	+	+	-
Histidine	+	-	-
Albumin (Trp or His)	?	-	
Lisozyme (Trp and activity)	+	-	-
Cholesterol	-	-	-

The study of TrpEE photosensitisation by chlorpromazine was complicated by the formation of a chlorpromazine photodegradation product which presents a broad and intense fluorescence emission in the 340 nm region, thereby overlapping the tryptophan emission spectrum (maximum at 360 nm). It was thus necessary to use small drug concentration in order to follow the decrease of TrpEE concentration in ethanol. The enzyme activity of lysozyme was strongly reduced after irradiation in presence of chlorpromazine.

The generation of photoproducts that interfere with the assay also prevented us from assessing the effect of chlorpromazine toward histidine, both as a free amino acid or incorporated into the albumine polypeptide chain.

No effect of chlorpromazine was observed toward cholesterol.

4 CONCLUSIONS

All the drugs screened by us proved to be markedly photolabile, with the possible exception of chloramphenicol. Only chlorpromazine generated a long-lived triplet state with a good efficiency. Hence, this drug was able to generate highly reactive singlet oxygen in a diffusion-controlled process. Since oxygen strongly inhibited the photodegradation of UV/visible light-irradiated chlorpromazine, it appears reasonable to hypothesise the photo-sensitised process by photodynamic pathways (i.e. oxygen-involving) and the photoinduced degradation represents competitive channels for triplet deactivation in the case of chlorpromazine. This may explain the observed inefficiency of chlorpromazine as a photosensitiser towards a variety of substrates, in spite of its ability to generate singlet oxygen. Only TrpEE is affected by photoexcited chlorpromazine, probably as a consequence of the high rate constant for the reaction of singlet oxygen with the indole side chain[20]; this rate constant is in the order of 10^8 s^{-1}; hence three orders of magnitude larger than that typical of the reaction between singlet oxygen and cholesterol, which was actually found to be insensitive to chlorpromazine photosensitisation.

It is likely that in the other drugs the photoinduced degradation and the photosensitisation processes (see nifedipine) occur via the lowest excited singlet state, whose short lifetime prevents a significant interaction with oxygen. Electron transfer processes leading to the generation of radical-type transient species could be mainly responsible for the observed effects. This hypothesis needs to be further checked by fast spectroscopic techniques and the use of radical scavengers.

Acknowledgments. This work was financially supported by the Istituto Superiore della Sanità (Rome, Italy) in the framework of the project 'Physico-chemical properties of drugs and their safety of use'.

References

1. C. H. Dicken, *J. Am. Acad. Dermatol.*, 1984, **11**, 541.
2. A. R. Shalita, R. B. Armstrong, J. J. Leyden, P. E. Pochi and J. S. Strauss, *Cutis*, 1988, **42**, 1.
3. C. E. Orfanos, R. Ehler and H. Gollnick, *Drugs*, 1987, **34**, 459.
4. G. J. Gendimenico and J. A. Mezick, *Skin Pharmacol.*, 1993, **6 suppl. 1**, 24.
5. S. M. Lippman, R. A. Heyman, J. M. Kurie, S. E. Benner and W. K. Hong, *J. Cell Biochem. Suppl.*, 1995, **22**, 1.
6. G. J. Kellof, C. W. Boone, J. A. Crowell, V. E. Steele, R. Lubet and C. C. Sigman, *Cancer Epidemiol. Biomarkers Prev.*, 1994, **3**, 85.
7. J. L. Palma Gamiz, *Cardiology*, 1997, **88 suppl. 1**, 39.
8. L. H. Opie, *Eur. Hearth J.*, 1997, **18**, A92.
9. E. F. Magann and J. N. Jr. Martin, *Obstet. Gynecol. Clin, North. Am.*, 1995 **22**, 337.
10. H. N. Hultgren, *Ann. Rev. Med.*, 1996, **47**, 267.
11. T. F. Luscher, R. R. Wenzel, P. Moreu and H. Takase, *Cardiol. Clin.*, 1995, **13**, 579.
12. D. J. Calcagnetti, B. J. Keck, L. A. Quatrella and M. D. Schechter, *Life Sci.*, 1995, **56**, 475.
13. P. M. Leary, *Drug. Saf.*, 1991, **6**, 171.
14. R. H. Levy, *Pharmacotherapy*, 1996, **16**, 125S.
15. N. T. Mathew, *Neurol. Clin.*, 1997, **15**, 61.
16. N. L. Friedman, *Pharmacotherapy*, 1996, **16**, 986.
17. G. R. Jones, *Med. Hypotheses*, 1996, **46**, 25.
18. G. H. Rabbani, *Dan. Med. Bull.*, 1996, **43**, 173.
19. Y. Harth and M. Rapoport, *Drug Saf.*, 1996, **14**, 252.

20. J. D. Spikes, in 'The Science of Photobiology', K. Smith, Ed., Plenum Press, New York, 2nd edition 1989, Chapter 3, p. 85.

21. E. L. Lozovskaia, E. N. Makareeva and I. U. V. Makedonov, *Biofizika*, 1997, **42**, 549.

22. A. Romani, F. Elisei, F. Masetti and G. Favaro, *J. Chem. Soc., Faraday Trans.*, 1992, **88**, 2147.

23. H. Görner, F. Elisei and G. Aloisi, *J. Chem. Soc., Faraday Trans.*, 1992, **88**, 29.

24. R. Schmidt, C. Tanielan, R. Dunsbach and C. *Wolff J. Photochem. Photobiol. A: Chem.*, 1994, **79**, 11.

25. H. Horinishi, Y. Hachimori, K. Kurihara and K. Shibata, *Biochim. Biophys. Acta*, 1964, **86**, 477.

26. A. W. Girotti, G. J. Bachowski and E. Jordan, *Lipids*, 1987, **22**, 401.

27. G. Jori, G. Galliazzo and E. Scoffone, *Biochemistry*, 1969, **8**, 2868.

Photostability of Drug Substances and Drug Products: A Validated Reference Method for Implementing the ICH Photostability Study Guidelines

Henry D. Drew
U. S. Food and Drug Administration, Center for Drug Evaluation and Research
Division of Testing and Applied Analytical Development
1114 Market Street, St. Louis, Missouri 63101-2045, U. S. A.

1 INTRODUCTION

The International Conference on Harmonization (ICH) Expert Working Group for Photostability Testing, partly on the basis of pilot studies conducted by the Japanese Pharmaceutical Manufacturing Association (JPMA), have proposed guidelines for photostability testing.[1] The U.S. Food and Drug Administration (FDA) and the United States Pharmaceutical Research and Manufacturers Association (PhRMA) conducted a collaborative study based on these guidelines and the JPMA pilot studies.[2] Consistent with the JPMA pilot studies and Option 2 of the ICH guidelines, this collaborative study used Cool White fluorescent tubes to simulate ambient indoor room lighting and UV-A fluorescent tubes to simulate the UV component of window-filtered sunlight. Similar sources were adopted by the American Society for Testing and Materials (ASTM) in the collaboratively developed "Standard Test Method for Accelerated Testing for Color Stability of Plastics Exposed to Indoor Fluorescent Lighting and Window-Filtered Daylight".[3]

Although there are numerous photostability studies of pharmaceutical drug substances and drug products in the literature, few articles have addressed procedures and protocols for performing such studies. Some principles and practices for drug photodegradation studies have been presented by Moore.[4] Anderson et al.[5] summarized photostability testing procedures used in the United Kingdom and Nema et al.[6] have provided an overview of pharmaceutical photostability testing considerations. This overview also includes a glossary of photochemistry terms, derived from the improper usage of terms used in describing pharmaceutical photostability as expressed by Piechocki et al.[7]

1.1 Collaborative Study Objectives

The primary goal of the FDA/PhRMA collaborative study was to confirm the photostability testing procedures proposed in the ICH Expert Working Group guidelines aimed at developing a Reference Method for Photostability Testing of Pharmaceutical Products. The collaborative study objectives included evaluation of a) the Cool White and UV-A fluorescent sources, b) the quinine actinometry system, c) the sample photolysis procedures applied to representative active drug substances and dosage form pharmaceutical products, and d) the ICH proposed minimum exposure limits for visible and near ultraviolet radiation.

The ICH proposed minimum exposure levels for the Cool White and UV-A fluorescent sources have been defined as 1.2 M-lux-hours and 200 watt-hours/square meter, respectively. The JPMA pilot study proposed a chemical actinometer system using a quinine solution to monitor the integrated exposure of samples with the UV-A fluorescent source. This system, with a modification of the actinometer photolysis cell, has been extended to the Cool White fluorescent source for the FDA/PhRMA collaborative study. The use of the quinine actinometer system for monitoring exposure to Cool White radiation has not been endorsed by the ICH Expert Working Group.

A series of papers on photostability issues has been published in the May-June issue of USP Pharmacopeial Forum (PF). These papers include a presentation of the ICH photostability process,[8] the FDA/PhRMA collaborative study,[9] and more fundamental studies on quinine photochemistry.[10] Another paper on the FDA/NIST collaborative study to calibration of the quinine actinometer system for monitoring UV-A fluorescent radiation will be published in the PF. In due course, a general chapter on photostability will be incorporated in the United States Pharmacopeia.

1.2 Collaborative Study Design

The collaborative study was designed jointly by the U.S. FDA and the PhRMA. The FDA Division of Testing and Applied Analytical Development (DTAAD - formerly the Division of Drug Analysis) served as the coordinating laboratory. The study was designed to determine quantitatively the loss of active drug substance after controlled exposure to photolytic radiation from Cool White and UV-A fluorescent sources. The quantitation procedures used stability-indicating chromatographic methods based on accepted, compendial procedures for the drug substances and formulated drug products involved in the study. The photolysed and control samples were assayed in triplicate to provide an assessment of intra- and inter-laboratory precision.

The photolysis apparatus consisted of standard fluorescent ceiling fixtures and Cool White and UV-A fluorescent tubes from a defined manufacturer. Each collaborating laboratory was requested to secure materials for the photolysis apparatus from the local marketplace.

Each collaborating laboratory was provided with a detailed study protocol specifying the experimental procedures for constructing the photolysis apparatus, performing the photolysis experiments, and quantitating the active drug substance. System suitability criteria were established for the photolysis apparatus and the assay procedures. The coordinating laboratory supplied the active drug substance, drug products, and quinine hydrochloride dihydrate investigated in the study as well as some commodities. Most commodities required in the study were supplied by each collaborating laboratory.

2 EXPERIMENTAL

2.1 General

2.1.1 Photolysis Apparatus. The photolysis apparatus used commercially available fluorescent ceiling fixtures and fluorescent tubes. The UV-A source consisted of two 24-inch, 20 watt tubes designated Sylvania "Black Light-Blue" Model F20T12/BLB fluorescent tubes. The Cool White source consisted of two 24-inch, 20 watt tubes designated Sylvania "Cool White" Model F20T12/CW fluorescent tubes. The lamp housing/power supply consisted of a standard 2-fluorescent-tube ceiling fixture with the tubes 6 cm apart.

2.1.2 Photolysis Samples. The drug substance and drug product photolysis samples were supplied by the coordinating laboratory. The drug substance samples included furosemide, megestrol acetate, metronidazole, and nifedipine. These powder samples were sized to less than 250 F by sieving through a No. 60 mesh screen.

Solid drug formulation samples, obtained from commercial suppliers, were furosemide (80 mg), megestrol acetate (40 mg), and metronidazole (500 mg) tablets and nifedipine (10 mg) soft shell capsules. Injection solution samples were furosemide (10 mg/mL) pre-loaded syringe product with a brown plastic primary wrapping and metronidazole (5 mg/mL) in a plastic intravenous (IV) pouch.

2.1.3 Quinine Actinometry Solution. A 2% (w/v) solution of quinine hydrochloride dihydrate was prepared in water and maintained protected from light. Standard 1 cm quartz spectrophotometer cells were used as actinometer cells. These cells were completely filled with quinine actinometry solution, all air eliminated, and the cells tightly stoppered.

2.2 Photolysis Procedures

The photolysis studies were conducted as follows:

2.2.1 Photolysis Apparatus System Suitability. The photolysis apparatus system suitability test consisted of a linearity study of the quinine actinometer absorbance change at 400 nm versus photolysis exposure time. The photolysis apparatus was assembled and new lamps burned in for at least 10 hours before use. After burn-in the lamps were turned on and equilibrated for about 30 minutes before use. Three quinine actinometer cells were employed: two actinometer cells exposed to radiation and the third actinometer cell, tightly wrapped in aluminum foil, use as a control cell. The two active actinometer cells and control cell were placed under the photolysis source, centered between the fluorescent tubes at least 10 cm from the end of the tubes. Periodically, the actinometer and control cells were removed from the photolysis apparatus, the optical surfaces cleaned, any small bubbles on the optical surface of the cell dislodged, and the absorbance of each cell at 400 nm recorded. The actinometer cells were replaced under the photolysis source and photolysis continued with periodic absorbance measurement until the desired quinine absorbance change was realized.

2.2.2 Photolysis of Drug Substance. A portion of the ground powder, accurately weighed, was placed in the provided plastic Petri dish to a uniform thickness of about 2 mm. The Petri dish without the cover was placed under the fluorescent source centered between the two fluorescent tubes. Two quartz actinometer cells with secure stoppers, previously filled with quinine actinometer solution, were placed on either side of the Petri dish centered on the two fluorescent tubes and parallel with the fluorescent tube ends. The distance of the drug substance photolysis sample was adjusted so that it was the same distance from the fluorescent tubes as the top of the quinine actinometer cells. A similar weight of powdered sample, a control sample, was placed in aluminum foil and wrapped tightly to exclude all radiation. The control sample was placed under the fluorescent tubes in close proximity to the photolysis sample and quartz actinometer cells. Periodically, the absorbance of the quinine actinometer solutions at 400 nm was recorded until the desired absorbance change was realized.

2.2.3 Tablet and Capsule Samples. Ten of the dosage units, tablets or capsules, were placed in a closely packed (but not touching) array centered under the fluorescent tubes. Two quartz actinometer cells with secure stoppers, previously filled with quinine actinometer solution, were placed on either side of the sample array centered on the two fluorescent tubes and parallel with the fluorescent tube ends. The distance of the tablets or capsules was adjusted to the same distance from the fluorescent tubes as the top of the quinine actinometer cells. Ten additional dosage units, wrapped in aluminum foil to exclude all radiation, were placed under the fluorescent tubes in close proximity to the photolysis sample. Periodically, the absorbance of the quinine actinometer solutions at 400 nm was recorded until the desired absorbance change was realized.

2.2.4 Furosemide Injection Sample. The Furosemide Injection formulation was a pre-filled clear glass syringe system with each unit individually wrapped in brown plastic. Two units of this formulation were photolysed: the first remaining in the plastic wrapping and the second with the plastic wrapping removed. During the photolysis of each unit a control sample, wrapped tightly in aluminum foil, was placed under the photolysis source. Two quinine actinometer cells were placed, one at each end of the syringe, under the fluorescent tubes. The distance of the syringe and the quinine actinometer cells was adjusted such that the top surfaces were equidistant from the fluorescent tubes. A third quinine actinometer cell, a control solution wrapped tightly in aluminum foil, was placed under the fluorescent tubes near the syringe sample. Periodically one of the quinine actinometer cells was removed and the absorbance at 400 nm recorded until the desired quinine absorbance change was observed.

2.2.5 Metronidazole Injection Sample. The Metronidazole Injection formulation was a 5 mg/mL, 100 mL IV plastic pouch system. Photolysis of the injection solution in the plastic IV pouch and in a quartz spectrophotometer cell was performed.

The plastic IV pouch was placed under the fluorescent tubes and bracketed with two quinine actinometer cells. The distance of the quinine actinometer cells or the IV pouch was adjusted such that they were equidistant from the fluorescent tubes. A third quinine actinometer cell, a control solution tightly wrapped in aluminum foil, was placed under the tubes near the pouch sample as was a control IV pouch tightly wrapped in aluminum foil.

Periodically, one of the quinine actinometer cells was removed and the absorbance at 400 nm recorded. Photolysis was continued until the desired exposure was achieved. Similarly, photolysis of the metronidazole IV pouch solution placed in quartz spectrophotometer cells was performed.

3 RESULTS AND DISCUSSION

3.1 Analysis of the Primary Photolysis Data

Each collaborating laboratory was requested to perform a total of 24 photolysis experiments and assay both the photolysed and control samples from each experiment in triplicate, a total of 72 assays per collaborating laboratory. Compliance with the testing protocol was exceptional, with 125 of 144 experiments reported. Of the 19 experiments not reported, 16 experiments were not performed by collaborating laboratory 6.

As with any collaborative study, some extraneous results are expected. Five collaborator results were rejected for the following reasons: a) specific comments from the collaborating laboratory about the validity of the results [two rejected values]; b) reported data were inconsistent with the defined study protocol (an erroneous sample weight) [one rejected value]; and c) outlier results based on the Grubbs test [five rejected values]. The Grubbs test[11] was calculated at the 95 % confidence level. The Grubbs test assumes a homogeneous sample. Because the sample preparation protocol requires compositing the solid photolysis samples to ensure homogeneity, the Grubbs outlier test is appropriate. Sampling or assay difficulties were evident in the results rejected by the Grubbs test. The average inter-laboratory standard deviation for these rejected results was 9.0%. The average interlaboratory standard deviation for all retained experimental results is less than 2.0%. The statistical analysis of the collaborative study will not be reported here.

3.2 Photolysis Apparatus System Suitability

The photolysis apparatus system suitability results are presented in Tables 1 and 2 for the UV-A and Cool White photolysis sources, respectively. These results are summarized as the linear regression analysis of the quinine actinometer cell absorbance at 400 nm versus exposure time. Two quinine actinometer cells were photolysed simultaneously.

A comparison of the duplicate results for the quinine actinometer cell indicates that, within each collaborating laboratory, the quinine actinometer provides a reproducible monitor of integrated radiation intensity for a given apparatus. This is indicated by the consistency in slope (AU/hour), which differs for the UV-A sources by less than 4% for each apparatus. For the Cool White photolysis assembly, the interlaboratory slope difference indicates a variation of less than 2%.

The photolysis slopes observed between laboratories for both the UV-A and Cool White photolysis sources are considerably different. This is not unexpected in view of variations in a) individual lamp housing power supplies, b) the spectral intensity of individual lamps, c) the operating voltage applied to the lamp assembly, and d) the distance of the quinine actinometer cells from the lamp source. Collaborator 3 assembled four separate Cool White

Table 1 *Summary of photolysis apparatus system suitability results for UV-A photolysis source.*

Lab	Cell	Slope (AU/hr.)	Intercept (AU)	Cor. Coef.
1	A	0.0351	0.0346	0.9834
	B	0.0361	-0.0062	0.9672
2	A	0.0293	0.0662	0.9996
	B	0.0282	0.0690	0.9994
3	A	0.0149	0.0543	0.9997
	B	0.0149	0.0542	0.9997
4	A	0.0278	0.0633	0.9988
	B	0.0270	0.0619	0.9985
5	A	0.0259	0.0382	0.9970
	B	0.0263	0.0367	0.9916
6	A	0.0190	0.0517	0.9995
	B	0.0198	0.0488	0.9993

Table 2 *Summary of the photolysis apparatus system suitability results for Cool White photolysis source.*

Lab	Cell	Slope (AU/hr.)	Intercept (AU)	Cor. Coef.
1	A	0.00248	0.0832	0.9975
	B	0.00251	0.0816	0.9961
2	A	0.00272	0.0820	0.9918
	B	0.00267	0.0800	0.9906
3	A	0.00150	0.0575	0.9980
	B	0.00152	0.0620	0.9965
4	A	0.00268	0.0907	0.9908
	B	0.00269	0.0900	0.9904
5	A	0.00424	0.0407	0.9939
	B	0.00419	0.0406	0.9914
6	A	0.00206	0.0711	0.9971
	B	0.00203	0.0692	0.9968

photolysis systems and reported that the lux meter readings at equivalent positions and distance from the lamp ranged from 10,000 to 17,000 lux. If it is assumed that the spectral power distribution of these lamps remains constant, the absolute intensity obtained from each assembly is not critical but does demarcate the exposure time necessary to achieve the desired target exposures.

Collaborator 4 reported a negative curvature in quinine absorbance versus photolysis time for the Cool White assembly. Curvature was not reported by the other collaborating laboratories. A possible cause of this curvature could be the reduction in lamp intensity during the first several hours of use (i.e. a burn-in period). This burn-in period is expected with fluorescent sources (private communication, Atlas Electric Devices, Inc., Chicago, IL). The ICH guidelines do not endorse the quinine actinometer system for monitoring photolysis with the Cool White photolysis source.

3.3 Drug Substance, Tablet, and Capsule Photolysis

3.3.1 Drug Substance Photolysis. The results of the bulk drug substance photolysis are presented in Figures 1a and 1b for the UV-A and Cool White photolysis systems, respectively. Furosemide, megestrol acetate, and metronidazole do not degrade on exposure to either UV-A or Cool White radiation. Nifedipine undergoes significant photodegradation with UV-A and Cool White radiation.

It is interesting to note that nifedipine photodecomposition is more extensive with Cool White, visible radiation (57.3% of nifedipine decomposed) than with UV-A, near ultraviolet radiation (12.6% of nifedipine decomposed). Nifedipine absorbs ultraviolet radiation (UV-A) more extensively than visible (Cool White) radiation. Presumably the Cool White radiation penetrates farther into the powder particles than does near ultraviolet radiation, thereby exposing more nifedipine to damaging photolytic radiation.

3.3.2 Tablet and Capsule Photolysis. The UV-A and Cool White photolysis studies of the tablet and capsule formulations are presented in Figures 2a and 2b, respectively. As with the drug substance photolysis studies, the furosemide, megestrol acetate, and metronidazole tablet samples do not photodegrade with either UV-A or Cool White radiation exposure.

Based on the interlaboratory average, the Nifedipine soft gel capsule formulation appears to undergo some photodegradation from UV-A exposure and even more with Cool White exposure. These results are consistent with the observed sensitivity of the nifedipine drug substance to UV-A and Cool White radiation, indicating a greater sensitivity of the Nifedipine soft gel capsules to Cool White radiation.

Photolysis of solid state samples, drug substance, tablets, and hard shell capsules is a surface phenomenon. Photolysis radiation does not penetrate beyond the surface layer of material. In addition, as the surface layer undergoes photolysis, the photoproducts effectively absorb the incident radiation and thus prevent further penetration of radiation and photodegradation. Thus, solid state photolysis is a self-limiting process; the extent of photolysis is limited not by the photolysis radiation but by the depth the photolysis radiation penetrates and the total exposed sample surface area.

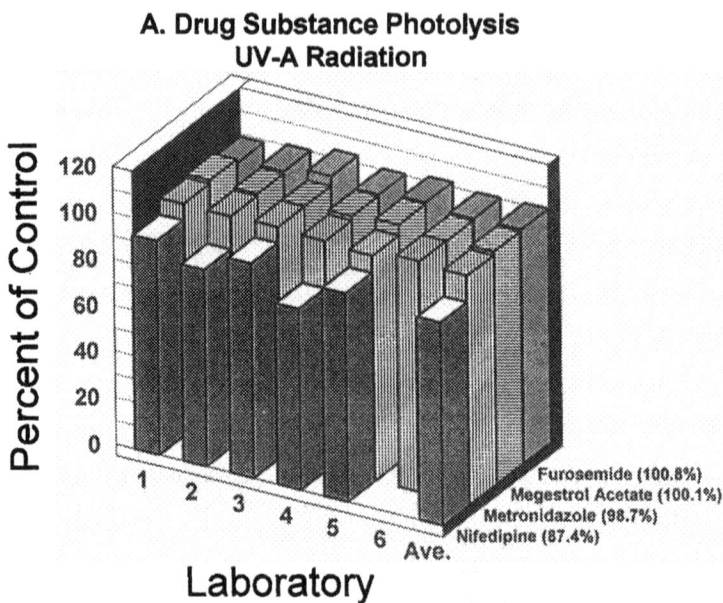

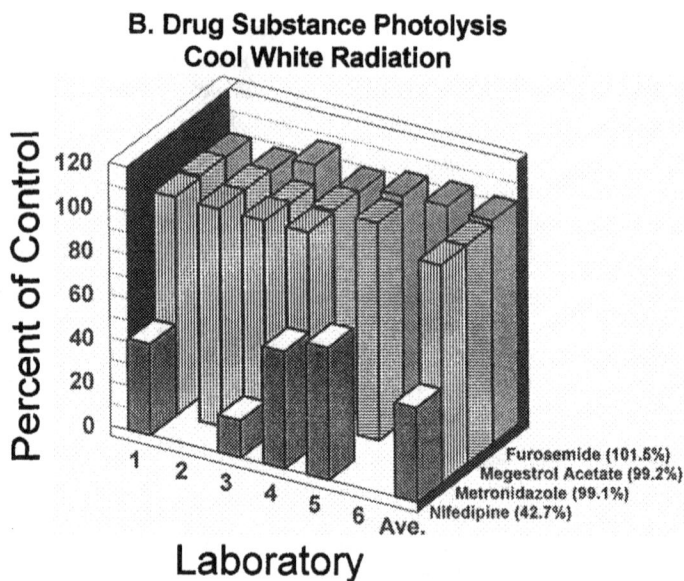

Figure 1 *Summary of bulk drug substance photolysis with UV-A (a) and Cool White (b) radiation. The entries in parentheses are the intralaboratory average.*

A. Tablet/Capsule Photolysis UV-A Radiation

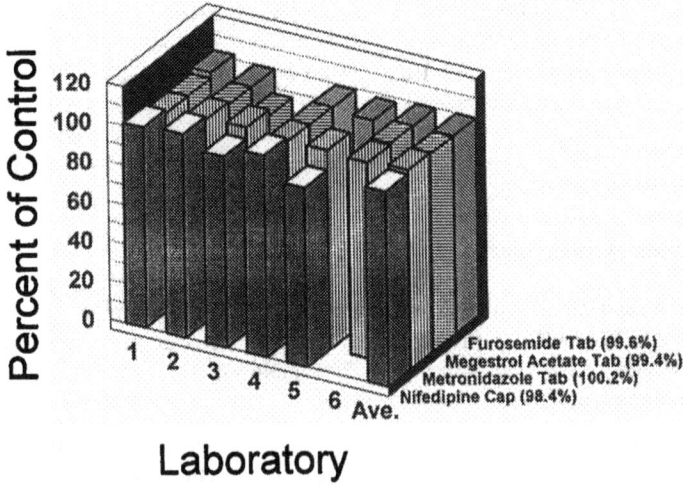

Furosemide Tab (99.6%)
Megestrol Acetate Tab (99.4%)
Metronidazole Tab (100.2%)
Nifedipine Cap (98.4%)

B. Tablet/Capsule Photolysis Cool White Radiation

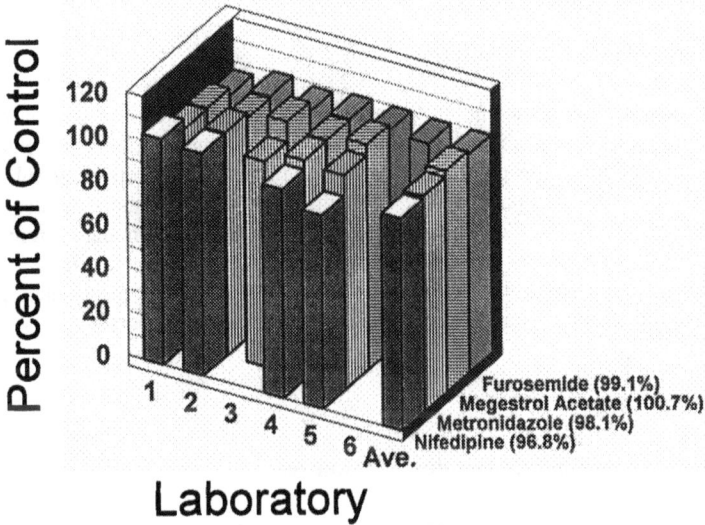

Furosemide (99.1%)
Megestrol Acetate (100.7%)
Metronidazole (98.1%)
Nifedipine (96.8%)

Figure 2 *Summary of tablet and capsule formulation photolysis with UV-A* (a) *and Cool White* (b) *radiation. The entries in parentheses are the intralaboratory average.*

For powder samples, the presentation of the sample to the photolysis radiation is critical. The powder layer should be thick enough to ensure that no incident radiation is transmitted. Preliminary photolysis studies with nifedipine powder samples indicated that the powder particle size is a significant factor. With a constant total exposed surface area (5 cm diameter Petri dish) and powder layer thickness (2 mm), the extent of nifedipine photodegradation increased as the powder particle size decreased. At a particle size of 250 F (60 mesh) or less, the extent of nifedipine photodegradation remained constant. Apparently, with large particles some incident radiation is transmitted through the powder layer and not absorbed. The drug substance powder samples used in the collaborative study were sized by screening to ensure a particle size of 250 F or less.

3.4 Furosemide Syringe Injection Photolysis

The UV-A and Cool White photolysis results for the furosemide syringe injection formulation with the primary brown plastic wrapper both in place and removed, are presented in Figures 3a and 3b, respectively. The photolysis results for the furosemide drug substance and tablets are included for comparison.

Based on the interlaboratory average, the primary brown plastic wrapper protects the furosemide injection solution from photodecomposition from either UV-A or Cool White radiation. Removing the wrapping produces considerable photodecomposition with Cool White irradiation (32.3% of furosemide decomposed) and extensive photodecomposition with UV-A irradiation (65.3% of furosemide decomposed). All collaborators reported that the injection solution was clear before photolysis but turned a straw brown during photolysis. Some collaborators also noted the formation of a precipitate in the syringe assembly upon photolysis with the brown wrapping removed.

3.5 Metronidazole Injection Photolysis

The UV-A and Cool White photolysis results for the metronidazole injection formulation solution in the plastic IV pouch and in a quartz spectrophotometer cell are presented in Figs. 4a and 4b, respectively. The photolysis results for the metronidazole drug substance and tablets are included for comparison. These results indicate that some photodecomposition occurs with both UV-A and Cool White irradiation whether the formulation remains in the plastic IV pouch or is photolysed in a quartz spectrophotometer cell. Photolysis is more extensive when the formulation solution is photolysed in the quartz spectrophotometer cell, indicating that the plastic IV pouch offers some protection from exposure.

4 CONCLUSION

4.1 Photolysis Apparatus and Quinine Actinometer

Quinine solution as an actinometer system for pharmaceutical photostability studies for the UV-A photolysis was proposed by JPMA. The ICH Expert Working Group has endorsed

A. Furosemide Photolysis UV-A Radiation

Drug Substance (100.8%)
Tablets (99.6%)
Injection Wrapper (99.5%)
Injection No Wrapper (34.7%)M

B. Furosemide Photolysis Cool White Radiation

Drug Substance (101.5%)
Tablets (99.1%)
Injection Wrapper (99.1%)
Injection No Wrapper (67.7%)

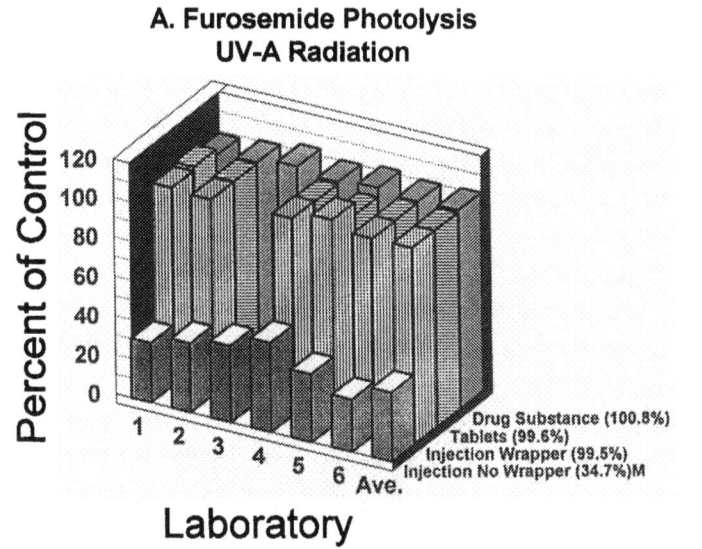

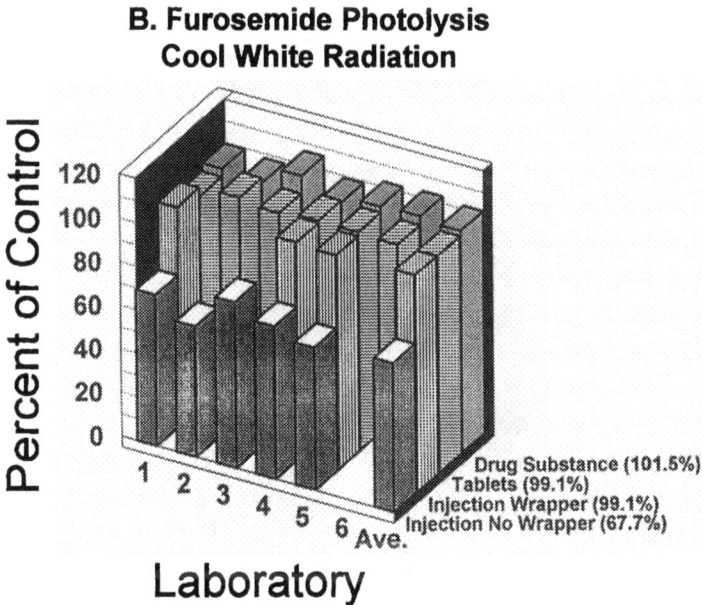

Figure 3 *Summary of Furosemide drug substance and formulated product photolysis with UV-A (a) and Cool White (b) radiation. The entries in parentheses are the intralaboratory average.*

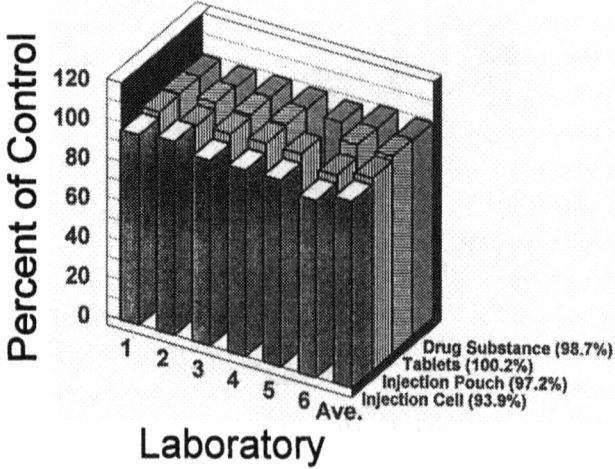

A. Metronidazole Photolysis
UV-A Radiation

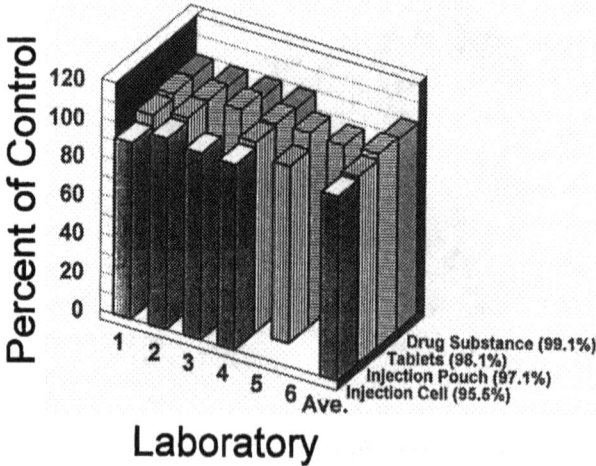

B. Metronidazole Photolysis
Cool White Radiation

Figure 4   *Summary of Metronidazole drug substance and formulated product photolysis with UV-A* (a) *and Cool White* (b) *radiation.   The entries in parentheses are the intralaboratory average.*

the quinine actinometer system for UV-A photolysis. The efficacy of this system for both UV-A and Cool White photolysis is demonstrated by comparing the interlaboratory standard deviation for samples that undergo photolysis with the variation in lamp intensities as reflected in the photolysis apparatus system linearity studies (Tables 1 and 2). The variation in Cool White and UV-A source intensities, as indicated by the photolysis slope (AU/hr, Tables 1 and 2), differs by a factor of about two with a percent relative standard deviation of 27.4% and 33.2% for the UV-A and Cool White photolysis sources, respectively.

The function of the quinine actinometer system is to monitor photolytic exposure and thereby normalizing the inherent intensity variations due to lamps, applied power, sample positioning, etc. To demonstrate the efficacy of the quinine actinometer system, the photolysis results for the Furosemide and Metronidazole Injection formulations with UV-A radiation are tabulated in Table 3.

The Furosemide Injection results indicate that the brown wrapper provides excellent protection from UV-A radiation. The inter- and intralaboratory standard deviation from the wrapped results for the Furosemide Injection reflects the standard deviation of the assay procedure. The Furosemide Injection results with the wrapping removed clearly indicate substantial photodecomposition. Although each laboratory confirms that the Furosemide Injection solution is susceptible to UV-A photolysis, the results show considerable variation between laboratories.

In the Metronidazole studies, sample presentation was controlled for the samples photolysed in the quartz cell. Note that the inter- and intra-laboratory standard deviation for the Metronidazole photolysis in the quartz cell (2.8) is comparable to the standard deviation due to the assay method alone (2.2). Thus, when the sample presentation is properly controlled, it appears that the quinine actinometer system provides reproducible monitoring of UV-A radiation.

Reproducible sample presentation will significantly affect the reproducibility of photolysis studies. In testing formulations in their primary or secondary packaging, adequate control of the sample presentation may not be possible. Such is the case with the pre-loaded Furosemide syringe formulation. Replicate photolysis with the same photolysis apparatus offers only a modest improvement in reproducibility as compared to the precision observed between laboratories in the collaborative study. Unfortunately, drug formulations are not ideal optical devices.

The ICH protocol recommends photolysis studies of the formulation directly and, if necessary, further studies in the primary and secondary marketing pack. Direct photolysis of formulated solutions in a quartz spectrophotometer cell, as with the Metronidazole solution in the FDA/PhRMA collaborative study, provides a well controlled sample presentation to the photolytic radiation.

4.2 Sample Photolysis Procedures

In general, the sample photolysis procedures for the drug substances and dosage forms in this study appear adequate. In photolysis studies, solid samples, drug substances, and solid oral dosage formulations are not expected to undergo extensive photolysis, given that the

Table 3 *Inter- and intralaboratory variation in UV-A photolysis of injection formulations.*

		Percent Drug Substance Remaining			
		Furosemide Injection		Metronidazole Injection	
Lab	UV-A Lamp[a] Intensity (AU/hr)	Wrapper	No Wrapper	Pouch	Quartz Cell
1	0.0356	100.1 (0.9)	29.8 (0.2)	98.5 (1.9)	95.6 (0.5)
2	0.0288	99.1 (0.3)	34.8 (0.7)	96.2 (1.6)	97.5 (SD)
3	0.0149	---- ----	38.5 (0.4)	98.0 (0.5)	93.3 (0.5)
4	0.0274	99.5 (0.4)	45.0 (1.4)	98.6 (0.3)	93.7 (0.3)
5	0.0261	103.8[b] (13.0)	34.0 (7.9)	98.6 (0.7)	94.0 (0.3)
6	0.0194	99.4 (0.4)	26.3 (0.1)	93.2 (0.5)	89.2 (0.6)

Interlaboratory Correlation:
Mean: 0.0254   99.5   34.   97.2   93.9
Std Dev (n): 0.0073(6)   0.4(4)   6.6(6)   2.2(6)   2.8(6)

a - The UV-A lamp intensity as determined in the photolysis apparatus system suitability test in units of quinine absorbance at 400 nm per hour of UV-A exposure (AU/hr).
b - This result is not included in the interlaboratory mean and standard deviation calculation based on the Grubbs test.
SD - Single determination.

photolysis of these samples is a surface phenomenon. Consequently, the precision of the assay procedure for determining the drug substance is a major factor in elucidating statistically significant photolytic decomposition.

Although photolysis of the tablet and capsule samples in general did not indicate significant decomposition of drug substance, all collaborating laboratories noted photolytically induced cosmetic changes. It is possible that these photolytic changes may alter product performance through damage to film coatings or modifications of other tablet or capsule attributes.

The procedures for photolysis of the injection solution formulations also appear to be adequate. As expected, these solution formulations can undergo considerable photolytic decomposition. Care should be taken that the samples are presented to the photolytic radiation in a consistent manner. When no photolysis occurs, sample presentation is not a concern. When extensive photolysis occurs (Furosemide Injection solution without the wrapper) the variance due to sample orientation will not hinder reliable conclusions about the photostability of the formulated product. In the equivocal case, such as the Metronidazole Injection study, the variance due to sample presentation may hamper definitive conclusions as to the photostability of the formulation.

The UV-A and Cool White exposure levels proposed in the ICH guidelines appear appropriate. Under these exposure conditions, drug substances and products of known photosensitivity (e.g. nifedipine) undergo significant photodegradation. Less photosensitive drug substances exhibit significant photodecomposition only in formulated solutions (e.g. Furosemide and Metronidazole Injections).

The adverse effects of heating during photostability studies is a difficult factor. The samples are irradiated with high energy, high intensity radiation. The majority of this electromagnetic radiation that is absorbed will result in increased sample heating. In classical photochemistry studies, these thermal effects could be largely controlled with water-jacketed photolysis cells and alternate means. Photostability studies of pharmaceutical products, particularly in their primary and secondary marketing packs, provide a significant challenge in maintaining isothermal conditions during testing. Photolysis sources more intense than the fluorescent tubes used in these studies (e.g. xenon, metal halide, etc.). We have found the quinine actinometer control cell to be an efficient monitor of excessive thermal heating. No thermal effects, noted by an absorbance change at 400 nm for the quinine actinometer control cell, were observed with either the Cool White or UV-A fluorescent sources. Preliminary studies with a Vita-Lite fluorescent sources did indicate some thermal decomposition of quinine. The lack of thermal heating with the Cool White and UV-A fluorescent sources was a major factor for utilizing the ICH photolysis source option.

Acknowledgment. The diligence and patience of the PhRMA collaborating laboratories are greatly appreciated. These laboratories include, in alphabetical order, Burroughs Wellcome Co.; Pfizer, Incorporated; R.W. Johnson Pharmaceutical Research Institute; Schering-Plough Corporation; and G.D. Searle and Company. The FDA Division of Testing and Applied Analytical Development served as the coordinating laboratory for this study. Particular thanks are due to James Brower, Larry Thornton, and William Juhl who served on the development team and to Jettie Travis-Bryant, the analyst who performed the collaborative study at DTAAD.

References

1. Int. Com. Harmonization, 'Tripartite Guideline for the Photostability Testing of New Drug Substances and Products', *Federal Registry*, 4th Rev. Draft, Dec. 15,1994.

2. S. Yoshioka, Y. Ishihara, T. Terazono, N. Tsunakawa, M. Murai, T. Yasuda, K. Kitamura, Y. Kunihiro, K. Sakai, Y. Hirose, K. Tonooka, K. Takayama, F. Imai, M. Godo, M. Matsuo, K. Nakamura, Y. Aso, S. Kojima, Y. Takeda and T. Terao, *Drug Dev. Ind. Pharm.*, 1994, **20**, 2049.
3. Am. Soc. Testing Materials; 'Standard Test Method for Accelerated Testing for Color Stability of Plastics Exposed to Indoor Fluorescent Lighting and Window-Filtered Daylight', *ASTM D 4674-4689*, 1989.
4. D. E. Moore, *J. Pharm. Biomed. Anal.*, 1987, **5**, 441.
5. N. H. Anderson, M. A. McLelland and P. Munden, *J. Pharm. Biomed. Anal.*, 1991, **9**, 443.
6. S. Nema, R. J. Washkuhn and D. R. Beussink, *Pharm. Technol.*, 1995, **19**, 170.
7. J. T. Piechocki and R. J. Wolters, *Pharm. Technol.*, 1994, **18**, 60.
8. N. Sager, F. R. G. Baum, R. J. Wolters and T. Layloff T, *Pharm. Forum*, 1998; **24**, 6331.
9. H. D. Drew, L. K. Thornton, W. E. Juhl and J. F. Brower, *Pharm. Forum*, 1998; **24**, 6317.
10. H. D. Drew, J. F. Brower, W. E. Juhl and L. K. Thornton, *Pharm Forum*, 1998; **24**, 6334.
11. J. K. Taylor, 'Statistical Techniques for Data Analysis', Lewis Pub., Chelsea, MI, 1990, p. 91.

The Elaboration and Application of the ICH Guideline on Photostability: A European View

Per Helboe
Danish Medicines Agency
378, Frederlkssundvej
DK-2700 Bronshoj, Denmark

1 INTRODUCTION

The development of the ICH guideline will be described. Photostability is an area not hitherto covered by guidelines in any of the three regions. In the talk, the previous situation without having any guideline will be discussed. The start of the ICH process will then be mentioned and it will be described how the guideline on photostability fits into the system on ICH quality guidelines. The content of the guideline will be described and it will be discussed how the new requirements will be implemented. Finally, in the conclusion the future in the EU within this field will be mentioned.

2 PREVIOUS SITUATION

Before the start of the ICH process the need for investigating photostability for active substances and medicinal products were briefly mentioned in the EU stability guideline. For the active substance, it was mentioned in the guide that among the conditions to be addressed when testing for stability of a new active substance were temperature, humidity and *light*. Also for testing the stability of finished products it was mentioned that the various test conditions used should be stated, and "The following may need to be considered - *light, either natural day light or defined artificial illumination*". The guidance given was so little and imprecise that one could not expect that a harmonised attitude would be applied throughout the European Union and, maybe, hardly between the assessors within the agency of one member state. Similarly for industry, the documentation received from various companies differed, and even from the same company you might receive dossiers where different perspectives in relation to photostability were applied. We could ask the question then: Has this lack of precise guidance resulted in not getting the information on light-sensitive substances and products? The answer is probably no for very sensitive substances. Such substances will at a very early stage of development be recognised as being photolabile and appropriate steps will be taken in relation to formulation of the product, manufacture, analysis and packaging. However, for substances and products where the light sensitivity is less pronounced you may run the risk that the problem is insufficiently investigated so that the patient may run into trouble with a slightly to moderately photolabile product.

3 START OF THE ICH PROCESS

The ICH process on harmonising technical requirement to medicinal products for licensing has been running since 1991. Expert working groups (EWG) in the fields of Quality, Safety, Efficacy and Multidiciplinary have been elaborating a series of guidelines to cover the various fields of expertise. Four major conferences were held, in Brussels in 1991, in Orlando in 1993, in Yokohama in 1995 and again in Brussels in July 1997. At each conference reports were given on the progress in elaborating guidelines in various fields. In the stability field the parent guideline on Stability of new active substances and medicinal products was finalised in December 1993 and the guideline is under implementation in a manner so that it will be fully implemented by January 1998. The guideline on Photostability and that on New Dosage Forms were both finalised in December 1996 and will be implemented together with the parent guideline by January 1998.

In the parent guideline the overall scope was to define a core stability data package to be acceptable in all three regions, US, Japan, and the EU. In that guideline the potential problems of photostability are only briefly addressed. Under Objective of the guideline it is mentioned that the stability should be investigated under the influence of a variety of environmental factors such as temperature, humidity and *light*. Furthermore, it is mentioned in the Glossary that *light testing* should be an integral part of stress testing. This means that photostability has been defined as stress testing and should not form part of the normal core stability data package covering long term data and accelerated studies.

Photostability was accepted by the Steering Committee in 1992 as an ICH topic. The wish to have such guideline came primarily from the Japanese industry and Japanese authorities. Much experimental work had been performed already at that time in Japan to be used in the future ICH discussions. The first meeting of a Photostability EWG was held in Japan in Kyoto in June 1992 and the first draft document was issued in October 1997. The photostability guideline took on board the basic concepts of the parent stability guideline, namely that light testing is an integral part of stress testing and that stress testing is normally carried out on one batch.

In the discussions in the EWG it was realised that a number of problems were related to the many variables involved and to the technical complexities of the testings to be performed. The testing variables were: daylight variations throughout the year and the variations within the geographical location. There could be different exposure levels to be recommended and also there was a substantial risk of having non-standardised conditions due to the different light sources used throughout the regions and even the different ways of measuring light. Therefore, in the proposed guideline should be included a definition of the type of light source(s) to be used, the number of batches to be tested, the exposure level to be used, how to measure the exposure and, last but not least how, to interpret the results.

4 CONTENT OF THE GUIDELINE

The guideline in its final version has taken into account the various problems mentioned above. Some readers of the guideline may feel that some of the problems are not definitively resolved. One problem which is still under discussion is the light sources proposed in the guideline. Obviously, the ultimate harmonisation would be to prescribe one light source only. In that way it would be more easy to achieve the same results in photostability testing all over the world. However, there were substantial differences in tradition for use of various types of light in the individual regions and it was found necessary to describe the possibility of using more light sources. One option is to use any light source giving an output similar to D65/ID65 such as a xenon or a metal halide lamp. Another possibility is to expose the sample to both a cool white fluorescent lamp, output similar to ISO 10977, and a near UV fluorescent lamp.

Even when taking into account the lack of total harmonisation as described above, the ICH guideline on photostability does describe a useful basic protocol for testing new active substances and medicinal products for first submissions, and this is to be regarded as substantial progress in harmonisation. The guideline also mentions the possibility of applying other alternative approaches if properly justified.

An important part of the guideline is the very illustrative decision tree (see the Fig. 1 on p. 67) which describes how to interpret the results of the stability testings. At several stages in the decision tree the conclusion ends up in a *no further action* statement. If the medicinal product outside the primary package shows a not more than acceptable change, the product is insensitive to light, i.e. no action is needed. The same is valid if the primary packaging or the final package shows sufficient protection to ensure that only acceptable change takes place. The term *acceptable change* did cause some discussions in the EWG. It was discussed whether we should rather require that *no change* should be the term. However, this was felt to be unsuitable and a too absolute requirement. Rather, the expression related to what is used in the parent guideline when assessing the results of accelerated testing should be used. Acceptable change means that a comparison should be done with the results from other, formal stability studies to assure that the product will be within specifications during the shelf life.

An important statement in the decision tree is seen in the case where a product fails to meet the requirements even when stored in the final primary and outer package. In such cases, the solution should not be that the manufacturer states a number of warnings to the user on the label such as *Store protected form light, Store in the dark, etc.* Instead, the manufacturer should go back and study the product once more considering to reformulate the product to make it better light resistant, or he should choose a package which can fully protect the product.

5 HOW WILL THE GUIDELINE BE IMPLEMENTED?

As mentioned above, the Photostability guideline was implemented in the EU by January 1998, a date to coincide with the full implementation of the parent stability guideline and the other additional guideline on stability studies for new dosage forms. The implementation should, as usual for CPMP guidelines, be understood as *studies commencing after*, i.e. the licensing authorities do not expect that all dossiers received after that date necessarily are in full compliance with the new guideline if the studies contained were performed before the implementation date.

Labelling requirements are not specified in the ICH guideline. Within the EU work is in progress on a general guideline on labelling requirements in relation to storage conditions. As mentioned above it is not found to be a good idea to accept statements on the package such as *Store in the dark, Store protected from light, etc.*, since the manufacturer should ensure that either the formulation or the package is able to protect the photolabile compounds of a preparation sufficiently. The statements to be recommended in the new CPMP guideline would rather be *Store in the original container, Store in the carton, etc.*

As usual for ICH guidelines, also the photostability guideline addresses specifically new active substances and new medicinal products. It is now to be discussed within the EU to what extent the content of the photostability guide should also be applied to generic products, abridged applications. Work has not yet been started but a guess on the content could be that for active substances it would be acceptable to base the documentation on information from literature if sufficient information is available. For the finished product, however, there may be more use of the recommendations of the guideline to check if the formulation shows any problems when exposed to light, e.g. due to sensitivity caused by one or more of the excipients used.

Two problems will probably need to be further discussed in the future with a view to a possible updating of the guideline. As mentioned earlier, it could be seen as problematic that different types of light source are given in the guideline without being totally equivalent. Further discussions may be difficult since it could be said that the ultimately ideal solution is yet to be described. The problems of the existing light sources are related to e.g. the spectral distribution of the light, the ageing of the lamps, *etc.* Also, the calibration towards a chemical actinometer may cause some problems. It has been shown (cf. a poster presented at the conference) that more control of the temperature may be needed since the actinometer based on quinine solutions can cause problems due to a temperature effect on the light sensitivity which may not be fully compensated by the dark control prescribed in the guideline.

6 CONCLUSIONS

From a regulatory point of view it is very useful that the photostability guideline has now been published to address a hitherto insufficiently covered area of the stability studies necessary to characterise new active substances and products fully. It is not clear whether or not a harmonised attitude towards generic products will be applied in the three regions, since they are not covered by the guideline. Future improvements of the guideline may include clarifications on light sources and on the application of chemical actinometry.

Selecting the Right Source for Pharmaceutical Photostability Testing

J. T. Piechocki
579 Sunshine Way
Westminster
MD 21157, U. S. A.

1 INTRODUCTION

In the past few years there have been several papers published recommending the proper source to use for pharmaceutical photostability testing. Little has been written as to how these sources vary or why they should be used. It is important to use the term "source" rather than lamp. By "source" I mean the lamp and all of its accessories including power supply, ballast, fixture and filter(s). Many of these items, which can affect the "spectral power distribution" (SPD) of the lamp selected are seldom appreciated and rarely reported in publications.

The reports and recommendations of Kerker,[1] Thoma,[2,3] Thoma and associates,[4-6] Tønnesen and associates,[7,8] Matsuda and associates[9] and Matsuo and associates[10] were developed based on experimental work with various lamps. Other recommendations, such as those of Piechocki,[11] Nema and associates[12] and Riehl and associates[13] were based on literature reviews.

All of these reports have put forth different ideas as to which source should be used. Some have clearly demonstrated why to use their recommended source while others have postulated as to why certain sources should or should not be used. Still others have merely listed the recommendations of others.

In this chapter I will attempt to add to the knowledge of proper lamp selection based upon these works and other works. At the end of this chapter it is hoped that you will have a better appreciation of many of the factors involved in proper lamp selection not mentioned by the cited authors. The effect of manufacturing and regional factors that serve to confuse the researcher and complicate the development of this field will also be discussed.

The literature is replete with many non-reproducible and non-relevant photostability studies. Quite often studies are run on alcoholic solutions and extrapolated to solids. It is not unusual to attend a scientific meeting today and have someone present studies, carried out using a bare unfiltered, mercury lamp. It is also quite common for such results to be extrapolated to solar conditions. Readers should be careful to understand exactly what sources were used and how the dosages were measured before they interpret much of the current photostability literature.

Table 1. *Objective of pharmaceutical photostability studies*

• Material Characterization
• Product development
• Shelf-life stability
• Environmental fate
• Pre-clinical photoactivity determination
• All of the above

The first action in any anticipated study is to define one's objective. Pharmaceutical photostability testing is no exception. The choice of photon source, exposure conditions used, etc. will define the equipment selected and actual conditions used. Table 1 lists some of the possible choices. The final selection is often based upon less than ideal conditions because one must often consider what resources are available and what can be achieved within a given time frame.

2 HISTORICAL

The effects of pharmaceutical photoinstability have been with us for many centuries. The ancient Egyptians may not have known why certain effects took place but they were wise enough to be able to see their positive effects and exploit them for the treatment of known maladies. Examples supporting this statement are the use of crude psoralens by ancient Egyptians to treat vitiligo, a fact well documented by Abdel M. Mofty[14] of the University of Cairo in the 1940's and during more modern times the 1903 Nobel Prize in Medicine, given to, Niels R. Finsen for his finding that skin lesions of tuberculosis often resolved after exposure to ultraviolet light.[14]

Table 2 is a list of the companies who supported the work of Arny and his associates[15, 16, 17, 18] in this field in 1926. Some of these companies no longer exist, many have been merged together but a few still exist, under the same name. The point I wish to make is that this is not a new field and the industry has been aware of it for many decades.

There are two very basic questions which I think the list presented in Table 1 brings to mind and are of greatest interest to those seeking to do photostability studies. First, can one source provide all of the information required, regardless of the application. Secondly, which source will produce the results desired in the shortest time and is the safest source to use. I would define "safest", from a development standpoint as that selection which will yield the maximum amount of data and be less likely to lead to post-marketing surprises. Post-marketing surprises are an embarrassment to both companies and regulatory authorities.

To use this approach is to adopt the "worst-case scenario". The value of this approach is in that it assures that all possible pharmaceutically relevant photochemical problems have been considered. There is no guarantee that in vitro photochemical data, as presently practised, will always correlate with in vivo performance but there is some indication that some of the studies might.[19] The ability to correlate in vitro with in vivo performance will depend greatly on whether very similar sources are used for both tests, as they were in this instance.

Table 2. *Sponsors of the research of Arny and associates.*[15, 16,17, 18]

• Dow Chemical Co.	• Sharpe and Dohme
• Drug Products	• Dr. William J. Schieffelin.
• Hynson, Wescott & Dunn	• Squibb & Sons
• Lehn & Fink	• Frederick Sterns & Sons
• Eli Lilly & Co.	• The Upjohn Company
• William S. Merrill	• Warner Co.
• Merck & Co.	• Perdue Frederick
• Chas. Pfizer & Co	• Corning Glass

The work of Cole and associates with psoralens and UV-A lamps,[19] reproduced in Figure 1, is a good illustration of a positive correlation between in vitro and in vivo photoreactivity. These workers were able to demonstrate that a measured decrease in vivo phototherapeutic performance was caused by a change in the SPD of the lamp used. The in vitro problem had been previously reported by Forbes and associates.[20]

The measured change by Cole and associates[19] was roughly of the same magnitude in vitro as in vivo. To obtain these data they plotted the effectiveness of both lamps by weighing their output by the published action spectra for 8-methoxypsoralen and plotting these values against wavelength, Figure 1. "Integrating the energy beneath the curves yielded a 2.2 to 1 ratio of estimated effectiveness of the BL-O compared to the BL-N lamp". Similar data regarding the effectiveness of certain wavelengths of electromagnetic radiation have also been published for the treatment of hyperbilirubinemia by Tan and his associates.[21]

These studies point out several important points to the pharmaceutical photostability chemist. The first point is the necessity of using a source which has a SPD matching or at least irradiating the entire "activation spectra" of the test substance. With an sample in a variety of matrices, i.e., solid, solution and cream, for example, only a broad-band continuous source with no spectral gaps can assure you of complete irradiation of the sample.

An "activation spectrum" (sometimes referred to as action spectrum) is defined as "the reciprocal of the number of incident photons required to produce a given effect compared with the wavelength of the radiation employed".[22-24] The importance of activation spectra lies in the fact that they represent the most active photochemical wavelengths of a test substance/system. While the absorption spectrum may often correlate directly with the activation spectra, this is not always assured. Searle[22-24] has written extensively on this particular topic.

Bathochromic or hypsochromic shifts in absorption spectra are possibly due to any one of a number of factors including, most often, changes in matrices. Likewise the test substance

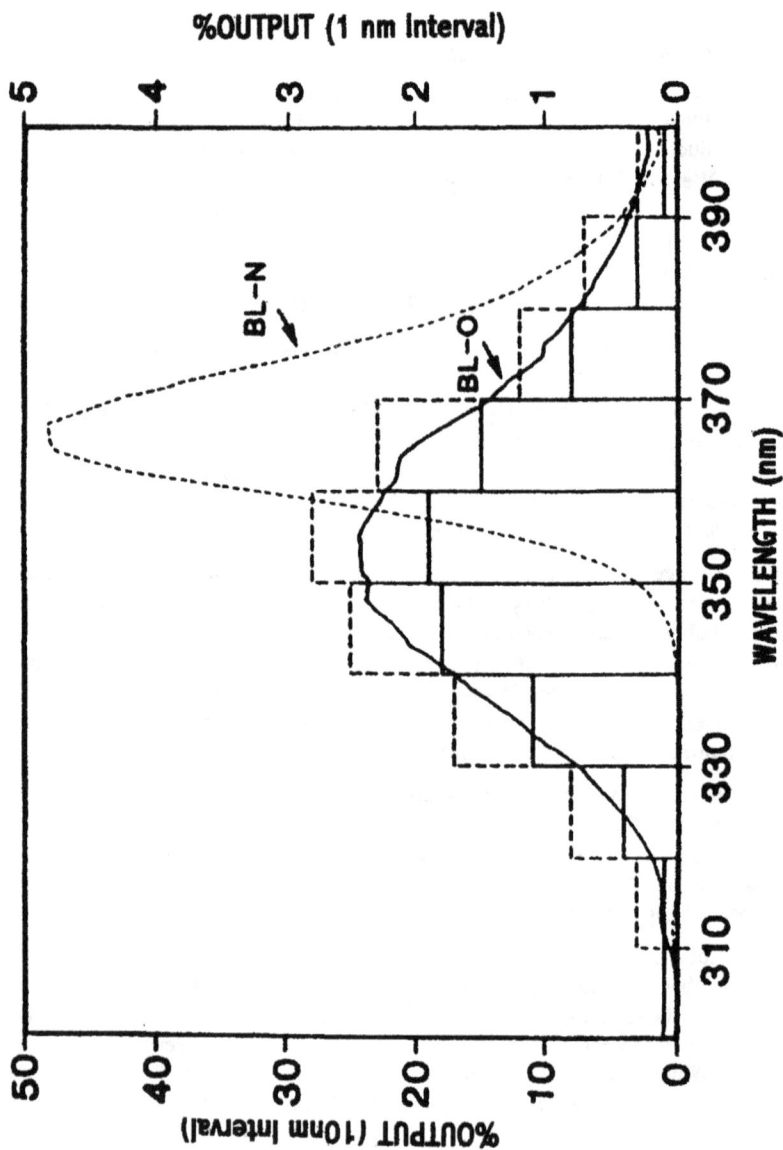

Figure 1. *Spectral power distribution of BL-0 and BL-N blacklight lamps plotted in relative output per nanometer (right ordinate). The histogram represents the limits of FDA-approved spectral distribution of the irradiation sources for PUVA therapy (from Oxsoralen package insert), presented in relative output per 10-nm interval (left ordinate) (Cole and associates).[19]*

may be degraded by a photoactivator present in the system as a contaminant or additive. Quinine is a good example of this latter case. To cover all these unknown circumstances it is best to utilize a broad-band, continuous source for all studies. Only this approach can assure that all wavelengths will be irradiated. Broad-band sources have also been recommended by Tønnesen and Moore,[25] Tønnesen and Karlsen,[8] and Riehl and associates.[13]

3 LAMP PROBLEMS

An important consideration in choosing a lamp/source is the lack of control one has when using mass marketed commercial lamps as opposed to specialty lamps such as the long-arc Xenon. This factor is currently much in evidence in the US where because of the putting into effect of the Energy Policy Act of 1992[26] and the American National Standard Institute (ANSI) and Illuminating Engineering Society of North America (IESNA) "Recommended Practice for Photobiological Safety for Lamps and Lamp Systems" RP-27.1-96[27] and RP-27-3.96.[28] The first law has already resulted in the removal from the mass market of certain lamps and substitution of SPD non-equivalents.

The ANSI standards cited recommend classifying lamps as to their UV hazard and strongly recommend to all manufacturers that they label all lamps as to their UV emissions and also attach UV warning labels for consumers. The Committee Internationale de l'Eclairage (CIE) has already established a photobiology working group for the purpose of reviewing the ANSI/IESNA standards. It is anticipated that they will also adopt these standards, or a modification thereof, as a world-wide standard.

Changes have already taken place in the market place, unbeknownst to many. The glass used to manufacture the tungsten-halogen lamps used in many desktop lamp fixtures is now doped to reduce their high UV content. There have been many reported cases of sunburn from users of the older lamps.

Based on all of the previously cited works, one can easily see that selection of a radiation source is the most critical part of any photostability study. Kerker[1] and Thoma and associates[4, 5, 6] have confirmed this fact by studying a series of UV and VIS absorbing materials, using lamps of different spectral power distributions. Lamps with different spectral power distributions yielded differing amounts of decomposition products.

While this fact may not seem important to some individuals, it is important for standards setting. How can one set limits for impurities or degradation products without a defined set of reproducible conditions? Even a simple limit test must be capable of detecting a definite quantity of the specific entity. In this case, if same wavelength and equivalent number of photons in a given bandwidth is not reproducible from source to source, the amount of degradation produced will vary. A "go" vs "no-go" method may suffice for a packaging decision but not for a stability or kinetic study.

The actual source selected will often depend on the results desired, cost of equipment, ease of equipment use, time required per test and overall applicability of the test. An important factor to consider is whether or not one's reference will be constant or subject to change over time. Since there is no standard "room", "pharmacy", "manufacturing area" or "hospital room", how can they be simulated on a world-wide basis?

4 TERMINOLOGY

Terminology has been a problem in optics and photostability from its inception. Many authors continue to use the term "light" to refer to both the VIS and UV electromagnetic spectral regions.

Most books define light as the 380-770 nm region of the visible spectrum or as "radiant energy capable of exciting the retina and producing a visual sensation".[29] To speak of light in any other terms is to be incorrect.

There are many different common terms that have been used in the lighting industry. These terms, often do not represent a definite standard but rather a firm's interpretation of the term. All cool white lamps manufactured, in the past, were identical because they all used the same single halophosphate for the fluorescent coating material. With the advent of the new laws limiting power consumption and the UV radiation given off by these lamps, these lamps are becoming less available.

Current lamps are most likely triphosphor lamps of a slightly different SPD, some having a slight pinkish color. It is not uncommon for many fixtures to now have no diffusers, only bare bulbs.

The designation "cool white" for lamps still remains in use. This term originated because people who entered a room illuminated with these lamps thought light emitted gave them the sensation that the room was "cooler" and the light less harsh.

A more complex situation exists for "daylight" fluorescent lamps. One can pick up any manufacturer's catalog and generally find two or three different varieties of "daylight" lamps, all with different SPDs.

Another frequently used term today is "equivalent". Aside from the dictionary, which generally defines it as meaning "..equal in force, amount, or value" what acceptable definition is there? For our purposes does this mean that the SPDs of equivalent lamps must agree within 10%, 20%, 30% or even more? If the dictionary definition indeed were true, there would be no need for this chapter.

The major argument for continuing the usage of these terms, including the misuse of "light" and "UV light" is that they are in common usage. Scientists, most of all should not accept this argument. Scientists use precise terms that are or should be unambiguous. There should be no place for wrongly used or imprecise terms in our terminology.

Much scientific work eventually reaches the legal arena. The existence of "common usage" terms in such places only serves to confuse not only others in the field but the general public, juries and judges.

5 ROOM LIGHT

The most often cited exposure situation for the final drug product is that of the "drug store", also designated as the equivalent of normal "room light". Indeed, this approach was taken by Lachman and associates[30,31] in their often cited landmark papers on evaluating the stability of pharmaceuticals. They included in Table 1 of their paper a listing of the "Light Intensity of Typical Retail Store Locations". While this information is useful it may easily give one the wrong impression as to how to proceed with developing a method to be applied on a global scale. An understanding of global practices is necessary before a simple, all encompassing solution can be reached.

There is no information reported in Lachman's and associates[30,31] paper or many published papers regarding the types of lamps, fixtures, diffusers, ballasts, detectors or meters used for these types of measurements. Failure to consider these items can lead to significant errors in the data obtained. This particular topic was clearly presented by Lape and Ryder[32] and Klibansky.[33]

Cole and associates[34] have perhaps done the most comprehensive study of what is room light. Important items pointed out in this study were the different lamps used, the effect of wattage on the SPD, the effect of diffusers on the SPD and that UV-C was emitted from bare bulbs.

Room light is a highly variable quantity. Perhaps it is most correctly described for a windowed room as "a combination of window-glass filtered daylight and artificial light." Since no one has ever, to my knowledge, defined the proportion of these two sources nor does a single source exist that can duplicate it, I recommend that we abandon this concept and term.

The term "room light" has changed throughout history, is changing with the increased use of more efficient sources such as the tungsten halide lamps and will surely again change in the future. One very efficient and intense source currently on the horizon is the "micro-wave discharge" lamp that is currently being introduced into the market place.[35,36] Will this be the "room light" of the future?

6 IMPORTANCE OF SPECTRAL POWER DISTRIBUTION (SPD)

The photostability studies of Kerker,[1] Tønnesen and associates,[7,8] and Matsuo and associates[10] include practically all lamps currently used for pharmaceutical photostability testing. The works of Thoma and associates[4-6] clearly demonstrated the effect that the SPD of a source can have on the data obtained.

The importance of the source SPD is easily demonstrable in everyday experience. How many have not experienced the problem of trying to match the color of two articles of clothing and been fooled? The fluorescent lamps, of a lower color temperature and lower blue light and UV content, give one a different color match than does daylight, essentially fooling the eye.

Riehl and associates[13] have dismissed this problem as only a color matching problem when it really indicates a significant difference in the SPD of the lamps. The lamp color index is a rough indicator of lamp temperature. Lamps with high color index numbers have higher color temperatures and more closely resemble the CIE D65 standard. Wyszecki[37] has done the most comprehensive study of this problem and developed a formula for the CIE by which lamps of various SPDs can be compared for their goodness of fit to a given standard.

A spectral power distribution presents not only an energy distribution diagram but also the power of the source being studied or used. To specify exposure limits in terms of power alone without reference to both spectral range to be used and the spectral sensitivity of the detector is to be scientifically imprecise. A 1.2 megaLux exposure obtained using "cool white" and "daylight" lamps would be quite different in terms of actual total energy. Lux measurements are generally taken using a meter having a "phototopic" response. All of the energy below approximately 500 nm will not be detected or measured by this instrument.

A newer, linear radiometer, capable of measuring all energies equally, will show a considerable increase in energy for the "daylight lamps" over the "cool white." This difference in power received by the sample is due to the difference in SPDs of the lamps used and also the spectral sensitivity of the detector used. If we truly mean to measure the total VIS energy of a source the use of the term Lux should eliminated since it only refers to a portion, not all of the VIS spectral region. Both the UV and VIS regions energies should be specified as all other standards do, in terms of a spectral range and the watts/meter-squared that should be present in that given region.

7 FILTERS

One item frequently neglected, when researchers report on the "room light" used for testing, is whether the fixture did or did not have a diffuser. This one item can affect the SPD of the radiation source very dramatically as illustrated in Figure 2, taken from a publication of Cole and associates.[34] Many have probably personally experienced this same phenomena. The absorption of UV rays by the plastic sandwiched between the glass of windshields, accounts for why photochromic lenses do not work in automobiles. A similar problem might be found in using windowsill or more properly called "energy-efficient window-glass filtered daylight" for testing in some modern laboratory buildings.

Filters are prescribed in the ICH[38] guideline, Option 1 for use with only Xenon (more correctly called long-arc Xenon [LAX]) and metal halide lamps. Cole and associates[34] clearly demonstrated that filters should also be used for fluorescent lamps. In their study they detected UV-C radiation coming from bare "cool white" fluorescent lamps. The importance of this problem is well demonstrated in the adoption of the previously cited ANSI/IESNA RP 27-1[27] and 27-3[28] standards.

Filters are used to isolate unwanted radiation from the system. They are generally defined by various parameters such as, narrow or wide band, sharp cut-off, narrow band pass and cut-off wavelength. ISO 10977,[39] ISO 4892-2[40] and the currently proposed ASTM G 3.03[41] standards currently define the glass to use relative to the SPD of a Xenon lamp.

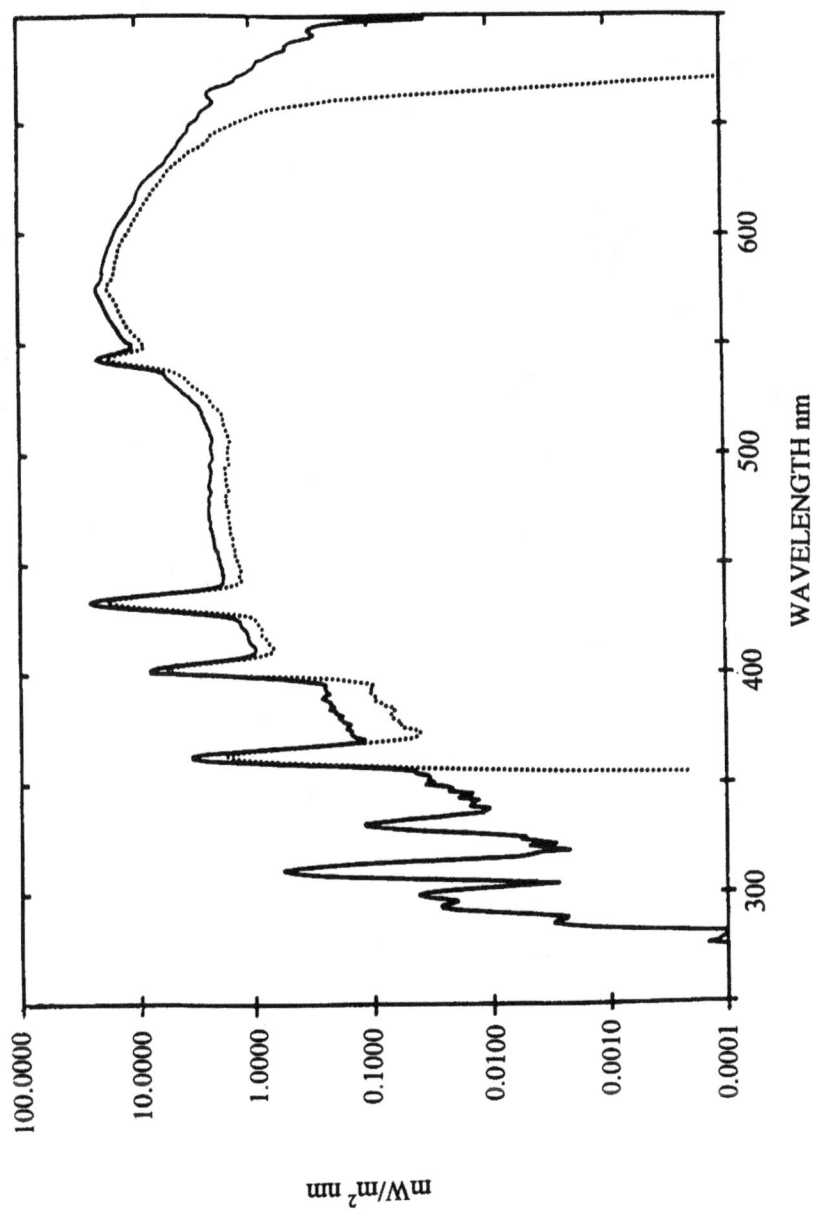

Figure 2. *Spectral irradiance of F40T12 WW RS WM with and without K-12 acrylic diffuser (Cole and associates).[34]*

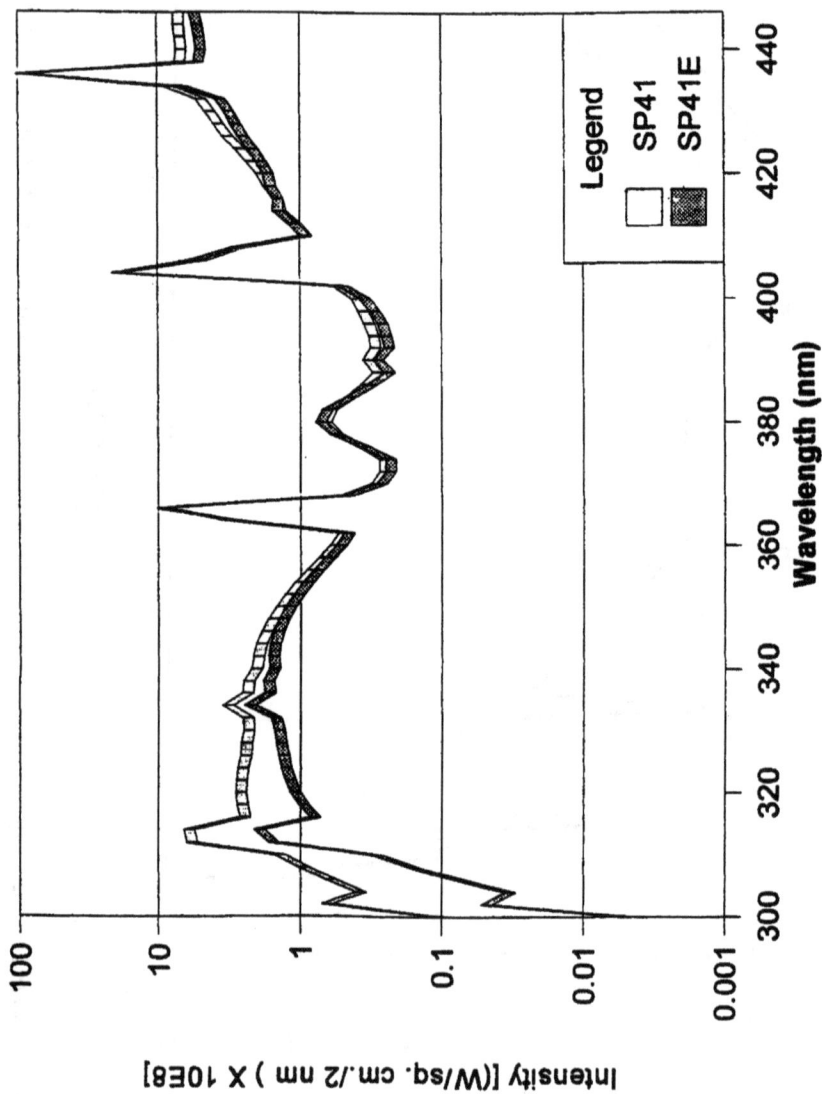

Figure 3. *UV spectral irradiance at 1000 lux for two fluorescent 32W T8 SP41 fluorescent lamps. SP41 - US glass, SP41E - European glass (Bergman and associates).* [42]

The spectral characteristics of the glass are never specified. This is a serious problem for developing a single standard.

Just how the glass used to manufacture fluorescent lamps can affect the lower UV cut-off wavelength is well demonstrated in Figure 3, taken from a paper presented by Bergman.[42] Both of these lamps are manufactured by GE, one in the US and the other in the UK. This difference has been known among lighting engineers for a long time and is taken advantage of when illuminating art objects, to reduce the deleterious effects of the UV radiation on them.

The absorption noted is due to the presence of very small quantities of iron in the European glass. Similar absorption can be obtained by doping the glass with cerium, as was reported by Bergman.[42] Several manufacturers are now using this technique to reduce the UV emissions given off by desk-top lamps which use tungsten halogen lamps.

Mullen and associates[43] and Searle and associates[44] have studied the aging characteristics of several lamp types and filters. Their work has clearly demonstrated the "solarization" of glass filters by intense sources. Solarization is the darkening of glass which occurs when it is exposed to intense electromagnetic radiation. This results in a shift of the lower cut-off frequencies to longer wavelengths and affects the higher energy, UV-B and lower UV-C regions the most. Searle and associates[44] clearly demonstrated that the observed "solarization was a function not only of the glass but also of possible surface contaminants and the atmosphere. Similar aging observations have been made by Ketola and associates[45] for various plastics and polystyrene Petri dishes.

ISO 4892-2[40] cautions the user to discard filters "... after 4 000 h of use or in accordance with the manufacturer's specifications". Some manufacturers are less specific in their recommendations.

8 CURRENT REGULATORY REQUIREMENTS

Table 3 lists the current requirements of the ICH guideline.[38] There are four different options a researcher can choose from. Three different sources under Option 1 and one under Option 2. I list Option 2 as a single source because both lamps are to be used either together or in separate studies, in order to complete one test.

Table 3. *ICH Guideline lamp options.*[38]

Option 1 (D65/ID65)
•   Combination UV + VIS fluorescent
•   Xenon
•   Metal halide
Option 2
Cool white fluorescent lamp (ISO 10977:1993(E) and UV fluorescent (max. between 350-370 nm)

Other organizations have also been doing photostability studies for many decades and served as a basis for the development of current pharmaceutical photostability practices. This point was brought out by Lachman and his associates.[30,31] Table 4 lists some of the current ASTM, International Standards Organization (ISO), German Institute for Standardization (DIN) and the American Association of Textile Chemists and Colorists (AATCC).

Table 4. *Similar phototesting procedures from allied sciences.*

ASTM G 3.03	Proposed Standard Practice for Exposing Nonmetallic Materials in Accelerated Test Devices That Use Laboratory Sources. (in revision)
ASTM G 24	Practice of Conducting Exposure to Daylight Filtered Through Glass
ISO 877	Plastics -- Methods of Exposure to Direct Weathering, Indirect Weathering Using Glass Filtered Daylight and Indirect Weathering Using Fresnel Mirrors
ISO 10977 [JIS* Z8902-1984]	Photography - Processed photographic colour films and paper prints - Methods for measuring image stability
AATCC** 16C-91	Colorfastness to Light - Daylight
ISO 105-B01	Textiles -- Tests for Colorfastness
JIS* L 0841	Test for Colourfastness of Textiles

    *   Japanese Industrial Standard
    **   American Association of Textile Chemists and Colorists

All of these organizations are in the process of revising their source standards and most of the focus has been on using the procedure found in DIN 53 387[46] and adopted in ISO standards 4892-1,[47] -2[40] and -3,[48] titled, "Plastics - Methods of exposure to laboratory light sources". These standards specify not only the total irradiance (watts per meter squared) but also the percentage of the total distribution or irradiance which must appear in each wavelength band range specified. The ISO standard specifies the acceptable variance in the irradiance levels and also specifies that the irradiance at the specimen surface shall not vary more than ± 10% between any two points over the sample surface.

9 PHOTOSTABILITY SOURCES

Table 5 lists the three most popular sources currently used for photostability studies. All can be artificially generated but some pharmaceutical researchers still insist on using natural sources as reported in the surveys of Anderson and associates[49] and Thoma.[2] There have been no published surveys for the US.

Table 5. *Most commonly sources used*

- Daylight
- Window-glass filtered daylight
- Room light

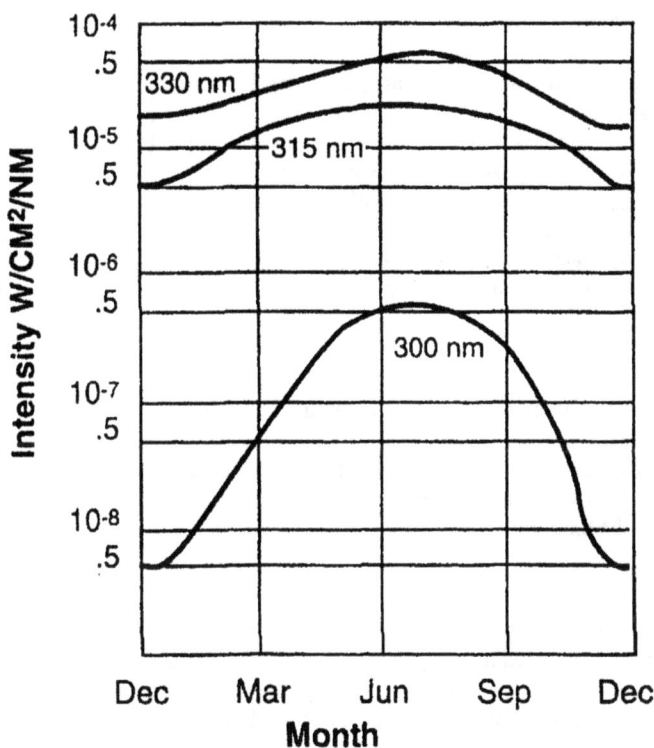

Figure 4. *UV variation at different wavelengths in Darvos, Switzerland at noon time, over the period of one year (with permission of Q-Panel Co.)*

9.1 Daylight

Daylight is defined by the ICH as that specified in ISO 10977:1993(E)[39] for outdoor daylight CIE Illuminant D65. This standard in turn cites the CIE Publication No. 15.2, Colorimetry.[50] Other standards, including those of the ISO[40] and ASTM G3.03[41] cite more appropriate CIE standards such as No. 20, "Recommendations for the integrated irradiance and the spectral distribution of simulated solar radiation for testing purposes" or No. 85 - 1989, "Technical Report - Solar spectral irradiance".

Daylight, for most photostability testing laboratories is defined as the CIE D65 standard which means that radiation which is equivalent to that produced by a black body heated to a temperature of 6500 K. To achieve this SPD, globally agreed upon by all CIE members, requires the use of glass filters and high intensity lamps such as xenon-arc or metal halide lamps.

9.2 Natural daylight

Natural daylight is not a reliable source for sample irradiation. It is well known that the amount of solar radiation not only varies with many positional and atmospheric factors but that the actual bandwidth is seasonally sensitive and affects the UV-B and short wavelength UV-A, much more severely than other regions of the spectrum. Figure 4 clearly illustrates this problem. The difference in both the radiation intensity and wavelength cut-off frequency at these two very different times of the year is readily apparent from this figure. Since the UV bandwidth changes with the time of the year the use of "natural daylight" for controlled scientific studies is not recommended.

9.3 Artificial daylight

There are currently three different types of artificial "daylight" photon sources available. These three different types of lamps are fluorescent, Xenon and metal halide lamps. From the surveys of Thoma[2] and Anderson[49] it would appear that the long-arc Xenon lamp is preferred in Europe; the use of the metal halide lamps would be second and fluorescent "daylight" lamps would be the third most commonly used sources.

9.3.1 Fluorescent lamps. Fluorescent lamps[51] are perhaps the most complex of all the lamps listed. They consist of a filament and/or cathode, anode, conducting gas, non-conducting gas, phosphor(s) and glass bulb. The conducting gas is mercury vapor provided by a drop of mercury inserted into each lamp. For smaller wattage lamps the mercury may be added amalgamated to the cathode. Some lamps may also contain an additional shield to reduce the black deposits of metal ions which occur during usage, at the cathode end of the tube.

All of these lamps will have SPDs which will contain the mercury emission lines at approximately, 254, 313, 365, 405, 436, 546 and 587 nm. The intensity of these lines will depend upon the amount of mercury added, which is difficult to control and the thickness of the phosphor layer. Not all published SPDs reproduce these lines although they are always there.

The spectral power distribution of a number of the commonly used fluorescent and other lamps is presented in Figure 5. This plot, supplied by Dr. Masaaki Matsuo, of the Tanabe Seiyaku Co., Japan, clearly shows the various SPDs of all of the lamps currently allowed by the ICH guideline are not all the same.

One should also be aware that published spectra are taken using a large integrating sphere and represent the average of a batch, not a particular lamp. The intensity and spectral distribution can vary across the length of a lamp.

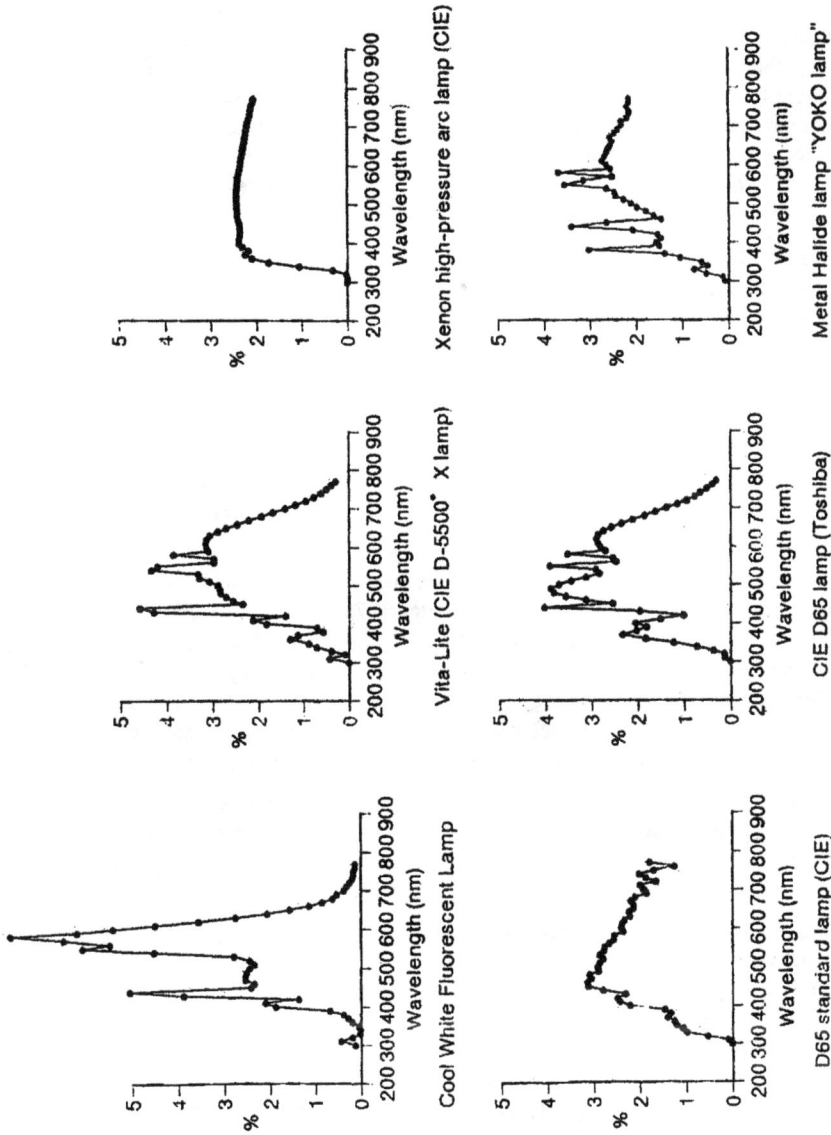

Figure 5. *Spectral power distribution of various lamps permitted for use by ICH guideline (courtesy of Dr. M. Matsuo, Tanabe Seiyaku Co., Japan).*

It is difficult to control the amount of mercury added during production. Only a very small drop is added but it is difficult to control this amount. For this reason, in particular, the intensity of these lamps will generally show a large variance. This is especially true of the "hardware store", mass produced, home use, monohalophosphate lamps. The multiphosphor lamps will generally show less of these problems but of course cost 5-20 times as much.

Additionally, fluorescent tubes are coated by first filling them with a slurry of the phosphor(s), then letting them drain before putting the tubes on a slightly inclined roller bed to allow them to dry without too much settling. These tubes are then put into an oven and fired to convert the slurry into a fine coat on the inside of the tubes. One outcome of this process, used particularly for the low price commercial market, e.g., "cool white" tubes, is shown in Figure 6. The differences in intensity noted, not readily evident in this black and white picture are actually color differences along the length of an individual tube.

This color difference, ranging from white to yellow to dark blue, results from difference in coating thickness of the lamp from one end to the other and the intensity of the fluorescence activating UV radiation. The thinner the coating the darker the color. The thicker the coating the whiter the color. Triphosphor and other more thickly coated lamps will not have this same problem. This is a problem specific to the thinly coated, mass produced and therefore cheap "cool white" lamps.

One advantage of fluorescent lamps is the very low amount of infra-red radiation. This advantage is not always found in practice. Since these lamps are low in intensity, it is not uncommon for several to be used at the same time. This practice leads to heat buildups that require cooling fans or cabinets.[30,31]

One should also be cognizant of the fact that not all fluorescent "daylight" lamps are equivalent. This problem is well illustrated in Figure 7. Here I have plotted the Relative Percent Energy of the CIE D65 standard, the ICH "cool white" lamp, two US and one European "daylight" lamp versus wavelength. The Relative Energy (%) was obtained by dividing the individual measured energy at one sample point by the total amount of energy over the 300-700 nm range and multiplying this value by 100. This procedure normalizes the data. If all of the lamps were equivalent, all of the curves would be superimposed.

The data obtained were plotted only up to the 580 nm data point to emphasize the differences in the more active and relevant 500 nm and below region of the spectrum. It is readily apparent that fluorescent "daylight" lamps are not good replicators of the D65 standard.

9.3.2 Xenon lamps. The power of Xenon, high-discharge lamps[52-54] to provide very intense white radiation led to their development as a radiation source in the 1930 - 1940s. The ability of this source to reproduce solar like spectra at high illuminance levels was recognized from the outset. Development was rapid due to their use in the war effort as lamps for search lights. Over the years the major problems that thwarted their adoption in testing laboratories, ozone formation and bulb explosion have been eliminated. Ozone formation was eliminated by the use modern UV absorbing glasses and the explosion problem (for high pressure short-arc lamps) greatly reduced with the development of stronger glasses. LAX (long-arc Xenon lamps) lamps are medium to low pressure lamps.

Figure 6. *Color variation along the length of single halophosphor "cool white" fluorescent lamps (personal communication).*

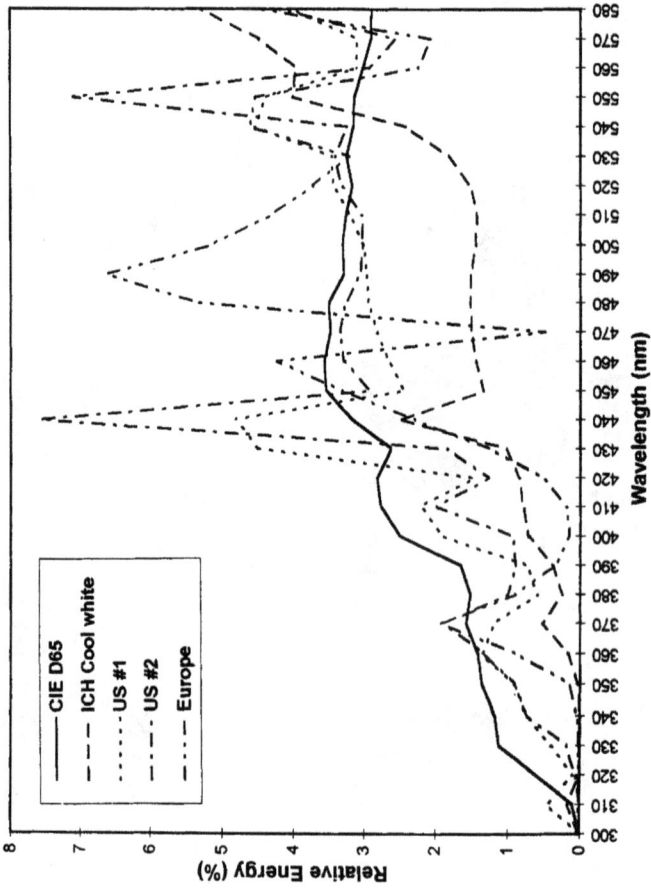

Figure 7. *Spectral power distributions of different fluorescent lamps. US #1, US #2 and European lamps advertised as daylight lamps.*

Xenon lamps consist of two electrodes enclosed in a quartz bulb or tube, which contain either mercury and Xenon or Xenon only as a fill gas. When ionized, SPDs of xenon + mercury or xenon only, are produced. Short-arc xenon, xenon + mercury lamps are used primarily for Sun Protection Factor determinations. The mercury is added when a more intense source is desired. However, the addition of mercury does add the additional mercury spectral lines and give data that do not correlate as well with natural sunlight.

The short-arc lamps have a smaller area of illumination and are more subject to intensity variations due to more significant changing arc lengths with time. Additionally these lamps emit line spectra from volatilized electrode metals and their impurities. The volatized metals also deposit on the cooler glass surface, resulting in a darkening of the glass inner surface and a reduction in the emission intensity.

Long-arc xenon lamps (LAX) (5 mm or more in arc length) have been used for several years and are the most common source used with the appropriate filters to artificially produce the CIE D65 SPD. They are, as noted, simple in construction, available world-wide, well characterized and accepted by practically all photostability testing groups such as the ASTM, ISO, DIN, JIS, etc.

These lamps are available in several wattages ranging from 150 to 6500 watts. Lamps larger than about 1500 watts generally require water cooling to reduce the large amount of infra-red energy emitted. The water cooled lamps emit a reduced amount of infra-red radiation. It is common to use these higher wattage lamps under conditions where the environment is controlled,[55] therefore the heat generated is not a significant problem.

Other methods of reducing this xenon lamp heat burden have been applied to currently available photostability instruments such as the use of cooled air and dichroic filters.

LAX lamps have a large area of illumination, emit a pure xenon gas spectra, change intensity less with age and show little darkening or metal deposition with age. Xenon lamps are also the basic reference lamps used for international standards of daylight, D65, or ID65 (window-glass filtered daylight).

With the concurrent development of the long-arc Xenon (LAX) low to medium pressure lamp the usage of this lamp was extended to other areas such as studio lighting and areas of photostability testing requiring large areas of illumination. The use of these lamps at higher wattages, i.e., greater than 1500 watts, required water cooling of the quartz envelope. This additional cooling did much to eliminate the infra-red radiation present in all xenon lamps.

Currently available lab top photostability testing equipment uses air-cooled LAX lamps of a wattage limit of 1500. Higher power lamps such as that used by Forbes and associates[55] require water cooling but have the advantage of reduced infra-red radiation at the sample.

9.3.3 Metal halide lamps. Metal halide lamps, first discovered at the end of the 18th century, did not come into practical use until the early 1960 s.[56-59] These lamps have been widely used to light large industrial areas, stages, studios and as street lights, i.e., sodium and mercury vapor lamps. The metal used, usually comes from the Group III/Period 6 rare earths. Those most commonly used for photostability testing metals are dysprosium, thallium and holmium because of their high color rendering index and high luminous efficacy which means they are good replicators of daylight spectrum, in particular, D65.

There is one lamp which does not fall into this category, the Dr. Hönle[60] lamp which is a proprietary lamp reportedly based upon iron as the active metal. Little has been published about this lamp, to date, and there is but one manufacturer, Dr. Hönle.

The purpose of the halide, usually iodine or bromine, in these and tungsten-halide lamps is to react with the volatilized metal, converting it into a volatile metal halide. This process results in less deposition of the metal on the lamp glass/quartz surface, which if not removed would cause a rapid change in the output of the lamp. Regeneration of both the metal and the halide also occurs when the volatile metal halide collides with the hot electrode. Both of these reactions prolong the life and intensity of the lamp.

Since these lamps are short-arc lamps they are subject to all of the properties of short-arc lamps, such as rapid changes in output and contamination of the emitted SPD by metals volatilized from the electrodes and shorter lifetimes then most other lamps.

As can be seen in Figure 8, taken from the IESNA Handbook, there are several lamps which can be found in this category and one should be careful to select the proper lamp and the proper filter. Eckhart[56] gives examples of many other metal halide lamps that have been developed but which are specific for special applications but not applicable to a general pharmaceutical photostabilty test.

9.4 Room light

Room light is a term, as cited previously, that is undefinable. Many people would consider it to be simply the light provided by a "cool white" fluorescent lamp. As previously discussed and documented in the work performed by Cole and associates,[34] large variances can be found in real room light situations.

Current fluorescent lamps, i.e. post 1997 in the US are most likely to be triphosphors, many not having same SPD of the pre-1997 lamps and of a lower intensity because of their thicker phosphor coating. To compensate for this latter item it is not unusual to see many situations where the diffusers, used in the past, have been eliminated, thereby increasing the possibility of exposure to UV-C.

The International Standards Organization (ISO) has developed a standard based on a Toshiba "cool white" fluorescent lamp, ISO10977:1993(E) "Photography - Processed photographic colour films and paper prints - Methods for measuring image stability". To date I have been unable to confirm that Toshiba does produce a lamp guaranteed to comply with this particular standard. I have also been informed that Toshiba fluorescent lamps are not sold in the US because they are not cost competitive. These items present problems for those who would want to use this particular source.

10 STANDARDIZATION

As previously pointed out, Drs. Thoma and Kerker,[1] Thoma and Klimek[2] and Matsuda,[9] Matsuo and associates[10] in particular, have studied the effect of several different radiation sources on the photostability of a series of UV and VIS photosensitive drug substances and

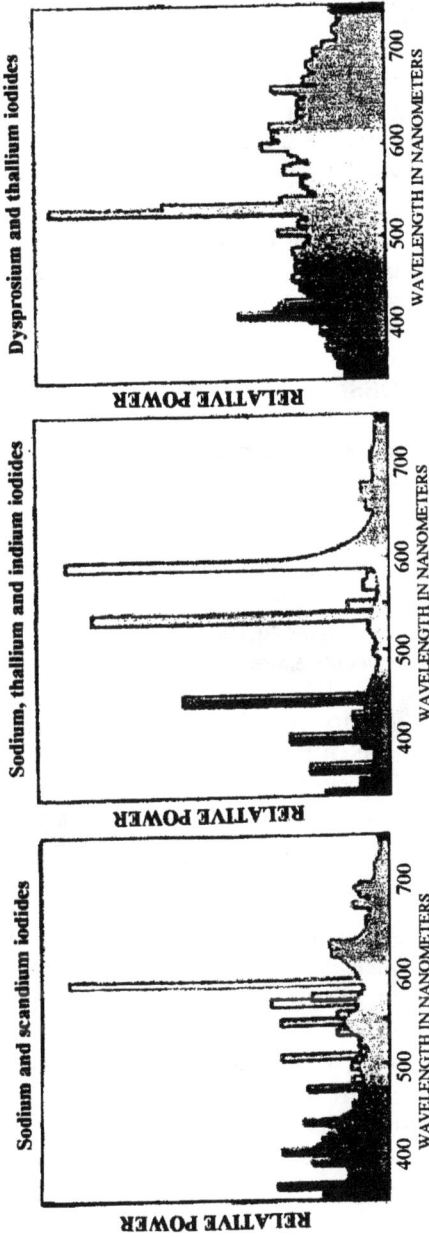

Figure 8. *Spectral power distributions of some commonly available metal halide lamps (permission of IESNA).*[29]

products. Their work and the works of others, clearly demonstrates the difficulty of obtaining similar results if the photon source is not standardized.

More recently Thoma and Kübler[61] pointed out this problem when they commented on how drug substance of high photochemical stability in their studies, Ketoconazole, was treated in the US and German Pharmacopoeias. The German pharmacopoeia requires the drug substance be labeled for "light protected storage", whereas the US pharmacopoeia does not.

If the speed of development and approval and cost reduction are the objectives of harmonization efforts, limiting the testing procedures, whenever possible, to a minimum and using the most inclusive procedure would seem to be the most logical approach to use. This approach assures covering most, if not all the cases, thereby protecting the public, innovator and regulatory authorities.

The need to standardize on a source and other issues has been pointed out by Yoshioka and associates,[62] Brower and associates,[63] Kester and associates[64] and Baertschi[65] and others.[66-68] Without standardization there cannot be reproducibility.

11 RECOMMENDATIONS

In Table 6 are listed the most important points to consider when selecting a photon source for pharmaceutical photostability testing. These points were assembled from many of the papers cited.

Table 6. *Factors to consider when selecting a photon source.*

• Does it have a reproducible SPD
• Does it cover the entire wavelength range
• Is it stable
• Does its SPD match that of The International Standards Organization (ISO) has developed a standard based on a Toshiba "cool white" fluorescent lamp, ISO D65
• Is the radiation emitted uniform over the entire sample area
• How much infra-red is emitted and can it be controlled

The use of a single, wide-band, reproducible source capable of illuminating the entire electromagnetic spectrum of interest at the same time gives the most all inclusive and applicable data in a single test. Extrapolations of this information to specific conditions should be straight forward.

Additionally, if photochromism were to occur, as it already has been reported for a drug substance by Sanvordeker,[69] its effects will be more correctly assessed. At the present time the Xenon and metal halide lamps, properly filtered, are the only two sources capable of meeting this requirement. These two sources do not show the regional variations found in fluorescent lamps and their SPDs are not likely to be affected by legislation or marketing pressures, since they are specialty lamps.

Use of a long-arc Xenon lamp, the most reproducible of all the options cited, provides a faster result, assures universal uniformity between testing sites and the least costly option. Since it is the option which best mimics the CIE D65 standard, the possibility of the results obtained correlating with in vivo performance are higher.

Acknowledgments. I am deeply indebted to all of the authors of the papers cited, many of whom have been helpful in supplying material and commenting on various phases of this paper. Thanks is also given to the U.S. Food and Drug Administration, who through their Professional Development Program made much of this work possible. However, the Agency's support should not necessarily be interpreted as approval of any statements in this paper.

References

1. R. Kerker, Ph D Thesis, Ludwig-Maximillians-Universität, München Germany, 1991. Available from Dissertatations und Fotodruck Frank GmbH, 8000 München 2, Gabelsbergerstrasse 15, Germany.
2. K. Thoma, *AAI Photostability Seminar*, Arlington, VA, Feb. 24-25,1997.
3. K. Thoma, in: H. H. Tønnesen, Ed., 'Photostability of Drugs and Drug Formulations', Taylor and Francis, London, 1996, p. 111.
4. K. Thoma, *10th Conf. Pharm. Techn.*, Shirakabako, Japan, July 1985, p. 1.
5. K. Thoma and R. Kerker, *Pharm. Ind.*, 1992, **54**, 169.
6. K. Thoma, *Pharm. Ind.*, 1992, **54**, 287.
7. H. H. Tønnesen, *Pharmazie*, 1991, **46**, 263.
8. H.H. Tønnesen and J. Karlsen, *Pharm. Eur.*, 1995, **7**, 137.
9. Y. Matsuda, *Pharm. Tech. J.*, 1994, **10**, 739.
10. M. Matsuo, Y. Machida, H. Furuichi, K. Nakamura, and Y. Takeda, *Drug Stab.*, 1996, **1**, 179.
11. J. Piechocki, *APQ Symp. 8th Ann. Meet. Expos.*, Lake Buena Vista, FL, Nov. 14-18, 1993.
12. S. Nema, R. J. Washkuhn, and D. R. Beussink, *Pharm. Technol.*, 1995, **19**, 170.
13. J. P. Riehl, C. L. Maupin, and T.P. Layloff, *Pharm. Forum*, 1995, **21**, 1654.
14. R. L. Edelson, *Sci. Am.*, 1988, **85**, 68.
15. H. V. Arny, A. Taub and A. Steinberg, *J. Am. Pharm. Assoc.*, 1931, **20**, 672.
16. H. V. Arny, A. Taub and A. Steinberg, *J. Am. Pharm. Assoc.*, 1931, **20**, 1014.
17. H. V. Arny, A. Taub and A. Steinberg, *J. Am. Pharm. Assoc.*, 1931, **20**, 1153.
18. H. V. Arny, A. Taub and R. H. Blythe, J. Am. Pharm. Assoc., 1934, **23**, 672.
19. C. Cole, P. D. Forbes and R. E. Davies, *J. Acad. Derm.*, 1984, **11**, 599.
20. P. D. Forbes, R. E. Davies, L. C. D'Alisio, and C. Cole, *Photochem. Photobiol.*, 1976, **24**, 613.
21. K. L. Tan, *J. Pediatr.*, 1989, **114**, 132.

22. N. D. Searle, *Int. Conf. .Adv. Stabil. Control. Degrad. Polym.*, Luzern, Switzerland, May, 1985.

23. N. D. Searle, *43rd Ann. Techn. Conf. (ANTEC)*, Washington, D.C., April 29, 1985.

24. N. D. Searle, *Sunspots*, 1984, **33**, 14.

25. H. H. Tønnesen and D. E. Moore, *Pharm. Technol. Int.*, 1993, **3**, 27.

26. Energy Policy Act of 1922, U.S. Congress, PL 102-486. Washington, DC: U.S. Government Printing Office.

27. 'RP-27.1 - Photobiological Safety for Lamps and Lamps Systems-Risk Group Classification and Labeling', revised in accord with the Illum. Engin. Soc. North America Photobiol. Com., January, 1966.

28. 'RP-27.3 - Photobiological Safety for Lamps and Lamps Systems - General Requirements', Illum. Engin. Soc. North America Photobiol. Com., January, 1966.

29. M. S. Rea, Ed. 'Lighting Handbook, Reference and Application', 7th Ed., of the Illum. Engin. Soc. North America, 1993.

30. L. Lachman and J. Cooper, *J. Am. Pharm. Assoc.*, 1959, **48**, 226.

31. L. Lachman, C.J. Swartz, and J. Cooper, *J. Am. Pharm. Assoc.*, 1960, **49**, 213.

32. P.B. Lape and D.V. Ryer, 1968 IEEE Northeast Electr. Res. Engin. Meet., November 6, 1968.

33. M. Klibansky, *AAI Sem. Photostab.*, Arlington, VA, February 24-25, 1997.

34. C. Cole, P. D. Forbes, R. E. Davies, and F. Urbach, *Ann. NY Acad. Sci.*, **453**, 305.

35. J. T. Dolan, M. G. Ury, and C.H. Wood, *6th Int. Symp. Sci. Technol. Light Sources*, Budapest, September 2, 1992.

36. Light Drive™ 100 lamp available from Fusion Lighting, Inc., 15700 Crabbs Branch Road, Rockville, MD 20855, 301-251-0300 FAX 301-279-0578.

37. G. Wyszecki, *Farbe*, 1970, **19**, 43.

38. ICH, 'Guideline for the Photostability Testing of New Drug Substance and Products' (see p. 66).

39. ISO 10977:1993(E), 'Photography - Processed Photographic Colour Films & transmittance, solar direct, total solar energy transmittance and ultraviolet transmittance, and related glazing factors', Int. Org. Standard., Genève, Switzerland, 1993.

40. ISO 4892-2:1994(E), 'Plastics - Methods of exposure to laboratory light sources - Part 2: Xenon-arc sources', Int. Org. Standard., Genève, Switzerland, 1994.

41. ASTM G3.03, 'Proposed Standard Practice for Exposing Nonmetallic Materials in Accelerated Test Devices That Use Laboratory Light Sources' Am. Soc. Test. Mat., Philadelphia, PA, to be published.

42. R. S. Bergman, T. G. Parham and T. K. McGowan, *IES Conf.*, Miami, FL, August 7-11, 1994.

43. P. A. Mullen, R. A. Kinmonth and N.Z. Searle, *J. Test. Eval.*, 1975, **3**, 15.

44. N. Z. Searle, P. Giesecke, R. Kinmonth and R. C. Hirt, *Appl. Opt.*, 1964, **3**, 923.

45. W. D. Ketola and J. S. Robbins, in 'UV Transmission of Single Strength Window Glass', W. D. Ketola and D. Grossman, Eds., Am. Soc. Test. Mat., Philadelphia, PA, 1993.

46. DIN 53 387, 'Artificial Weathering and Ageing of Plastics and Elastomers by Exposure to Filtered Xenon Arc Radiation', Dtsch. Inst. Norm., 1986.

47. ISO 4892-1:1994 (E), 'Plastics - Methods of exposure to laboratory light sources -', Part 1: General guidance', Int. Org. Standard., Genève, Switzerland, 1994.

48. ISO 4892-3:1994(E), 'Plastics - Methods of exposure to laboratory light sources -, Part 2: Fluorescent UV lamps', Int. Org. Standard., Genève, Switzerland, 1994.
49. N. H. Anderson, D. Johnston, M. A. McLelland and P. Munden, *J. Pharm. Biomed. Anal.*, 1991, **9**, 443.
50. CIE, No. 15 (E-1.3.1), 'Colorimetry, Official Recommendations of the International Commission on Illumination', Com. Int. Ecl., Paris, 1971.
51. "Toshiba Lamps", Catal. H8-8, Toshiba Light. Techn. Co., Shingawa-ku, Tokyo 140, Japan.
52. J. N. Aldington, *Trans. Illum. Eng. Soc.*, 1949, **14**,, 19.
53. R. Phillips, Ed., 'Sources and Applications of Ultraviolet Radiation', Academic Press, London, 1983, Chapter 11.
54. M. Rehmet, *IEE Proc.*, 1980, **127A**, 121.
55. P. D. Forbes, R. E. Davies, F. Urbach, D. Berger and C. Cole, *Cancer Res.*, 1982, **42**, 2796.
56. K. Eckhardt, *Lichttechnik*, 1975, **27**, 407.
57. R. Phillips, Ed., 'Sources and Applications of Ultraviolet Radiation', Academic Press, London, 1983, Chapter 9.
58. D. E. Work, *Light. Res. Tech.*, 1981, **13**, 143.
59. T. M. Lemons, *Light Des. Appl.*, 1978, **8** (8), 32.
60. Dr. K. Hönle GmbH, Fraunhoferstr. 5, D-82152 Planegg bei München, Germany.
61. K. Thoma and N. Kübler, *Pharmazie*, 1996, **51**, 885.
62. S. Yoshioka, Y. Ishihara, T. Terazono, N. Tsunakawa, T. Murai, T. Yasuda, Kitamura, Y. Kunihiro, K. Sakai, K. Hirose, K. Tonooka, K. Takayama, F. Imai, M. Godo, M. Matsuo, K. Nakamura, Y. Aso, S. Kojima, Y. Takeda and T. Terao, 1994, *Drug Dev. Ind. Pharm.*, **20**, 2049.
63. J. F. Brower, H. D. Drew, W. E. Juhl and L. K. Thornton, *1st Int. Meet. Photost. Drugs*, Oslo, Norway, June 8-9, 1995.
64. T. C. Kester, S. Zhan and D. H. Bergstrom, *Ann. Meet. Am. Ass. Pharm. Sci.*, 1996.
65. S. Baertschi, *Drug Stab.*, 1997, **1**, 193.
66. N.D. Searle, 'Comments on Docket No. 96D-0010, Notice in Federal Register, Vol. 61, No. 46, Thursday, March 7, 1996, p. 9310-9313', Docket Management Branch (HFA-305) 12420 Parklawn Drive, Rockville, MD 20857, Item code C4.
67. J. Piechocki, 'Comments on Docket No. 96D-0010, Notice in Federal Register, Vol. 61, No. 46, Thursday, March 7, 1996, p. 9310-9313', Docket Management Branch (HFA-305), 12420 Parklawn Drive, Rockville, MD 20857, Item code C6.
68. T.J. Wozniak, 'Comments on Docket No. 96D-0010, comments on the Notice in Federal Register, Vol. 61, No. 46, Thursday, March 7, 1996, p. 9310-9313', Docket Management Branch (HFA-305), 12420 Parklawn Drive, Rockville, MD 20857, Item code C10.
69. D.R. Sanvordeker, *J. Pharm. Sci.*, 1976, **65**, 1452.

Design and Validation Characteristics of Environmental Chambers for Photostability Testing

Jörg Boxhammer and Christine Willwoldt
Atlas Material Testing Technology
PO Box 1842
D-63558 Gelnhausen, Germany

1 INTRODUCTION

Photostability tests, as discussed since a couple of years in the pharmaceutical industry, have been carried out for many years in the most various fields of industrial application. The main objectives for tests under simulated environmental conditions in laboratories are to conduct the tests under more controlled, as well as accelerated, conditions compared to real time exposure.

The reproduction of real world effects on the one hand, as well as the precision and the speed on the other, are the key factors of a good accelerated photostability test. A great deal of time and money are continuously spent by industry in scrutiny of these points in an effort to improve the quality of test design. Concurrently, there has been an accompanying strong effort by instrument manufacturers concerning improved and evolving instrument technology.

A more general description and discussion of today's photostability test methods and the characteristics of testing chambers might give valuable impetus as well to decisions on a valid testing and instrument technology for photostability testing of new pharmaceutical substances and products.

2 ENVIRONMENTAL CHAMBERS FOR PHOTOSTABILITY TESTING – PRINCIPLES

For more than half a century, laboratories have been dealing with photostability testing of most various kinds of materials. For conducting those tests in instruments three typical configurations are spread worldwide as shown in Figure 1. They essentially consist of an artificial radiation system facing a sample area/volume where the materials, substances or products are exposed. The configuration may be a horizontal, stationary sample tray, irradiated in most cases by a likewise horizontally positioned radiation system installed above,

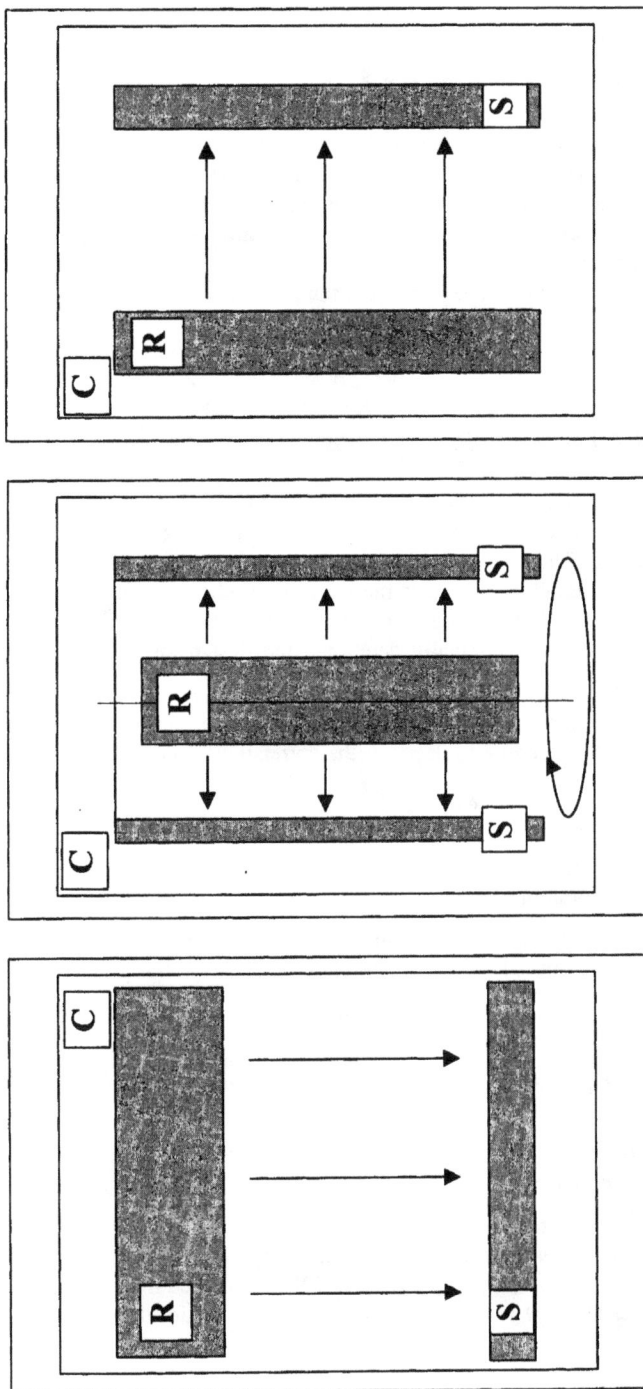

C - Chamber
R - Radiation system
S - Sample area/volume

Figure 1. *Photostability exposure chambers - Principles*

or a vertically positioned radiation system with a sample area rotating around resp. with a vertical, stationary sample tray facing it.

The size of the sample area may be from about one square centimeter[1] up to some square meters,[2] dependent on the purpose of the test. Normally the radiation system does not only consist of the lamp (specific type, one or more, may be a lamp array), but includes at least one or more of the following components: specific power supply, filter elements, mirrors, lenses, reflector elements, diffusors and a cooling unit (as part of the radiation system). Therefore, based on the spectral radiant flux of a specific lamp, the spectral distribution of radiation as well as the irradiance level and the uniformity of radiation at the location where the specimens/substances are exposed (the sample area/volume) are not only determined by the type and technical quality of the lamp itself; they may widely be influenced by the above listed system elements and their technical quality (e.g. tolerances, temperature shift, ageing stability).

3 PHOTOSTABILITY TEST SPECIFICATIONS

The main objectives for tests under simulated environmental conditions in laboratories are to conduct the tests under more controlled as well as accelerated conditions compared to real time exposure. Those test conditions may vary depending on the purpose and the specific objectives of the tests. The latter must therefore thoroughly be defined as the basis for the described and specified testing procedure as well as the required precision data. The contents of the test specification finally determine the design and essential characteristics of the testing chamber(s).

From all experiences in the general field of photostability testing in equipment there are some key factors tending to decrease the correlation/repeatability and reproducibility of test results.

Most important are

- the use of ultraviolet radiation shorter than that occurring in actual use environment/variations in the "defined UV-cut off"

- the use of a spectral distribution that differs widely from the one in actual use or shows significant differences in spectral distribution

- the use of extremely high irradiance levels compared to actual use conditions resp. extremely different irradiance levels

- high sample surface temperatures or significant differences in sample surface temperatures.

Therefore, these factors especially have to be thoroughly defined and specified as precisely as possible.

There were various national standards in the past with detailed descriptions of specific instrumentation, based on commercially available chambers. At international level (and now more and more adopted during revision of the mentioned national standards) it has been the intention since many years (with some exceptions) to develop so-called performance based standards that allow the user to conduct a test using non-specific instrument parameters, specified as test procedures. This standardization "strategy" is shown as a scheme in Figure 2. The manufacturer of any type of testing chamber has to ensure that the (equipment related) requirements are met within the specified tolerances at the location where the specimens are exposed.

Some of those most important International Standards in the general field of material photostability testing are listed in Figure 3. The main objective of these standards is to conduct tests under simulated sun radiation (outdoors and behind window glass), except for ISO 10977. The latter also includes room illuminants as a possible reason for photoreactions; as these rather different conditions are similar to the usual environment of pharmaceuticals, this may be the reason why the ICH-Guideline especially refers to this standard.

The essential contents of ISO 4892 for testing plastics may serve as an example for the "structure" of those standards and are detailed as follows:

- Definition of a "reference spectrum" as a basis for simulation

- Specification of "spectral functions" to be met on sample area within tolerance bands (from 300 nm to 800 nm in steps)

- Recommendation of level of irradiance on sample area to be used

- Requirements on uniformity of spectral irradiance over the entire (defined) sample area

- Requirements on maximum sample surface temperature

- Requirements on ambient (test chamber) relative humidity

- Others (depending on the purpose or the test)

- Description of devices for measuring irradiance and radiant exposure as well as maximum sample surface temperature (on sample area); application recommended - partially required.

It is obvious that for photostability testing in equipment the characterization of radiation incident on specimen surface is of highest importance. The definition of a "reference spectrum" depending on the objectives of the test (e.g. sunlight outdoors or behind window glass resp. specific room illuminants) serves as a real basis for simulation. The specification of a "radiation function" with as acceptable regarded tolerances based on the reference spectrum allows for judging the quality of light simulation in a testing chamber.

As an international criterion for comparing artificial light sources with sunlight outdoors, spectral data from the CIE publication No.85 :1989, Table 4, called global solar radiation, are worldwide accepted today. The spectral distribution is shown in Figure 4. Compared to outdoor measurements the curve is smoothed.[3] The "reference spectrum" for sunlight behind window glass has been defined by folding the curve with the spectral transmittance of a defined window glass (thickness 3 mm).

The necessary accuracy of simulating the reference spectrum by an artificial radiation system may depend on the objectives of the test and may therefore be regarded under several aspects, e.g. test purposes, technical and economical limitations, but nevertheless it should be defined. Figure 5 shows an example for a specified "radiation function" with as acceptable

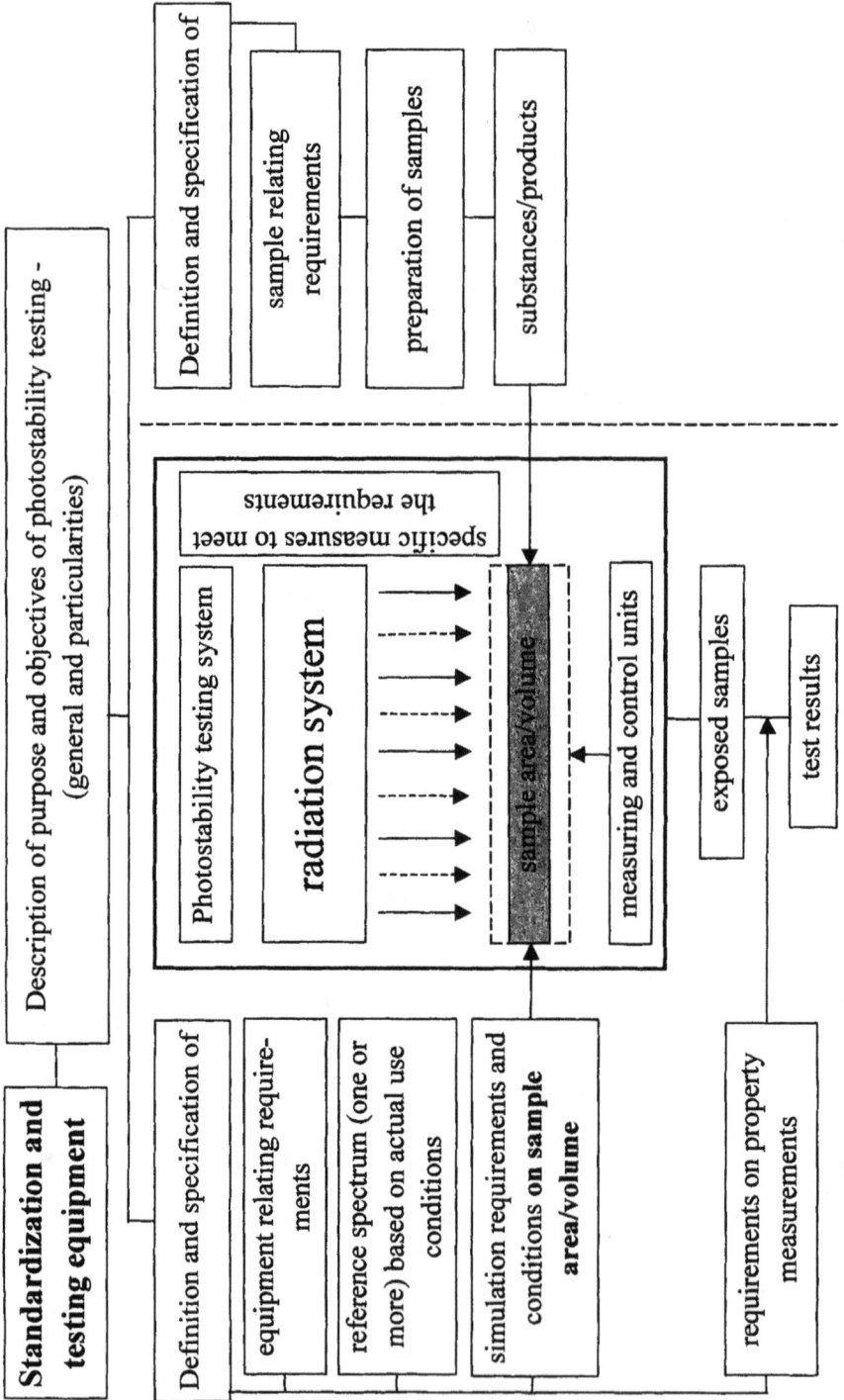

Figure 2. *Photostability testing in equipment - Basis*

Standard	Title	Edition
ISO 4892 part 1 part 2 part 3 part 4	**Plastics** - Methods of exposure to laboratory light sources General guidance Xenon-arc sources Fluorescent UV-lamps Open flame carbon-arc lamps	1994
ISO 11341	**Paints and varnishes** - Artificial weathering of coatings and exposure of coatings to artificial radiation - Exposure to filtered xenon-arc radiation	1994
ISO 11507	**Paints and varnishes** - Exposure of coatings to artificial weathering in apparatus - Exposure to fluorescent ultraviolet and condensation conditions	1997
ISO 10977	**Photography** - Processed photographic colour films and paper prints - Methods for measuring image stability	1993
ISO/DIS 105 part BO6 *)	**Textiles** - Tests for colour fastness Colour fastness and ageing to artificial light at high temperatures: xenon-arc fading lamp test	1997

*) Automotive applications

Figure 3. *International Standards in the general field of photostability testing in equipment*

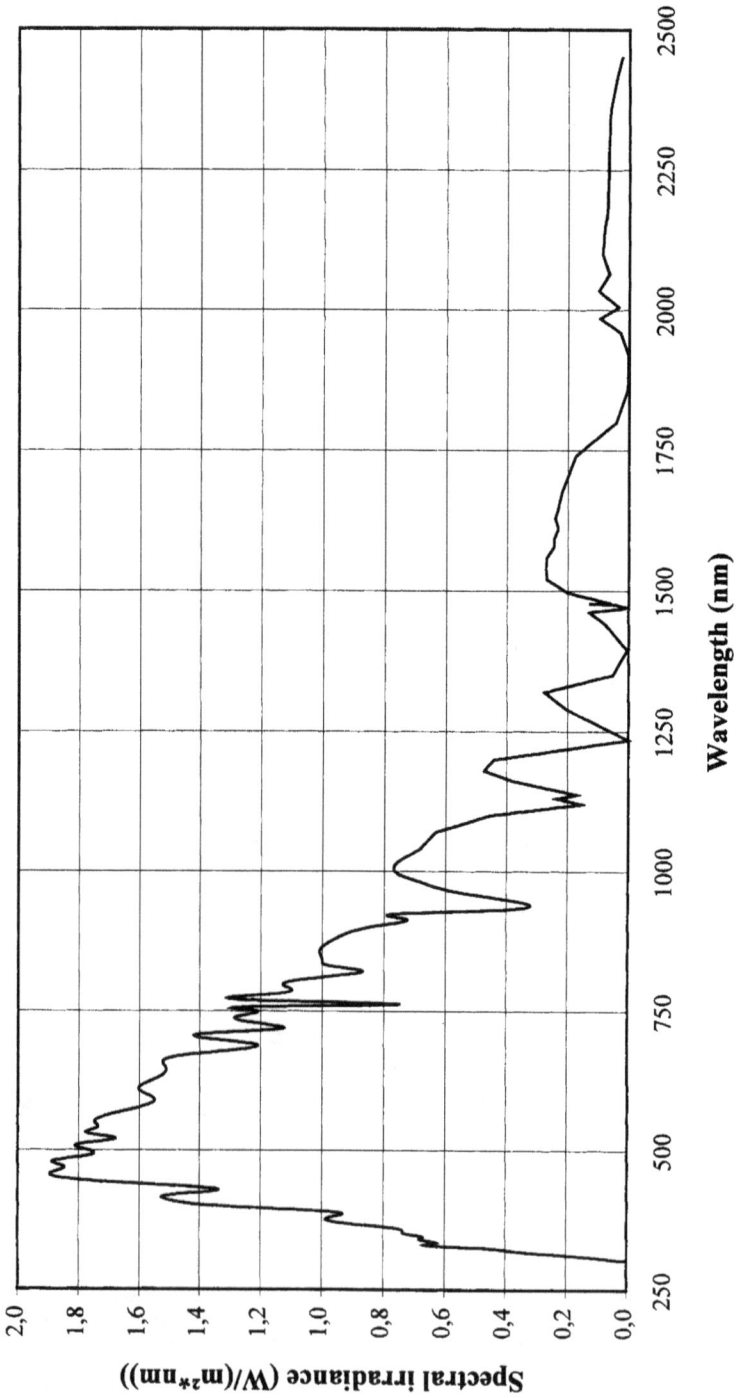

Figure 4. *Spectral energy distribution of global solar radiation according to publication CIE No. 85, Table 4 (1989)*

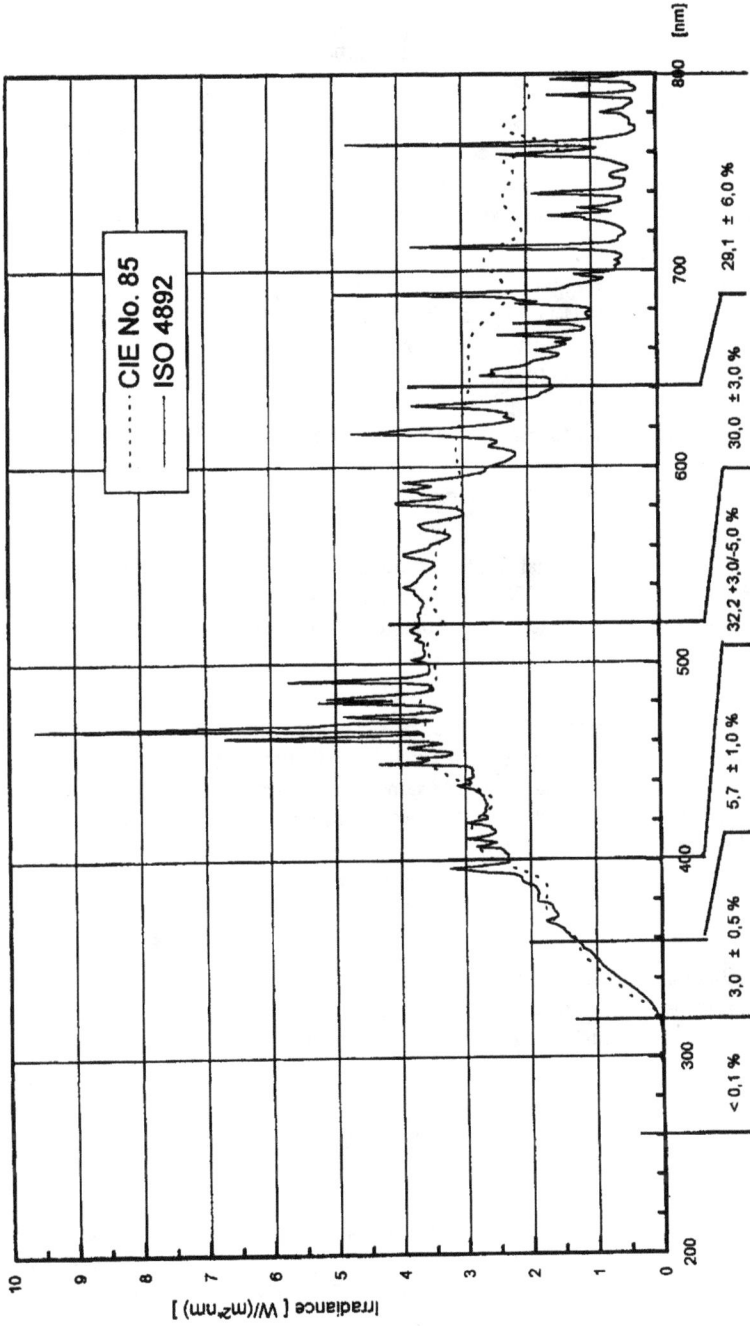

Figure 5. *Global solar radiation (CIE No. 85, Table 4/1989) behind window glass - radiation function ISO 4892 part 2 - method B (1994) and ISO 105 - B06 (proposal for revision - 1995).*

regarded tolerance bands based on the reference spectrum for simulating sunlight behind window glass as well as the spectral distribution of appropriately filtered xenon radiation measured on sample area in an equipment for comparison. It is obvious that the latter matches the requirements rather close.

It may not always be necessary to simulate the reference spectrum over the whole wavelength range of light emission, especially in those cases where detailed information on the spectral sensitivity or/and activation spectra[4] are available. But on the other hand, the most accurate simulation minimizes the risk of anomalous reactions that may not occur in actual use environment.

Comparing the described requirements on simulation of spectral distribution of radiation with the contents of the ICH-Guideline, it is obvious that in the latter, quantitative requirements are missing. It is only pointed out that the spectral output of the artificial sources should be similar to ID65 (defined in ISO 10977:1993, as the spectral distribution of indoor indirect daylight). Instead of specifying the spectral distribution of artificial radiation on specimen area, various quite different artificial radiation sources are listed as being acceptable for photostability testing with the only additional remark - and this one is really important - that the radiation below 320 nm must be eliminated by an appropriate filter system.

It may be of interest to note that ID65 is defined by the relative spectral distribution of D65 as specified in CIE No.15.2:1986 (a predecessor of the mentioned CIE No.85), folded with the spectral transmittance of soda lime float glass (thickness 6 mm). Based on the latter the relative spectral distributions compared to each other show only slight deviations (Figure 6).

Focussing only on the various listed types of radiation sources in option 1 of the ICH-Guideline, it is well known that their spectral distributions may be quite different. This may be true as well for specific light sources of the same general type, as shown in Figure 7 (UV-wavelength range) for filtered metal halide lamps. In consequence of the missing quantitative requirements there are no data given that could support a decision whether these differences may be acceptable or not. Nevertheless, in practice testing chambers with these different radiation sources are allowed to be used for conducting tests according to the Guideline and to compare test results.

The same situation as discussed concerning the spectral distribution of radiation is valid for other important test conditions, especially the level and uniformity of irradiance on sample area and the sample surface temperature under incident radiation.

4   VALIDATION CHARACTERISTICS OF PHOTOSTABILITY TESTING CHAMBERS

To meet thoroughly defined and precisely specified testing conditions (as described) within narrow tolerances at sample area, where the specimens or substances are exposed, various measures are needed to solve the technical problems for realizing a validated system. Those

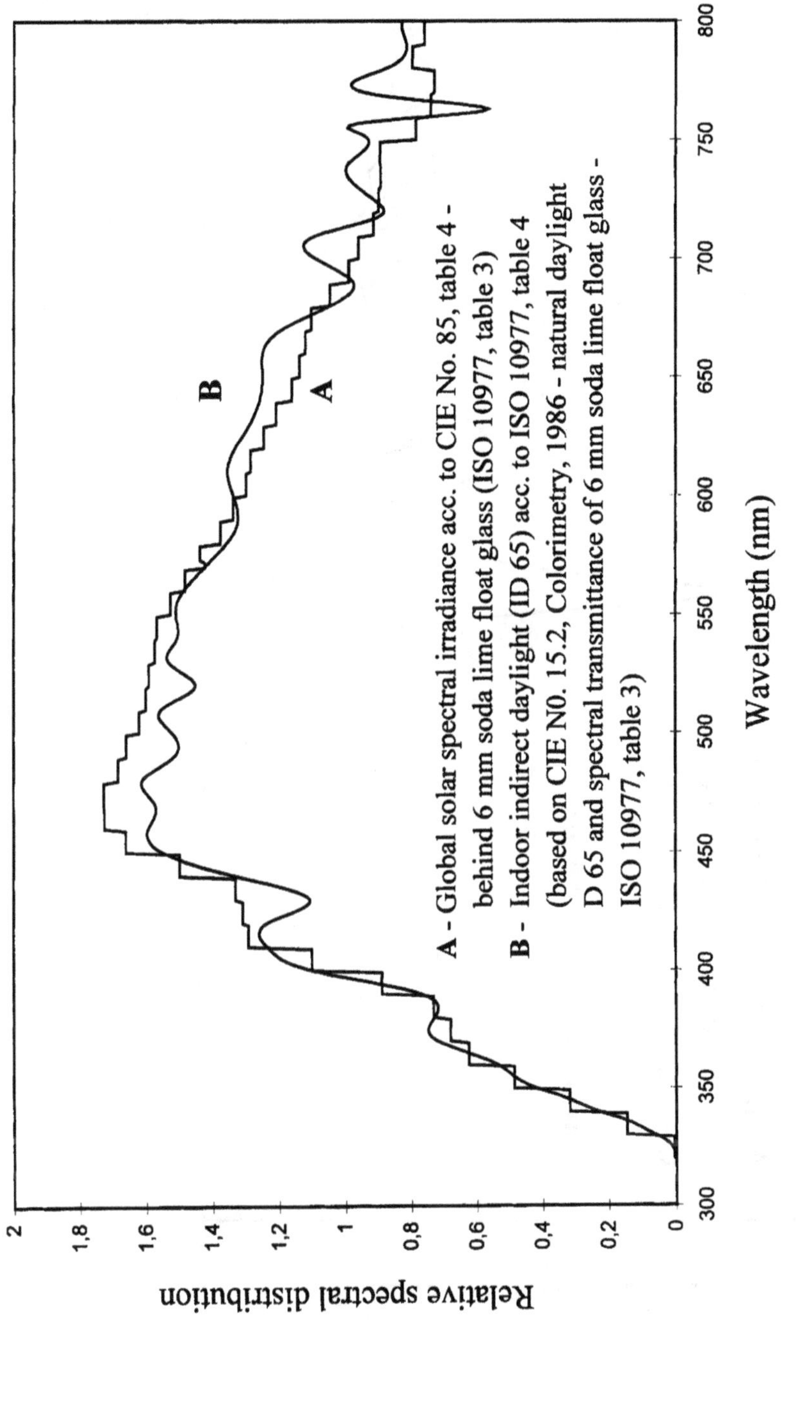
A - Global solar spectral irradiance acc. to CIE No. 85, table 4 - behind 6 mm soda lime float glass (ISO 10977, table 3)
B - Indoor indirect daylight (ID 65) acc. to ISO 10977, table 4 (based on CIE N0. 15.2, Colorimetry, 1986 - natural daylight D 65 and spectral transmittance of 6 mm soda lime float glass - ISO 10977, table 3)
A
B
Relative spectral distribution
2
1,8
1,6
1,4
1,2
1
0,8
0,6
0,4
0,2
0
300
350
400
450
500
550
600
650
700
750
800
Wavelength (nm)

Figure 6. *Global solar irradiation (+6 mm window glass) and indoor indirect daylight - comparison of relative spectral distribution.*

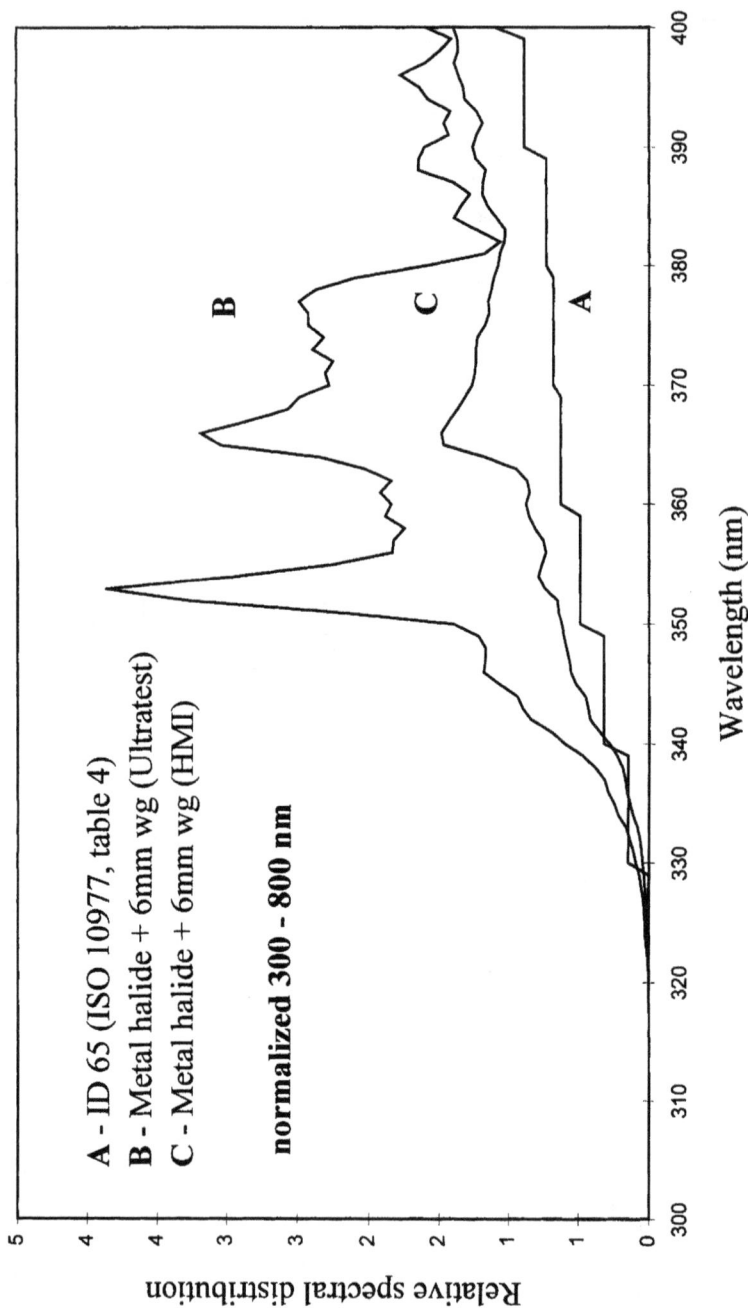

Figure 7. *Relative spectral distribution of different metal halide lamps (+6 mm window glass) - UV-wavelength range (300 - 400 nm).*

which are most important and may partially vary depending on the type of testing system are discussed below.

1.   The simulation of a "reference spectrum" within specified limits needs the use of thoroughly selected specific kinds of radiation system components.

2.   Technical tolerances in radiation system components require the use of high quality elements and the guarantee for a permanent quality control.

3.   The same is valid concerning the ageing behaviour. In special cases a specific component preageing may be necessary to eliminate the first essential decrease in radiant flux and/or the change in transmission characteristics. Specific and efficient cooling may also be needed depending on the type of light source(s) and filter elements. Concerning the different types of light sources it is most important to know that the definition of "lifetime" is quite different for lamps normally used as room illuminants on the one hand and "technical radiation sources" on the other hand, developed especially for simulation purposes. In any case, the light source and specific types of filter elements have to be replaced after a definite time of operation following the manufacturer's instructions or, as realized in some types of modern instruments with controlled irradiance, when directly indicated and displayed by the test equipment itself.

4.   The most effective measure to ensure the constancy of spectral irradiance at sample level is measuring and controlling constant irradiance via variable lamp power; this should be done in the wavelength range which is known as being most effective for photodegradation of materials on the one hand as well as known for significant changes in spectral distribution due to ageing of components on the other hand. When using a combination of two lamp types this has to be done for both types of lamps.

As an alternative to a controlled constant irradiance, at least the radiant exposure should be determined rather than only the exposure time. Again, it should be noted that using a combination of two lamp types one has to perform the measurements for both lamps and to ensure that the measuring device is calibrated to both spectral distributions.

Since from the technical point of view, as well as for economical reasons, the material's absorption specific wavelengths are not regarded for measuring the irradiance/radiant exposure but wavelength bands; the additional use of an actinometric system which is sensitive to specific wavelengths is recommended.

5.   Keeping the uniformity of spectral irradiance at sample level within acceptable limits needs in general a specific design of the whole test chamber system. But moreover, specific lamp replacement schedules may be required and a repositioning schedule for the samples within the sample area may be needed, when taking into account the possible combination of different types of lamps as well as the different ageing behaviour of individual lamps.

6.   The application of precise means for determining irradiance resp. radiant exposure on specimen area, or at least calibrated to specimen area, is a real basic requirement for reliable photostability testing instruments.

Two approaches to the measuring of radiation (with emphasis on the UV-wavelength range) are widely spread. The first is the use of a physical standard (chemical actinometric system as well) where a reference material or substance is exposed which shows a change in property in proportion to the dose of incident radiation (UV or other specific wavelength bands). The generally preferred approach is the use of a radiometer which responds to a defined wavelength band. Nevertheless, the additional use of an actinometric system may provide further valuable information.

The International Standard ISO 9379:1997 provides guidance on the use of spectroradiometers as well as filter radiometers and on calibration (recalibration) requirements.

7. The level of irradiance on specimen area as used for conducting tests may vary in a wide range. For simulating sunlight outdoors or behind window glass the maximum irradiance of sun radiation at the earth´s surface is widely used as a reference (CIE-publication No.20:1972) in the general field of photostability testing on technical materials. Systematic investigations concerning the application of still elevated irradiance values are right now going on.[5] For specific purposes, e.g. testing of processed photographic colour films (ISO 10977), very low levels of irradiance (illuminance) are specified for tests under simulated indoor indirect daylight and fluorescent illumination.

The literature concerning the photostability testing of drugs reflects the most different lighting conditions in the respective actual use environments as well as the missing specified irradiance level and is reporting quite different applied irradiance (illuminance) levels, varying by a factor of about 150. Using such a wide range of spectral irradiances there is a high risk that the mechanisms of material degradation may completely be changed, even at an equal spectral distribution. Therefore, from general experiences, and really depending on the types of materials and the properties measured, results from tests conducted at significantly different irradiance levels should not be compared unless correlation has been established.

Nevertheless, there is a need that the irradiance on specimen area in testing chambers stays variable within a wide range without the risk to change the spectral distribution of the radiation. For specific radiation sources, e.g. xenon arc systems, this may be done by varying the lamp power. In other cases the distance between the radiation system and the sample area has to be varied.

8. Owing to the incident total irradiance the surface temperature of irradiated sample/substances increases above the ambient temperature. Based on a specific photochemically effective spectral irradiance (UV- resp. (UV+Visible)-wavelength range) the total irradiance may differ widely, depending on the emitted wavelength range of the used light system (lamp + filter element); this may cause very different surface temperatures. Significant increased temperature levels may be expected, especially for those light sources containing IR-radiation, e.g. similar to sun radiation. Not exceeding the IR-level of the sun or staying below needs most effective filter elements resp. a modified lamp technology as e.g. already available for xenon-arc systems today.

It is obvious that the surface temperature of exposed samples can be reduced significantly by a specific cooling system, but the real cooling effect depends on whether the exposed surfaces may be directly affected by the cooling air stream (e.g. substances in containers) – or not.

Thus, the heating effect may be the reason for the necessary specification of an upper limit of the applied irradiance - independent of the available irradiance operating range in any photostability testing instrument.

In general, dark controls are no suitable substitutes for controlling or limiting surface temperatures during exposure. Whereas reactions initiated by light are accelerated by increase in temperature, the temperature will have only a small effect on materials shielded from light. Therefore, differences in temperature cannot be compensated for by a dark control.

Thus, the surface temperature at sample level has to be measured and controlled constant. As it is not practicable to monitor the surface temperature of individual exposed samples, precisely calibrated measuring devices (so-called black standard thermometers) are used in practice,[6] for characterization of the maximum surface temperature that may occur under total incident radiation.

9. As a conclusion from the preceding discussion the principle design and the essential validation characteristics of environmental chambers for photostability testing are illustrated by the schematic graph in Figure 8. The configuration may vary as shown in Figure 1. The major system elements will always remain similar.

The radiation system itself should be as stable as possible by using high quality components. The radiant flux from the radiation system will be directed to the sample area, but radiation reflection and/or scattering from the surrounding elements have to be taken into account. The real stress conditions have to be measured and controlled at the location where the specimens resp. substances are exposed or, at least, they have to be calibrated to this area using precisely calibrated radiation measuring devices (RMD) - if available an actinic system (AS) - and temperature measuring devices like TMD (normally a black standard thermometer, installed at sample level) and CT (sensor for measuring the surrounding temperature, shielded from radiation). Moreover, already discussed system components serve for controlling constant the measured values respectively for adjusting the stress conditions to specific specified levels.

5 SUMMARY

It has been shown that the radiation system is only one influencing factor in a photostability test, even though one of the most challenging. Today's approach should be to take into account all factors contributing to the final test result: the test standard prescribing the test conditions and the equipment related requirements, the preparation and the treatment of samples before and after the photostability test and – last but not least – the technical performance of the test equipment itself.

Out of the analysis concerning requirements and problems related to photostability test chambers and the possible measures to solve these problems, the validation characteristics of a photostability test chamber have been discussed and summarized.

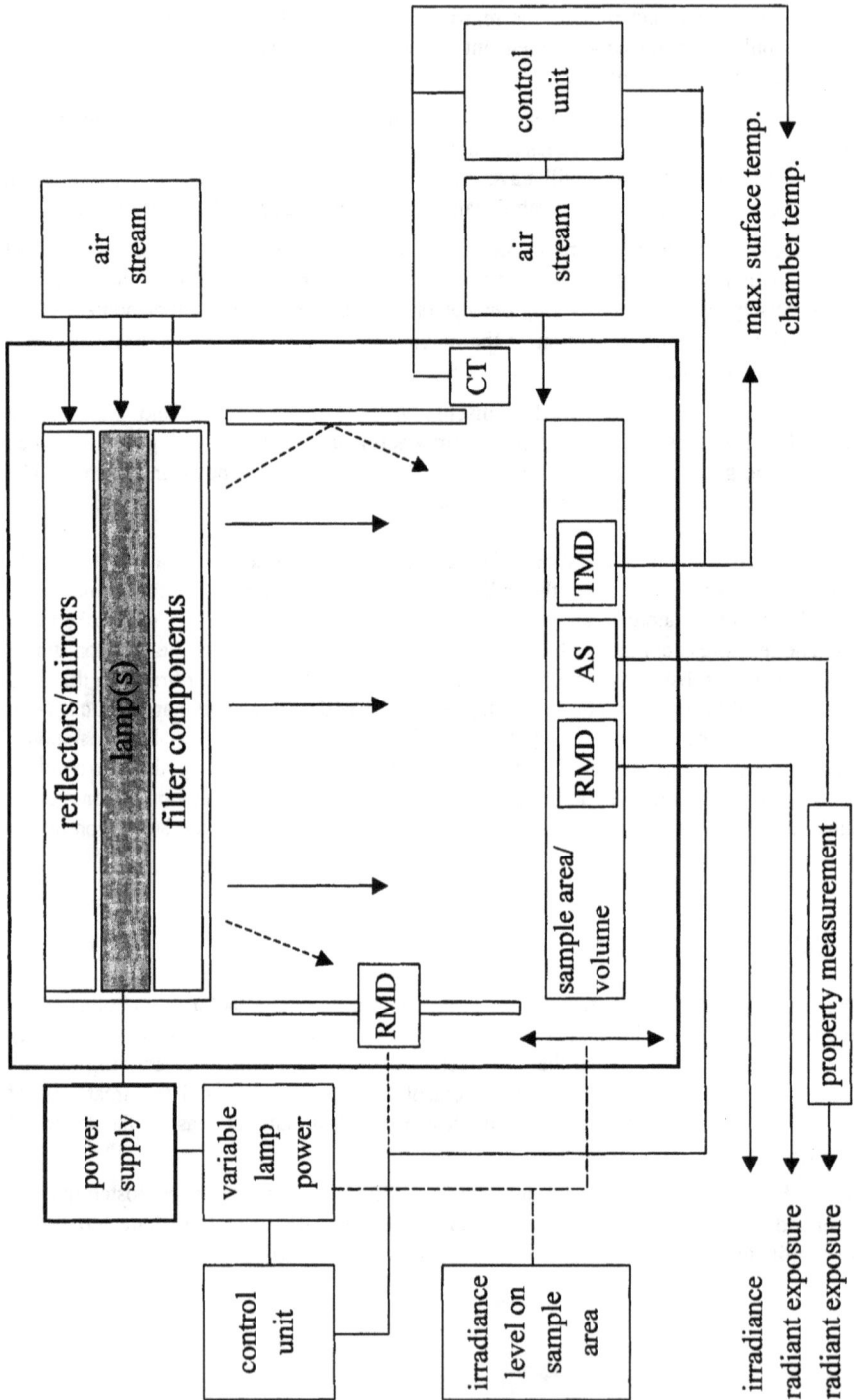

Figure 8. *Photostability exposure chamber - characteristics.*

References

1. D. S. Berger, *J. Investig. Dermatol.*, 1969, **53**, 192.
2. D. Kockott, *Symp. Bewitt. Kunststof*, Süddeutsches Kunststoffzentrum (SKZ); Würzburg (Germany), 1997, March 19.
3. J. Boxhammer, in 'Photostability of Drugs and Drug Formulations', H. H. Tønnesen, Ed., Taylor and Francis, London 1996, p.39.
4. D. Searle, *7th Int. Conf. Adv. Stabil. Control. Degrad. Polym.*, 1985, Luzern, May 22 to 24.
5. J. Boxhammer and K. P. Scott, *3rd Int. Symp. Weatherability*, Material Life Society, Tokyo, 1997, May 14 to 16.
6. J. Boxhammer, D. Kockott and P. Trubiroha, *Materialprüfung*, 1993, **35**, 143.

Design Limits and Qualification Issues for Room-size Solar Simulators in a GLP Environment

P. Donald Forbes
Argus Res. Lab., The Genzyme Transgenics Co.
905 Sheehy Drive
Horsham, PA 19044, U. S. A.

1 INTRODUCTION

Pharmaceutical photostability (PPS), viewed from one limited perspective, comprises a subset of issues related to chemical stability. From this perspective, the story ends when the drug product is administered. From another and equally limited perspective, PPS is simply a subset of issues related to phototoxicology (i.e., adverse biological consequences of one or more interactions among "light", chemicals and living tissue).

The disciplines behind each of these perspectives provide bodies of details that should be integrated if the goal of providing a framework for PPS is to be achieved. This paper offers details about techniques used in phototoxicology which may be beneficial to workers seeking the means to test photostability of their pharmaceutical products. Examples include producing and measuring the optical radiation used in the test system.

2 ROOM-SIZE SOLAR SIMULATORS

2.1 Solar Spectra and their Simulation

Although the natural solar spectrum encompasses cosmic radiation, gamma, UVR, visible, IR, and radio frequencies, our principle photochemical concerns are with what we might call terrestrial optical radiation (i.e., wavelengths 290 - 800 nm; see Figure 1). At the lower end of that range the solar energy distribution is strongly influenced by atmospheric absorption (including the influence of ozone) and solar angle. Harnessing a "sun in the laboratory" is a fairly elementary proposition when the area to be exposed is small, but becomes rapidly more difficult as the intended target expands. Limited numbers of devices are available for room-size solar simulation.

2.2 Monitoring Laboratory Optical Radiation Exposures

The essential tools include spectroradiometry (relative and absolute energy distribution) and radiometry (from hand-held meters to virtual instruments). Studies conducted under

Good Laboratory Practice (GLP) guidelines incorporate validation of data capture, display, and documentation.

Spectral Irradiance: Xenon Arc with 1 mm WG320

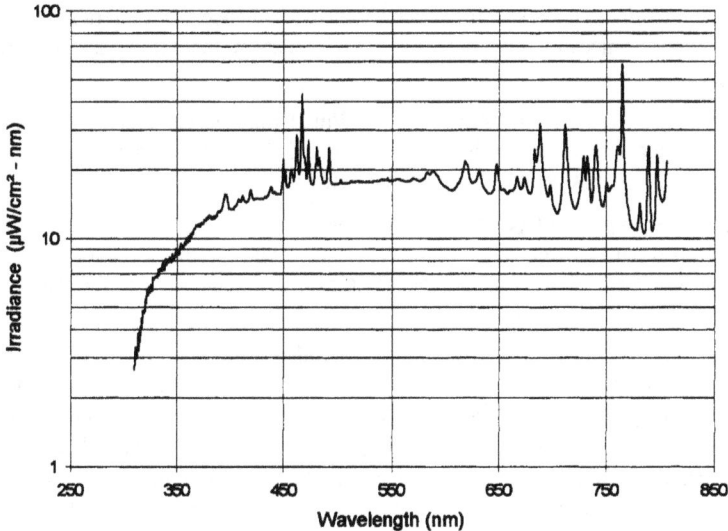

Figure 1. *Emission spectrum: Atlas Model RM 65 Xenon arc lamp (with WG320 filter), showing UVR and visible light regions.*

When describing the chemical and biological consequences of a relatively broad spectral band exposure, any reference to the total energy involved (e.g., *irradiance* in W/m^2 or *radiant exposure* in J/m^2) is of limited relevance. That is because the efficacy of photochemical interaction per incident quantum and the photobiological effects per unit radiant exposure vary substantially with wavelength. A quantitative plot of such spectral variation, usually normalized to unity at the most effective wavelength, is technically known as a "weighting function"; photobiologists frequently use the term "action spectrum".[1-3]

As one example, visual acuity is much greater at wavelength 550 nm than at 450. Thus, the weighting function for the part of the electromagnetic spectrum that activates the human retina ("visible light") assigns a greater value to photons around 550 nm. When the "photopic vision" weighting function is applied to the measured irradiance, the result ("brightness") is expressed as *luminance* (in lumen/m², i.e., lux). In brief, a measure of W/m^2 cannot tell us how bright an illuminated surface will look, but the number of lux can.

A second example involves ultraviolet radiation (UVR) and the skin response known as threshold erythema (the first perceptible pinkness following exposure to sunburning UVR). The *radiant exposure* producing the threshold response could be 200 J/m^2 or 2000 J/m^2, depending on the wavelength or mix of wavelengths involved. When the "erythema action spectrum" weighting function is applied to the measured or calculated radiant exposure, the result ("erythema effectiveness") is expressed as *"exposure dose"* in J/m^2_e (where J/m^2_e means "producing the response that would be associated with the same J/m at the peak of erythema effectiveness, wavelength 297 nm").[4,5] In brief, a broad-spectrum measure of J/m^2 cannot tell us whether a given surface dose will produce erythema in skin, but a figure of J/m^2_e can. (Note that under specific circumstances, "fluence" is preferable to "exposure dose". The circumstances include measurement of radiation, including back scatter, from all directions).[1,6]

The erythema-effective exposure dose (in J/m^2_e) may be calculated from a measured (unweighted) radiant exposure, or it may be estimated directly by a radiometer designed for the purpose (i.e., with a wavelength-dependent detector).

Wavelength-*independent* (thermal) radiometers incorporate detectors that have a flat spectral response over a wide range of wavelengths. Such thermal detectors operate on the principle that incident radiation is absorbed by a receiving element, and the temperature rise of the element is measured, usually by a thermopile or a pyroelectric detector.

Wavelength-*dependent* radiometers have spectral responses that vary widely depending on the types of detector and filters that may be incorporated. Detectors can be designed to have a spectral response that matches a particular action spectrum for a photobiological end-point. A case in point is the Robertson-Berger (Solar Light™) meter which incorporates optical filters, a phosphor and a vacuum phototube or photovoltaic cell.[7,8] This device measures wavelengths in the global UV spectrum with a spectral response that rises sharply with decreasing wavelength. The R-B meter has been used to monitor natural UVR continuously at several sites throughout the world, and also to monitor exposures to artificial sources of UVR in the laboratory.

2.3 Recording Exposure Data

The GLP environment favors on-line capture of radiation dosimetry data. Control CB™ is an example of accessory hardware and software that provides multi-channel display and recording of exposure data in real time. Virtually any type of electronic detector can be used with such software as part of a "virtual instrument" installation. In contrast, chemical actinometry has little to commend it in this application.

Spectroradiometry is used to characterize a source of UVR in terms of its spectral power distribution. The output, as indicated above, can be used to calculate photochemically or biologically weighted radiometric quantities. Emission spectra illustrated in this paper were determined with an Oriel™ multidiode array spectrograph linked to a computer running Instaspec II software; tabular output was imported into an Excel™ spreadsheet to generate presentation graphics.

2.4 Simulated Sunlight as a Source for PPS Testing

 . At least qualitatively, the xenon arc lamp is generally accepted as the best available source of simulated optical radiation.[9] Optical filters, such as those produced by Schott™, can be used to attenuate the shorter wavelength emissions and thus mimic the result of terrestrial exposures from various solar angles.[10] With a WG 320 filter of appropriate thickness, the xenon lamp provides UVR similar to mid-latitude noon summer sunlight, with good color balance in the visible spectrum (Figure 1). Without the WG 320 filter, the emission spectrum includes the entire far-UVR component characteristic of extra-terrestrial sunlight (Figure 2). With an added WG 345 filter (Figure 3), the UVR component is typical of that found in winter sunlight, or in sunlight through window glass.

Atlas RM 65 Xenon Arc Lamp at 2.25 Meters
(with and without WG 320 filter)

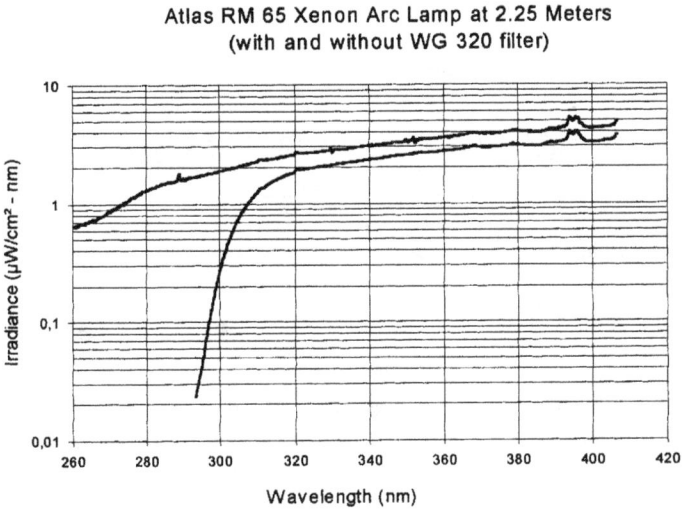

Figure 2. *Emission spectrum: Atlas Model RM 65 Xenon arc lamp (with and without WG320 filter; reading taken at 2.25 m from source).*

Table 1. *Watt-hour conversion to Joules (J). From this relationship the unweighted UVR needed to achieve the ICH "200 W-hr/m^2" would be 720 kJ/m^2.*

Watt-hour conversion to Joules (J)	
200 W-hr/m^2	= 200 X 3600 J/m^2
	= 720,000 J/m^2
	= 720 kJ/m^2
(example 1: ca. 9 hours c. bank of BL (TL10) lamps at 80 kJ/m^2 in 1 hour.)	
(example 2: ca. 72 hrs at 2m from Atlas model RM 65)	

The ICH draft guideline for PPS[11] (ICH Expert Working Group, 1997) expresses UVR and visible light exposures in terms of Watt-hours and lux-hours, respectively. For laboratory workers accustomed to using more conventional international units of measure, the Joule (J) is the appropriate energy unit for at least the UVR component (Table 1), provided that no effectiveness weighting is required.

**Atlas RM 65 Xenon Arc Lamp at 1 Meter
(with WG 345 filter)**

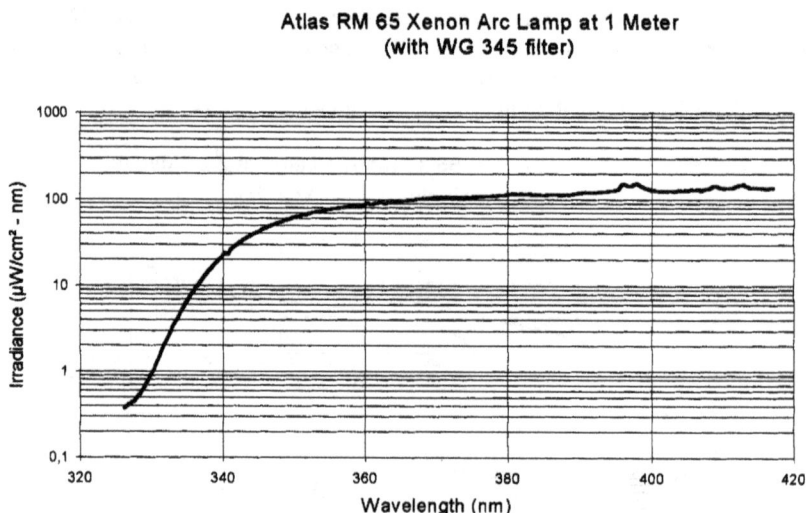

Figure 3. *UVR Emission Spectrum: Atlas Model RM 65 Xenon arc lamp (with WG345 filter to restrict emission to near ultraviolet and visible regions; reading taken at 1 m from source).*

Table 2. *Visible light intensities as a function of distance from Atlas RM 65 xenon arc lamp, compared with mid-day sunlight, shown with duration required to achieve ICH "1.2 million lux-hours".*

Atlas 6.5 kW xenon arc compared with mid-day sunlight (40° N. latitude)			
distance (meters)	ft-candles (lumens/ft²)	lux (lumens/m²)	duration (for ICH 1.2 million lux-hr) (hours)
2.25	370	3983	301.3
2	450	4844	247.7
1	1370	14747	81.4
0.5	4500	48438	24.8 e.g., 3 eight-hour days
sunlight	10000	107640	11.1

In contrast, the ICH draft guideline specifies the quantity of optical spectrum exposure as a biologically-weighted function (i.e., lux-hours, an estimate of its effectiveness in terms of human vision). The rationale for this specification is not obvious, since the photochemical sensitivity of retinal pigments should not be expected to represent all other compounds.

The intensity of optical radiation drops off as a function of target or work surface distance from the xenon arc source (Table 2). With a source of finite size, applications requiring the exposure of a relatively large surface area are limited to lower intensities.

Table 3. *Ratios of total UVR to visible light (lux) in Atlas Model RM 65 xenon lamp emission and in ICH guidelines. Using simulated sunlight to deliver the ICH-specified visible component results in simultaneously delivering 6000/1742, or approximately 3.5 times the ICH-specified UVR component.*

Ratios of UVR/lux: Solar Simulator vs. ICH Guideline		
UVR W/m²	Visible (lux)	
Atlas 2.78	4844	(Atlas total UVR = ca. 10 kJ/m²/hr @
1	1742	= 10 kJ/m² per 3600 sec
0.574	1000	= 2.78 W/m²)
ICH 200	1,200,000	(200 W-hr/m² and 1.2 X 10⁶ lux-hrs)
		6000
0.167	1000	

The UVR and visible components specified in the ICH document are not in the same proportions as those found in sunlight (Table 3). As a consequence, the use of a solar simulator as the radiation source for the visible component will result in disproportionate exposure to the UVR component. Tønnesen and Karlsen[12] have pointed out an analogous problem with another brand of xenon lamps. A re-evaluation of the specifications (including international units of measure), or at least the provision of a clearer rationale, would appear to be in order for future updates of this document.

References

1. J. Jagger, 'Solar UV Actions on Living Cells', Praeger Pub, New York, 1985, p. 159.
2. IARC Working Group, 'Solar and Ultraviolet Radiation' (IARC Monographs on the Evaluation of Carcinogenic Risks to Humans), WHO, Washington, 1992, Vol. 55.
3. A. F. McKinlay and B. L. Diffey, *CIE Journal*, 1987, **6**, 17.
4. D. F. Robertson, in 'Impacts of Climatic Change on the Biosphere' (CIAP Monograph 5), D. S. Nachtwey, Ed., Nat. Tech. Inf. Ser., Springfield, VA, 1975, Appendix B.
5. F. Urbach and P. D. Forbes, *Int. Cong. Photobiology*, Vienna, Austria, Sept. 1996.
6. C. S. Rupert and R. Latarjet, *Photochem. Photobiol.*, 1978, **28**, 1.
7. D. Berger, D. F. Robertson, R. E. Davies and F. Urbach, in ref. 4, Ch. 1-3, Appendix D.
8. D. S. Berger, *Photoch. Photobiol.*, 1976, **24**, 587.

9. COLIPA Task Force on Sun Protection Measurement, 'SPF Testing Method', L'Oreal RAD, Clichy, France, 1994, p. 1.

10. P. D. Forbes, R. E. Davies, F. Urbach, D. Berger and E. Cole, *Canc. Res.*, 1982, **42**, 2796.

11. ICH, 'Photostability Testing of New Drug Substances and Products', Draft, 1997.

12. H. H. Tønnesen and J. Karlsen, *Pharmeur.*, 1997, **9**, 735.

Actinometry: Concepts and Experiments

G. Favaro
Dipartimento di Chimica
Università di Perugia
via Elce di Sotto 8, I-06123 Perugia, Italy

1 INTRODUCTION

The aim of this report on actinometry is to meet the widespread demand of obtaining objective criteria which could be applied to evaluate photostability of drugs. In this context, the central point is to quantify the efficiency of the photodegradative processes that they may undergo so as to be able to reliably compare experiments carried out in different laboratories.

Determining the reaction quantum yield, that is, the rate at which molecules undergo a given event per photon absorbed per unit time, is a central problem to photochemistry. Since the rates of photochemical reactions are affected by the intensity of the source of irradiation, this intensity must be known. Actinometry provides the way by which intensity, that is, the number of photons in a beam, can be determined. The importance of actinometry therefore resides in its capability to quantify photochemical reactions.

From the expression of quantum yield (Φ):

$$\Phi = \frac{\text{number of molecules undergoing the process of interest per unit time}}{\text{number of photons absorbed by the photoreactive substance per unit time}} \quad (1)$$

it is seen that two distinct measurements are required: the analytical determination of the reactant consumed or product formed and the determination of the number of photons absorbed by the reactant. The first can be achieved by appropriate analytical methods, such as spectrophotometry, chromatography, etc., the second by using an actinometer, that is, a physical device or a chemical system by which the number of photons in a beam can be determined either integrally or per unit time.[1]

2 PHYSICAL DEVICES

Physical devices, such as photomultipliers, photodiodes and thermopiles, transform the energy of the incident photons into measurable quantities, that is, electric current or heat.

However, high perf rmance physical devices are very expensive and are currently found in only a few, highly equipped, laboratories. The majority of common commercial physical detectors have several disadvantages, the most important being the need for frequent recalibration, due to both aging and damages caused by high energy UV radiation. Moreover, sensitivity, which is generally very low in the UV, depends on wavelength. While the detectors are provided with calibration curves, the correction factors for wavelength dependence are not very precise and, in addition, calibration curves lose their validity during use. Finally, because of inhomogeneous spatial distribution of the sensitivity of the target, while the sample solution is homogeneous, the target and the sample respond differently to the same light stimulation.[2-4]

For these reasons, photochemists prefer using chemical actinometers and we will now focus our attention on these systems.

3 CHEMICAL ACTINOMETERS

Chemical actinometers are systems undergoing a photochemical process with an accurately known quantum yield: the photochemical conversion is directly related to the number of photons absorbed. Thus, the determination of absolute intensities, that is, number of photons incident onto the sample, is one of the main advantages of chemical actinometers. In addition, the detector is generally cheap allowing easy replacement in case of damage. There are no geometry problems: the actinometric system can mimic the experimental situation of the sample of interest, thus any experimental error due to differences in shape, surface, and physical arrangement can be easily avoided. Finally, no calibration is needed.

Even though any photoreaction whose quantum yield is known could be used as actinometer, actinometric systems should meet the following requirements:[3-5]

- the quantum yield value should be wholly reliable and easily reproducible;

- the quantum yield should be wavelength-independent or accurately known for a large number of wavelengths;

- the actinometer should be thermally stable to avoid complications due to dark reactions;

- the analytical methods should be simple, precise and not time-consuming;

- the actinometric material should be commercially available and cheap or, at least, it should be easy to synthesize and purify;

- absorption spectra of reactant(s) and photoproduct(s) should be well know; photoproducts should not absorb in the wavelength range of the actinometer (if they do, it is possible to take this into account);[3]

- the sensitivity of the system should be sufficiently large.

Only a few of the numerous proposed actinometers meet these criteria.

3.1 Kinetic Principles in Chemical Actinometry

In exposing the kinetic principles which govern actinometry,[3] we will first consider the simplest general photoreaction:

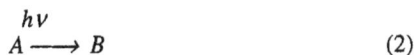

$$A \xrightarrow{h\nu} B \qquad (2)$$

The kinetic equation is:

$$- d[A]/dt = d[B]/dt = \Phi I_A \qquad (3)$$

where Φ represents the known quantum yield of the actinometer and I_A is the fraction of incident light absorbed by A per unit time which can be expressed as:

$$I_A = I_0(1 - 10^{-A_A}) = I_0(1 - 10^{-[A]\varepsilon_A}) \qquad (4)$$

where I_0 is the total incident light at the wavelength of irradiation, A_A is the absorbance of A at this wavelength and ε_A is the molar absorption coefficient of A at the wavelength of irradiation. By introducing eq. (4) in eq. (3), the following relationship is obtained:

$$- d[A]/dt = d[B]/dt = \Phi I_0(1 - 10^{-[A]\varepsilon_A}) \qquad (5)$$

The problem is to determine the incident light, I_0.

Three different absorption conditions may be met:

1) <u>concentrated solutions</u>, which totally absorb the incident light at the wavelength of irradiation ($A_A > 2$, total absorption). In total absorption conditions, I_A is equal to I_0, because the very large value of the negative exponent makes the exponential part in eq. (4) negligible. Thus, the kinetic equation reduces to:

$$- d[A]/dt = d[B]/dt = \Phi I_A = \Phi I_0 \qquad (6)$$

from where I_0 can be easily obtained from A (or B) concentration measurements before irradiation and after a certain irradiation time, t:

$$I_0 = ([A]_0 - [A]_t)/ \Phi t = ([B]_t - [B]_0)/ \Phi t \qquad (7)$$

2) <u>dilute solutions</u>, which partially absorb the incident light at the wavelength of irradiation ($A_A \approx 0.3 - 1$). In partial absorption conditions, no approximation is possible on the absorbed light and the rate equation is (5). From where, by integration:

$$I_0 = (1/2.303\varepsilon_A \Phi t) \ln (10^{[A]_t\varepsilon_A} - 1)/ (10^{[A]_0\varepsilon_A} - 1) \qquad (8)$$

3) <u>very dilute solutions</u>, which absorb a negligible fraction of the incident light at the wavelength of irradiation: $A_A < 0.03$. In negligible absorption conditions I_A can be approximated to $2.303 I_0 \, \varepsilon_A \, [A]$, after developing the exponential in Taylor series and neglecting all terms after the second. Thus the rate equation is:

$$- d[A]/dT = 2.303 I_0 \, \Phi \varepsilon_A \, [A] \qquad\qquad (9)$$

by integration:

$$I_0 = (1/2.303 \varepsilon_A \, \Phi t) \ln \, [A]_0 / [A]_t \qquad\qquad (10)$$

From the above relationship the incident radiation intensity per unit time, I_0 (Einstein cm^{-3} s^{-1}), can be obtained.

The first condition is by far the most convenient and is therefore recommended.

3.2 Experimental Quantum Yield Measurement

To set-up a model quantum yield experiment several arrangements are possible.[5] The simplest experiment consists of a reaction cell mounted on an optical bench in a fixed position relative to the light source. The irradiation wavelength is selected by means of a filter or a monochromator. A shutter, mounted in the optical path, is opened only for the duration of the photolysis. A known volume of test solution, V, is irradiated for a known time, t_1, and analysed. Then, the same reaction vessel is charged with the same volume of actinometer solution and irradiated for t_2 and analysed.

Possibly, but not necessarily, the absorbances of the test and the actinometer solution and the corresponding irradiation times should be very close each other. This means that, ideally, the light sensitivity of the two systems should be comparable. Since photoproducts may occasionally be present even before irradiation, a careful procedure requires chemical analysis not only after irradiation but also before irradiation (blank). For both the actinometric and the test solutions, product formation is easier to monitor than reactant loss, because the former can be measured at low conversion, while reactant loss measurement can involve a consistent experimental error because of the subtraction of two comparable large quantities.

By assuming total absorption conditions for both the actinometer and the test solution, I_0 can be obtained from the experiment on the actinometer:

$$I_0 = \frac{(C_{t_2} - C_{blank})V}{\Phi_{actin} \, t_2} \quad \text{(Einstein s}^{-1}\text{)} \qquad (11)$$

and its value is used to obtain the quantum yield of the system investigated from the experiment with the test solution:

$$\Phi_{test} = \frac{(C_{t_1} - C_{blank})V}{I_0 t_1} \tag{12}$$

3.3 Irradiation Sources

The most interesting wavelength range to which substances may be sensitive is that included in the solar emission. Solar emission on the Earth's surface extends from UV-A region (320 - 400 nm), which is attenuated by the atmosphere, to the visible (400 - 750 nm) and IR (λ > 750 nm) regions. UV-B (280 - 320 nm) and C (λ < 280 nm) are almost completely reduced by absorption in the atmosphere by gases (especially by O_3). Thus, only systems which absorb in the near UV and visible regions should potentially be affected by solar light. The most common, commercially available sources used by photochemists, are mercury and xenon lamps. While low-pressure mercury lamps display only the resonance 254 nm UV emission, medium pressure mercury lamps are enriched with emission lines towards the visible (λ = 313, 366, 405, 436, 546, 579 nm) and can be coupled with filters for selective irradiation. In the emission of high pressure mercury lamps, the resonance 254 nm line is missing and the visible spectrum broadens. An alternative for the near UV and visible regions are the Xe lamps which exhibit a continuous spectrum and, therefore, need to be coupled with a monochromator or an interference filter to select the irradiation wavelength. When a photochemical experiment has to be set up, some obvious criteria have to be followed in choosing light sources: the emitted light has to be absorbed by the reactant, its intensity should be sufficient to induce reaction in a reasonable time and be quite stable during photolysis. A point source is, generally, preferable for ease of collimation.[6]

3.4 Well-established Chemical Actinometers

A report on Chemical Actinometry was prepared between 1985 and 1988 by the IUPAC Commission on Photochemistry. It was coordinated by S. Braslavsky and M. Kuzmin and included representatives from about 15 countries. The results of the Commission's work appeared in Pure and Applied Chemistry in 1989.[4] This compilation lists photochemical systems which had been tested as actinometers. A few of them (7) were in solid phase, some other (19) were gaseous, the majority (49) were in liquid phase. They were characterised by wavelength range of operation, quantum yield at the irradiation wavelength, sensitivity and analytical method. The systems presented were also classified as "well established", "still under discussion", or "disproved", based on the confidence they gave. For eight of them, recommended procedures were described.

A list of the well-established actinometers in liquid phase is reported in Table 1. As can be seen, they cover a broad wavelength range, from the far UV to the red. The most used analytical method is spectrophotometry.

A few of them will now be described with greater detail.

Table 1. *Well-established actinometers in solution*

Current name	wavelength range (nm)	analytical method
cis-Cyclooctene	185	GC
Ethanol	185	GC
Uranyl oxalate	200-500	titration
Malachite green	225-289	abs. at 620 nm
Azobenzene (Actinochrome 2R)	230-460	abs. at 358 nm; HPLC
Actinochrome 1R	248-334	abs. at 572 nm
Potassium ferrioxalate	250-500	abs. at 510 nm (complex)
Aberchrome 540	310-375 435-535	abs. at 494 nm
Potassium Reineckate	316-750	abs. at 450 nm (complex)
Actinochrome N	475-610	abs. at 429 nm

3.4.1 Potassium Ferrioxalate (or Hatchard-Parker Actinometer)[7] is by far the most widely used actinometer system. The wavelength range of operation is between 250 and 500 nm, where it displays a wavelength-independent quantum yield ($\Phi = 1.25$) from 250 to 366 nm. The quantum yield slightly decreases in the visible ($\Phi = 0.93$ at 514 nm). In addition, the quantum yield is independent of temperature. The reaction (Scheme 1) substantially results in reduction of Fe^{3+} to Fe^{2+}, which is spectrophotometrically revealed by the red complex with *tris* 1,10-phenanthroline in acidic buffered solution (analysis wavelength, 510 nm).

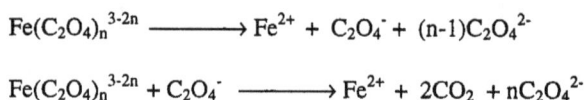

$$Fe(C_2O_4)_n^{3-2n} \longrightarrow Fe^{2+} + C_2O_4^- + (n-1)C_2O_4^{2-}$$

$$Fe(C_2O_4)_n^{3-2n} + C_2O_4^- \longrightarrow Fe^{2+} + 2CO_2 + nC_2O_4^{2-}$$

Scheme 1

This actinometer exhibits several desirable features:

- high sensitivity;

- high quantum efficiency over a wide wavelength range;

- operation over a wide range of intensities;

- temperature-independent quantum yield;

- wavelength-independent quantum yield over large wavelength ranges.

However, the analytical procedure, which must be carried out under red-safe light, is somewhat complicated and time-consuming, but the results are very precise.

3.4.2 Potassium Reineckate.[8] This actinometer extends the operative wavelength range further towards the visible (316 to 750 nm). The actinometric reaction ($\Phi = 0.3$) consists in the photoaquation of the complex $KCr(NCS)_4(NH_3)_2$ (Scheme 2); the thiocyanate ion released is determined by spectrophotometry on the absorption band of the photoproduced complex with Fe^{3+} ($\lambda = 450$ nm).

$$Cr(NCS)_4(NH_3)_2^- \longrightarrow Cr(NCS)_3(H_2O)(NH_3)_2 + SCN^-$$

Scheme 2

Advantages of this actinometer are:

- relatively high sensitivity;

- broad wavelength range of operation;

- simplicity of procedure;

- usability with lasers.

Despite several good qualities, this actinometer has some disadvantages:

- the quantum yield is slightly wavelength-dependent and this requires interpolation procedures for certain wavelength values;

- the quantum yield is temperature-dependent and its temperature-dependence is related to pH, thus a temperature and pH control is needed;

- the complex slowly decomposes in the dark, so fresh solutions should be used;

- the inner-filter effect, due to light absorption by photoproducts, needs some corrections or selected wavelengths should be used.

3.4.3 Actinochrome N.[9] has recently been suggested as an actinometer for the visible region. The reactant (meso-diphenylhelianthrene) undergoes self-sensitized photo-oxygenation forming an endoperoxide. The quantum yield for the endoperoxide formation in air-saturated toluene ($\Phi = 0.22$) is wavelength independent over the range 475 - 610 nm. The reaction is monitored by observing the growth of an absorption maximum of the endoperoxide at 429 nm, a wavelength where the reactant, which has a broad absorption band lying to the red of that of the photoproduct, absorbs only weakly.

This actinometer has several positive features:

- the commercial solution is directly usable as supplied;
- it exhibits high accuracy and is wavelength-independent;
- it may be used as a polychromatic quantum counter.

However, Actinochrome N is temperature sensitive owing to the temperature-dependence of the solubility of oxygen in toluene.

3.4.4   Photochromic Actinometers: Actinochrome 2R and Aberchrome 540. Actinometers have recently been developed, which are based on a photochromic reaction. The principal advantages of these "new generation" actinometers are ease of handling, accuracy and re-usability. Re-usability implies that the actinometric reaction is photo-reversible or thermoreversible (Scheme 3) and therefore the actinometer can be regenerated by irradiating or heating.

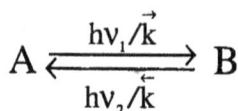

$$ A \underset{h\nu_2/\overleftarrow{k}}{\overset{h\nu_1/\overrightarrow{k}}{\rightleftharpoons}} B $$

Scheme 3

Azobenzene or Actinochrome 2R[10] undergoes thermal-reversible and photo-reversible *trans-cis* photoisomerization (Scheme 4). The thermal or photochemical back reactions may be used to regenerate the actinometer and hence makes it reusable, but it also requires that temperature must be kept as low as possible.

Both the forward ($\Phi_{trans \to cis} = 0.14$) and back ($\Phi_{cis \to trans} = 0.48$) photoreactions are used as actinometric reactions: the first in the 275 - 340 nm wavelength range, the second in the 350 - 440 nm wavelength range.

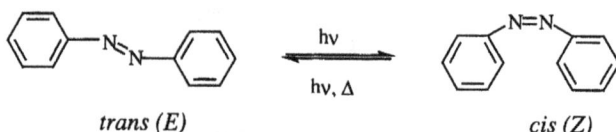

trans (E)   cis (Z)

Scheme 4

The analysis is carried out by spectrophotometric determination on the absorption maximum of the *trans* isomer ($\lambda = 358$ nm). However, different procedures are used for different wavelength ranges. While for the shorter wavelength range total absorption conditions are used, for the visible region a pre-irradiation is required to produce the *cis* isomer; the photostationary-state solution obtained results to partially absorb the visible light, which implies a more complicated procedure.[3]

The principal advantages of Actinochrome 2R are:

- availability in a sealed spectrophotometric cell ready for use;

- re-usability;

- accuracy and simplicity;

- usability with lasers.

The different procedures used for different wavelength ranges, corresponding to the forward and back photoreaction, cannot be considered a great disadvantage.

Aberchrome 540[11] has met with great success during the last decade. It is based on the photoreversible electrocyclic ring-closure reaction of a furylfulgide (Scheme 5). The operative wavelength range is 310 - 375 nm ($\Phi = 0.209$).

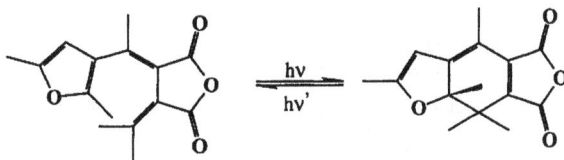

Scheme 5

The colored photoproduct is monitored by spectrophotometry at its absorption maximum, 494 nm. White light shifts the reaction to the left.

The back reaction is generally not used in actinometry because its quantum yield is both temperature- and wavelength-dependent.

The advantages of this actinometric system are:

- re-usability;

- sensitivity and simplicity of operation;

- temperature independence of the forward reaction;

- usability with lasers.

However, the re-usability of this actinometer has recently been criticized, because of the occurrence of photochemical *trans - cis* isomerization under UV irradiation. Since only the *trans* isomer cyclizes, the quantum yield apparently decreases after regeneration since visible light is only absorbed by the cyclic form.[12] A further source of error can be neglecting the occurrence of ring-opening by irradiation not only in the visible but also in the UV range.[13]

8 CONCLUDING REMARKS

In conclusion, the chemical actinometers described in this report, which is far from being exhaustive, fulfil to varying degrees the requirements mentioned above, but each system also has some disadvantages. The most appropriate actinometer system must be selected time by time depending on the type of experiment to be performed. Thus, when choosing an actinometer, great care should be taken to define the exciting wavelength, concentration, analytical method, temperature, etc. In this way, the quantification of photodegradation through reaction quantum yield should be very consistent.

References

1. 'Glossary of Terms Used in Photochemistry', *EPA Newsletter*, 1985, **25**, 13.
2. G. Gauglitz, *EPA Newsletter*, 1983, **19**, 49.
3. G. Gauglitz, "Photochromism of Organic Molecules", Ed. H. Dürr and H. Bouas-Laurent, Elsevier, Amsterdam, 1990, p. 314 and ref. therein.
4. H. J. Kuhn, S. E. Braslavsky, R. Schmidt, *Pure Appl. Chem.*, (1989) **61**, 187 and ref. therein.
5. N. J. Bunce, 'Handbook of Organic Photochemistry', Ed. J. C. Scaiano, CRC, Orlando, 1989, Vol. I, p. 241 and ref. therein.
6. P. de Mayo, H. Shizuka, 'Creation and Detection of the Excited States', W. R. Ware, Ed., Dekker, New York, 1976, Vol 4, p.139.
7. C. G. Hatchard, C. A. Parker, *Proc. Roy. Soc.*, 1956, **A 235**, 518.
8. E. E. Wegner, A. W. Adamson, *J. Am. Chem. Soc.*, 1966, **88**, 394.
9. H. D. Brauer, R. Schmidt, G. Gauglitz, S. Hubig, *Photochem. Photobiol.*, 1983, **37**, 595.
10. G. Gauglitz, *J. Photochem.*, 1981, **15**, 255.
11. H. G. Heller, J.R. Langan, *J. Chem. Soc., Perkin Trans. 2*, 1981, 414.
12. Z. Guo, G. Wang, Y. Tang, X. Song, *J. Photochem. Photobiol., A: Chem.*, 1995, **88**, 31.
13. E. Uhlmann, G. Gauglitz, *J. Photochem. Photobiol., A: Chem.*, 1996, **98**, 45.

trans-2-Nitrocinnamaldehyde as Chemical Actinometer for the UV-A Range in Photostability Testing of Pharmaceuticals

E. Bovina.[1] P. De Filippis,[2] V. Cavrini[1]* and R. Ballardini[3]

[1] Dept Pharmaceutical Sciences, via Belmeloro 6, 40126 Bologna-Italy
[2] Glaxo Wellcome, Medicines Research Centre, Pharmaceutical Development Dept., via A. Fleming 4, 37435 Verona-Italy
[3] Institute FRAE-CNR, via Gobetti 101, 40129 Bologna-Italy

1 INTRODUCTION

The study of drug substance degradation under the action of UV/visible light is important because it could influence the stability of drug formulations and it could lead to formation of toxic photodegradation products.[1,2]

In photostability testing of pharmaceuticals, light sources which simulate sunlight or indoor lighting[3,4] (according to D65 and ID65 standards) are employed; in these exposure tests a standardised method for calibrating the radiation intensity from light sources is required.[5]

In this effort, various forms of actinometers, both physical and chemical, are used to measure radiation intensity. Physical actinometers, such as radiometers or photometers, often yield discordant results depending on spectral characteristics; moreover, they require periodic recalibration and do not take into account fluctuations in light intensity over time.[6] Chemical actinometers, based on measurements of chemical changes occurring in a photosensitive standard, are instead considered generally more accurate and reproducible; they are quantitative, well established, stable with time and they accurately measure the total amount of radiation received.[6,7]

The Hatchard-Parker ferrioxalate chemical actinometer[8] is perhaps the most useful and popular system used for light intensity measurements: it absorbs a wide spectral range of light (250-500 nm), it is inexpensive, easily available and requires only the use of a spectrophotometer, an instrument available in every pharmaceutical laboratory. The major disadvantage which affects its use, in drug photostability studies with sunlight-simulating irradiation, is that the photoreaction is rapid in relation to most drug photodecomposition, when used with a wide range wavelengths.[9] If it can be assumed that the source intensity is constant, this would not affect the result but unfortunately this is not the case for Xenon-arc lamps.[9]

The International Conference of Harmonisation (ICH), with the aim to establish a standard protocol for photostability testing of pharmaceuticals, proposed quinine monohydrochloride dihydrate (2% solution in water) as a chemical actinometer (with a slower reactivity than ferrioxalate) for calibrating the UV radiation in sunlight-simulating conditions.[6] The quantum yield for quinine photodecomposition and the photodegradation mechanism have not been determined, but the proposal is for a unit of integrated intensity of radiation expressed as ΔA at 400 nm of the irradiated quinine solution.[10] Quinine absorption maximum is at 330 nm, falling off rapidly as wavelength increases; the relatively high concentration has been used as an attempt to extend the absorption range. However, quinine system might be satisfactory for drugs which absorb in the UV-B region, since its absorption spectrum does not cover the full UV-A range; a typical example can be the photodegradation of nifedipine (a model for photosensitive drugs), for which the quinine spectrum does not overlap well and the results are therefore unsatisfactory.[5,11]

In the present study, the potential use of *trans*-2-nitrocinnamaldehyde (NCA) as chemical actinometer in the UV-A range (320-400 nm) was evaluated. Its use has been suggested from the similarity to o-nitrobenzaldehyde, a well known solid-phase chemical actinometer,[12] and from its employment as a UV-sensitive "photoacid progenitor" in phototranschromic inks.[13] Moreover, *trans*-2-NCA seems to possess the right properties since its absorption spectrum covers the full UV-A range and its photoreaction occurs in the time of the ICH confirmatory studies. In particular, the increase in absorbance at 440 nm (A_{440}) of a 0.5% (w/v) methanolic solution, after UV exposure, was found to be proportional to the UV-dose emitted from light sources. A 0.5% methanolic solution was concentrated enough to show absorbance values greater than 2 over the studied spectral range (UV-A) so that it could be effectively used as a chemical actinometer.

2 MATERIALS AND METHODS

2.1 Materials

trans-2-Nitrocinnamaldheyde and quinine monohydrochloride dihydrate were purchased from Fluka Chemie AG (Switzerland) and Aldrich (Italy) respectively; potassium ferrioxalate was obtained according to the Hatchard-Parker method.[8] Solvents for chromatography were of HPLC grade from Romil LTD (U.K.); deionised double distilled water was used for mobile-phase preparation; all other chemicals were of reagent grade.

2.2 Equipment

For UV radiation exposure tests two different kinds of light sources were employed:

1) 150 W Xenon-arc lamp (solar simulator, model 68805, ORIEL Corporation USA) provided with a dicroic mirror (mod. 81045) to block VIS and IR and to minimise the

samples heating, an air-mass filter 1.5 (mod. 81090) to simulate solar conditions and a UV-B/C blocking filter. In these conditions only radiations in the UV-A range were emitted. The output beam was directed downward by a "beam turning assembly", which held the dicroic mirror;

2) 150 W medium pressure Mercury lamp Q400 from HANAU, characterised by a line output, equipped with interference filters to isolate λ=313 and 365 nm.

The UV-dose (mJ/cm^2) of the Xe-arc lamp was measured by a radiometer (Goldilux, mod.70127, from ORIEL Corporation, USA) assembled with two external interchangeable probes, for UV-B and UV-A respectively.

A UV-VIS spectrophotometer V530 (JASCO, Japan) was used for absorbance measurements.

An HPLC system with a photodiode array detector (DAD, Hewlett-Packard HP Ti series 1050) was used. Chromatographic analyses were performed on a reverse-phase Alltima 5μ, C18 (150 x 4.6 I.D.) column, using a mobile-phase of ammonium acetate buffer (pH 7.0; 0.01M) - acetonitrile 60:40 (v/v), at a flow rate of 0.8 mL/min.

Gas-chromatographic analyses were performed on two different chromatographic systems:

1) HRGC FRACTOVAP 4160 with a 410 temperature programmer (Carlo Erba Strumentazioni, Italy) and a VARIAN 4720 integrator;

2) Hewlett-Packard, model HP 5890 series II, with a mass selective detector HP 5971.

In both the systems a 25m bonded phase fused capillary column Methyl, 5% phenyl Silicone (0.32 mm I.D.; 0.25 μm film thickness) was used; the chromatographic conditions employed were: Helium (carrier gas) at 0.6 kg/cm^2, injection port at 280°C, oven temperature 50°C for 1 minute, 50°-250°C at 10°C/min.

2.3 Sample preparation and procedures

trans-2-Nitrocinnamaldehyde was dissolved in methanol to make a 0.5% (w/v) solution (in ethanol the solubility was too low to get absorbance values ≥2 over the full UV-A range); three mL of the solution were placed into quartz cells with a 1 cm path length and closed with teflon caps. Quartz cells were horizontally positioned and exposed to Xenon-arc lamp for increasing irradiation times, corresponding to increasing UV-A doses. After UV exposure, the solution absorbance was measured at 440 nm and plotted against UV-dose; similar experiments were carried out with 0.3%, 0.7%, 0.8%, 1,0% (w/v) solutions to observe the concentration effect on the response linearity.

Exposure tests were also performed with monochromatic lights of 313 and 365 nm, using 3 mL solutions continuously stirred during irradiation. The "micro-version" of potassium ferri-oxalate actinometry[14] was used to measure the photons emitted per minute (Nhν/min), so that it was possible to correlate the increase of A_{440} with the number of photons absorbed by 0.5% *trans*-2-NCA solutions.

Temperature effects on *trans*-2-NCA degradation were studied keeping a quartz cell at 38-40°C during UV-exposure to Xe-arc lamp; for comparison a sample completely shielded with aluminium foil was also irradiated in the same conditions to discriminate between UV-A and temperature effects.[15]

Photoexposed solutions were subjected to HPLC and GC analyses to detect photoproducts responsible of A_{440} increase; such analyses were also carried out on samples stored in the dark (monitoring of possible degradation products due to temperature or solvent interaction) to allow a periodical preparation of the NCA solution.

3 RESULTS AND DISCUSSION

3.1 Validation of *trans*-2-nitrocinnamaldehyde as actinometer

After UV-A exposure to 150 W Xenon-arc lamp (provided with a UV/B-C cut-off filter), the absorbance in the tail of the absorption band (visible region) of a *trans*-2-NCA (0.5% w/v) solution appeared to increase regularly with integrated UV-A intensity (Figure 1), determined by a UV-A radiometer calibrated at 365 nm. It was chosen to measure the absorbance increase at $\lambda = 440$ nm, because at this wavelength slight absorption of the unexposed solution and significant ΔA variation after irradiation were observed. At shorter wavelengths the ΔA increase is affected by the decrease in absorption of the starting solution; at higher wavelengths the ΔA values are smaller.

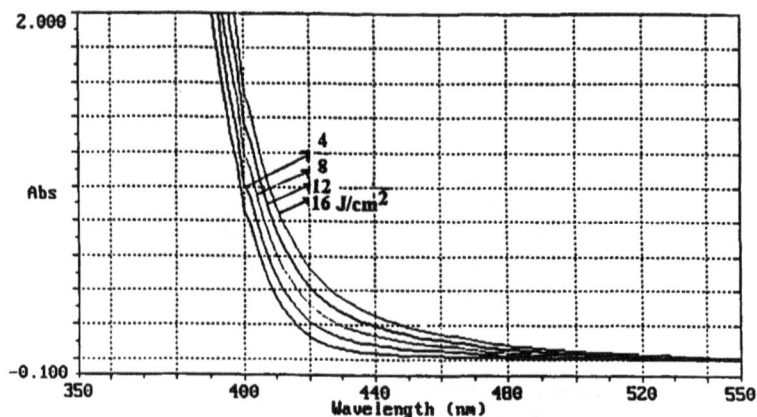

Figure 1. *Changes in absorbance of NCA 0.5% (w/v) methanolic solution exposed to Xe-arc lamp, as a function of UV-A dose.*

These experiments were carried out in several days, using several samples; A_{440} was plotted against UV-A dose, expressed as J/cm^2 ($1 J/cm^2 = 2.778$ Whr/m^2) and the linear regression lines (Figure 2), calculated for data obtained, showed high correlation coefficients ($r > 0.997$), as well as similar slopes and intercepts (Table 1). These results indicated that A_{440} increase was proportional to the integrated UV-A intensity and it could therefore be used as measure of UV-A dose of the lamp. Response linearity was maintained also for long exposure periods, corresponding to high UV-A doses (about 70 J/cm^2, approximately the 200 Whr/m^2 recommended by ICH Photostability Guidelines); such a device could be placed in close proximity to a sample during an ICH confirmatory photostability study, measuring in this way the entire amount of UV-A radiation incident on the sample.

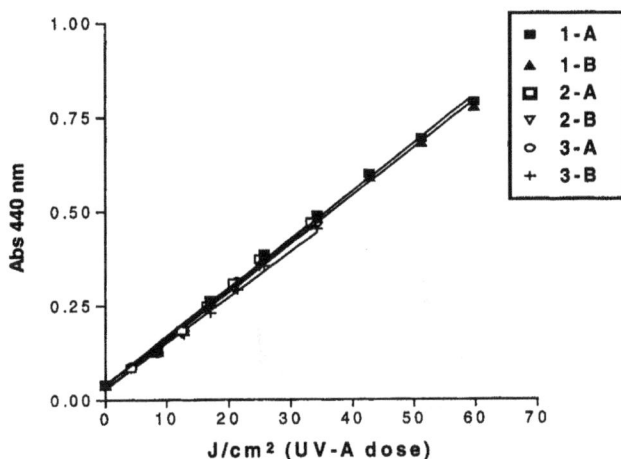

Figure 2. *Absorbance at 440 nm of NCA 0.5% solution as a function of UV-A dose ($1 J/cm^2$ = 2.778 Whr/m^2); experiments were carried out in several days (1, 2, 3) and using different solution lots (A, B).*

Table 1. *Regression lines of Abs vs UV-dose ($J/cm2$) from Xe-arc lamp*

Solution (Day-lot)	r	Slope x 10^{-2} $(J/cm^2)^{-1}$	sd x 10^{-2}	Y-int x 10^{-2}	sd x 10^{-2}
1-A	0	1.28	0.025	4.21	0.89
1-B	0	1.26	0.025	3.92	0.90
2-A	0	1.34	0.036	2.85	0.65
2-B	0	1.30	0.029	2.84	0.53
3-A	0	1.30	0.035	3.18	0.67
3-B	0	1.23	0.036	2.92	0.68

A *trans*-2-NCA 0.5% solution was concentrated enough to absorb all UV-A radiation received from light source, as it showed absorbance values ≥ 2 over the studied spectral range (320-400 nm), on the contrary of quinine monohydrochloride dihydrate solution (2% in water), the chemical actinometer proposed by ICH[5,6] for calibrating UV radiation, as shown in Figure 3.

Similar exposure tests were carried out with 0.3%, 0.7%, 0.8%, 1.0% (w/v) methanolic solutions as an attempt to extend the wavelength absorption range and to observe the concentration effect on *trans*-2-NCA photoreactivity. At high concentration, solubility problems were observed and slightly worse linear responses were obtained, confirming the 0.5% solution as optimal.

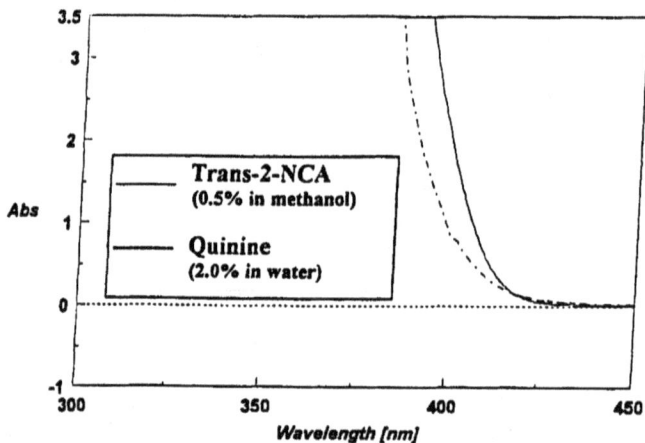

Figure 3 *Comparison between quinine monohydrochloride dihydrate (2% in water) and trans-2-NCA (0.5% in methanol) absorption spectra.*

Irradiation experiments were also performed with monochromatic lights at two different wavelengths, 313 and 365 nm respectively; A_{440} vs. absorbed photons (Nhv x 10^{-5}) for both wavelengths is reported in Fig.4, while regression lines were compared in Table 2. The data showed that *trans*-2-NCA absorbance increase was linearly correlated to the number of photons emitted from the light source, which were all absorbed from the actinometer solution, and was not depending on the irradiation wavelength. These results suggested that A_{440} increase of *trans*-2-NCA solutions exposed to different polychromatic UV-A lamps, should be dependent only from the total number of photons emitted from each lamp and not from their spectral distribution in the UV-A range.

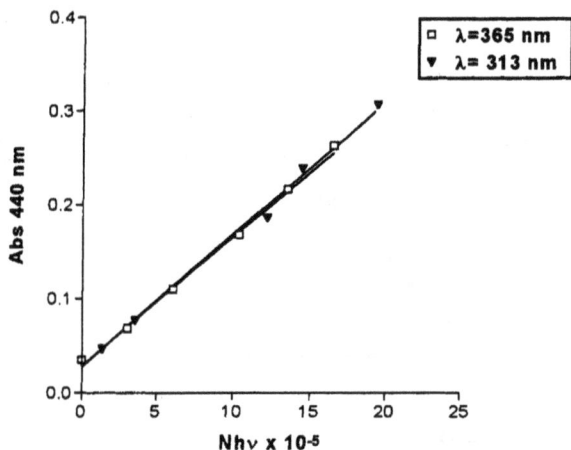

Figure 4. *Changes in absorbance at 440 nm of 0.5% solutions irradiated at 313 and 365 nm, respectively, as a function of cumulative number of photons (Nhν x 10⁻⁵).*

Table 2. *Regression lines of Abs vs Nhν x 10⁻⁵, from monochromatic lights at 313 and 365 nm.*

λ	r	Slope x 10^{-2} (Nhν.10^5)	sd x 10^{-2}	Y-int x 10^{-2}	sd x 10^{-2}
313	0.998	1.39	0.039	2.91	0.39
365	0.998	1.40	0.048	2.85	0.53

3.2 Temperature effects on *trans*-2-NCA photoreactivity

The effect of temperature on *trans*-2-NCA photoreactivity was studied, as described above, storing at 38-40 °C a test and a control sample wrapped in aluminium;[15] the behaviour of the test sample was identical to that observed for irradiation at room temperature (20-25 °C), while the control sample did not show any change in absorbance (Figure 5 and Table 3).

No decomposition products due to temperature were detected by HPLC analysis, both in the test and in the control sample, confirming that *trans*-2-NCA photoreactivity was not affected by temperature; this allows *trans*-2-NCA to be used as actinometer also in commercial light cabinets, provided with Xe-arc lamps without mirrors, where high temperatures (~ 40 °C) can be reached.[11]

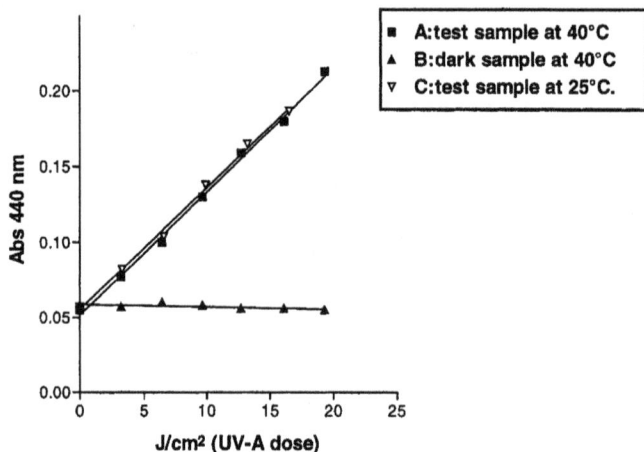

Figure 5. *Absorbance at 440 nm of NCA 0.5% solutions exposed to Xe-arc lamp at room temperature (C) and at 38-40°C (A), close to a control sample wrapped in aluminium foil (B).*

Table 3. *Regression lines of Abs vs UV-A dose (J/cm2) from Xe-arc lamp, at different temperatures.*

	r	Slope x 10^{-2} $(J/cm^2)^{-1}$	sd x 10^{-2}	Y-int x 10^{-2}	sd x 10^{-2}
25°C	0.998	0.82	0.021	5.10	0.25
40°C	0.998	0.81	0.024	5.52	0.24

3.3 Long-term stability

In order to study the long-term stability of the proposed solution, *trans*-2-NCA solution lots, stored in the dark at room temperature as well as refrigerated, were subjected to UV/Vis spectrophotometric and HPLC analyses. No absorbance changes or degradation products were detected in solutions stored up to 1 month, though an opalescence was observed, mainly in the refrigerated solutions, due to a decrease of solubility at low temperature. It was therefore suggested that the same lot of *trans*-2-NCA solution could be used as actinometer at least for two-three weeks, if stored in the dark and refrigerated (to limit the solvent evaporation); a possible precipitate is redissolved at room temperature.

3.4 Photodegradation mechanism

The photodegradation mechanism of *trans*-2-NCA, responsible for the absorbance increase in the visible region after UV exposure, was studied by HPLC and GC.

The analyses of photoexposed solutions showed the formation of a new chromatographic peak, at a retention time lower than that of NCA (peak 2 in Figure 6), whose area increase was found linearly correlated to the UV-A dose, up to ~ 30 J/cm^2. This photoproduct was identified as the NCA *cis*-isomer by GC-MS and ^{1}H NMR data; its mass spectrum was analogous to that of *trans*-2-NCA, while NMR spectrum showed different olefinic hydrogen coupling constants, according to the general rule of *cis-trans* isomerism.

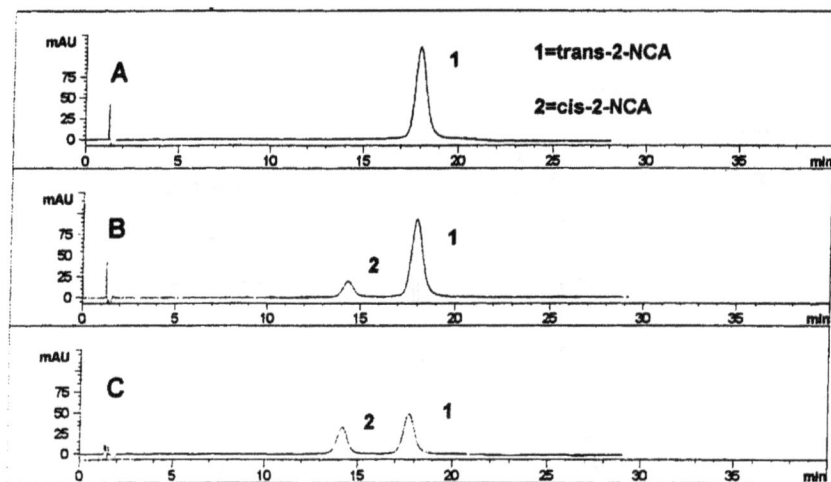

Figure 6. *HPLC chromatograms of trans-2-NCA solutions exposed to different UV-A dose levels: A) 0.00 J/cm2, B) 8.50 J/cm2, C) 25.5 J/cm2. Stationary phase: reverse phase C18, 5μm (150 x 4.6 I.D.); mobile phase: ammonium acetate buffer (pH 7.0, 0.01M)-CH3CN 70:30 (v/v); flow rate: 0.8 mL/min; UV detection at 240 nm.*

The quantum yield (φ) of this photoisomerisation was calculated according to the following equation:

$$\varphi = \frac{n° \text{ of moles of photoproduct (per unit time)}}{n° \text{ of moles of absorbed photons (per unit time)}}$$

and was found to be 0.15.

The number of moles of photons absorbed by the NCA solution was determined by the "micro-version" of potassium ferrioxalate actinometry (Fischer, 1984), at the excitation wavelength of 313 and 365 nm.

Two aditional photoproducts, at longer retention times (3 and 4 in Figure 7), were detected in significant concentration in solutions exposed to high UV-A doses (>30 J/cm^2).

Their GC-mass spectra were the same and suggested the compounds to be the *cis* and *trans* forms of 2-nitrosocinnamic acid methyl ester. The molecular ion peak (m/z 191) of 2-nitrosocinnamic acid methyl ester and a peak at m/z 161, due to the loss of nitroso radical (-NO), were observed in the mass spectrum. The characteristic methylcinnamate fragmentation pattern was also identified by two important peaks (m/z 130 and m/z 102), associated with the expulsion of the methoxyl radical (-OCH$_3$) and of the ester group (-CO$_2$CH$_3$), respectively.[16]

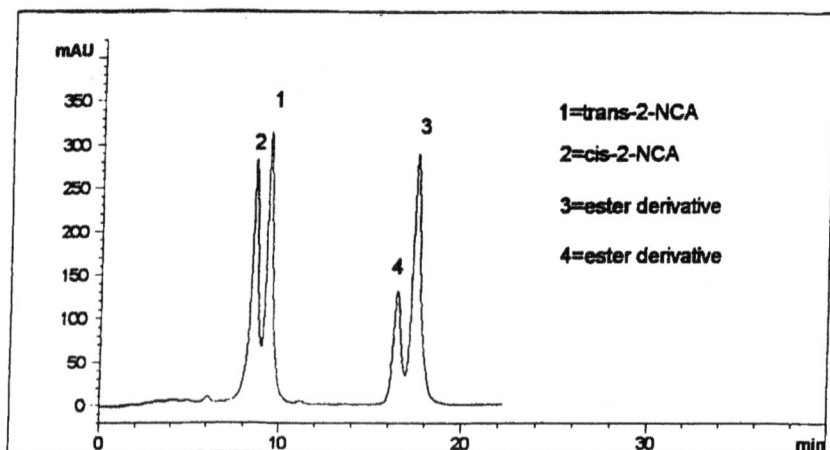

Figure 7. *HPLC chromatogram of a trans-2-NCA photoexposed solution (UV-A dose = 130.0 J/cm^2) Chromatographic conditions as in Fig. 6, with the exception of mobile phase composition: ammonium acetate buffer (pH 7.0, 0.01M)-CH$_3$CN 60:40 (v/v).*

Therefore, the data obtained suggested that a primary photoisomerisation to *cis*-2-NCA occurred, followed by a secondary photoisomerisation to o-nitrosocinnamic acid (both of *trans* and *cis*-2-NCA) and by the esterification with the solvent (methanol). Accordingly, when ethanol was the reaction solvent, mass spectra confirmed the ethyl ester formation. This photochemical pathway appeared to be similar to a carefully studied photoisomerisation of a well known solid phase chemical actinometer, o-nitrobenzaldehyde,[12] which undergoes a molecular rearrangement to o-nitrosobenzoic acid, due to UV exposure.

A consecutive reaction model, actually under evaluation, seems to explain the *trans*-2-NCA photoisomerisation to its *cis*-isomer and the following conversion to ester derivatives (Figure 8).

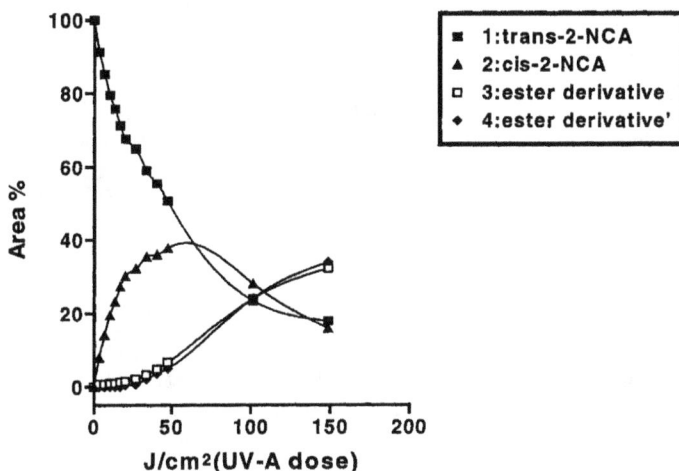

Figure 8. *trans-2-NCA photoisomerisation (1→2) and following formation of ester derivatives (3 and 4). Graph constructed from data obtained from GC chromatograms of photoexposed NCA solutions.*

4 CONCLUSIONS

trans-2-NCA is proposed as a new simple and reproducible actinometer for calibrating the intensity of UV-A radiation from light sources in photostability testing of pharmaceuticals, since it exhibits some interesting properties which allow its use in ICH confirmatory studies. In particular: its absorption spectrum covers the full UV-A range, its photodegradation quantum yield over the UV-A range is dependent only on the photons number and not on the radiation wavelength, its photodegradation mechanism has been elucidated and the photodegradation rate is in the time of the photostability testing. Finally, the methanolic solution is stable for periodical preparations.

Studies are in progress on the characterisation of some full-spectrum lamps by a spectroradiometer, in order to express the effective UV-A integrated intensity (for example the 200 Whr/m^2 recommended by ICH Photostability Guidelines for confirmatory studies)[4] in terms of ΔA_{440} of the *trans*-2-NCA actinometer solutions exposed to these lamps.

References

1. H. H. Tønnesen, *Pharmazie*, 1991, **46**, 263.
2. H. H. Tønnesen, "Photostability of Drugs and Drug Formulations", H. H. Tonnesen, Ed., Taylor & Francis, London, 1996, Chapter 1, p. 1.
3. D. E. Moore, in ref. 2, p. 9.
4. J. P. Riehl, C. L. Maupin and T. P. Layloff, *Pharm. Forum*, 1995, **21**, 6, 1654.
5. N. H. Anderson, in ref. 2, p. 305.
6. J. T. Piechocki and R. J. Wolters, *Pharm. Technol*, 1993, **17** (6), 46.
7. J. C. Scaiano, Ed., CRC Handbook of Organic Photochemistry, CRC Press, Boca Raton, FL, 1989, p.188.
8. C. G. Hatchard and C. A. Parker, *Proc. Royal Soc. London*, 1956, **A235**, 518.
9. D. E. Moore, in ref. 2, p. 63.
10. S. Yoshioka, Y. Ishihara, T. Terazono, N. Tsunakawa, T. Murai, T. Yasuda, Kitamura, Y. Kunihiro, K. Sakai, K. Hirose, K. Tonooka, K. Takayama, F. Imai, M. Godo, M. Matsuo, K. Nakamura, Y. Aso, S. Kojima, Y. Takeda and T. Terao, 1994, *Drug Dev. Ind. Pharm.*, **20**, 2049.
11. S. Nema, R. J. Washkuhn and D. R. Beussink, *Pharm. Technol.*, 1995, **19**, 172.
12. J. N. Pitts Jr, J. K. S. Wan and E. A. Schuck, *J. Am. Chem. Soc.*, 1964, **86**, 296.
13. P. M. Goman, Sirdesai, Sunil (Xytronyx, Inc.), "Aqueous printable phototranschromic inks for manufacture of UV dosimeters", *PCT Int. Appl.*, WO 94 01, 20 Jan 1994, 503; *Chem. Abstr.*, 1994, **122**: 163745y.
14. E. Fischer, *EPA Newsletter*, July 1984, **21**, 33.
15. H. H. Tønnesen and J. Karlsen, *Pharmeuropa*, 1995, **7**, 137.
16. E. M. Emery, *Anal. Chem.*, 1990, **32**, 1495.

Subject Index

www.ingramcontent.com/pod-product-compliance
Lightning Source LLC
Chambersburg PA
CBHW050522190326
41458CB00005B/1629